Antiviral Mechanisms

PERSPECTIVES IN VIROLOGY IX

Sir Christopher Howard Andrewes

THE GUSTAV STERN SYMPOSIUM

Antiviral Mechanisms

PERSPECTIVES IN VIROLOGY IX

Edited by MORRIS POLLARD

LOBUND LABORATORY
UNIVERSITY OF NOTRE DAME
NOTRE DAME, INDIANA

ACADEMIC PRESS New York San Francisco London 1975

A Subsidiary of Harcourt Brace Jovanovich, Publishers

ACADEMIC PRESS, INC.
111 Fifth Avenue, New York, New York 10003

United Kingdom Edition published by
ACADEMIC PRESS, INC. (LONDON) LTD.
24/28 Oval Road, London NW1

LIBRARY OF CONGRESS CATALOG CARD NUMBER: 59-8415

ISBN 0-12-560565-X

PRINTED IN THE UNITED STATES OF AMERICA

Contents

List of Invited Participants xi

Foreword *In Honor of Sir Christopher Howard Andrewes*
D. A. J. Tyrrell xxiii

Preface Morris Pollard xxix

Welcome Irene Stern xxxi

Chapter 1 **Perspectives in Virology: Vaticinations**
Edwin H. Lennette

Text 1
References 7

Chapter 2 **Detection and Identification by Immune Electron Microscopy of Fastidious Agents Associated with Respiratory Illness, Acute Nonbacterial Gastroenteritis, and Hepatitis A**
Albert Z. Kapikian, Stephen M. Feinstone, Robert H. Purcell, Richard G. Wyatt, Thomas S. Thornhill, Anthony R. Kalica, and Robert M. Chanock

I. Introduction 9
II. Detection of Rhinoviruses 11
III. Detection of Coronavirus Strain 692 14
IV. Identification of an Agent Associated with Acute Nonbacterial Gastroenteritis 19
V. Identification of an Agent Associated with Hepatitis A 33
VI. Conclusion 40
References 42
Discussion 45

Chapter 3 Current Understanding of Hepatitis B Virus Infection and Its Implications For Immunoprophylaxis

Robert H. Purcell

I.	Introduction	49
II.	Nomenclature	49
III.	Tests for HB_s Ag and anti-HB_s	50
IV.	Epidemiology	50
V.	Antigenic Diversity of HB_s Ag	54
VI.	Nature of Hepatitis B Antigens	56
VII.	Studies of Hepatitis Type B in Nonhuman Primates	61
VIII.	Prospects for Immunization Against Hepatitis B	63
IX.	Conclusion	66
	References	67
	Discussion	71

Chapter 4 Defective Interfering Particles As Antiviral Agents

Alice S. Huang and Eduardo L. Palma

I.	Introduction	77
II.	Continuous Infection of Cells by Vesicular Stomatitis Virus	78
III.	Nucleocapsid Accumulation	80
IV.	Molecular Basis of Interference	82
V.	Discussion	86
	References	88
	Discussion	89

Chapter 5 Synthetic Vaccines of the Future

Michael Sela

I.	Introduction	91
II.	Specificity Determinants	92
III.	Antigenic Competition	92
IV.	Synthetic Antigens and Role of Conformation	93
V.	Synthetic Antigens Cross-Reactive with Natural Products	94
VI.	Genetic Control of Immune Response	95
VII.	Built-in Adjuvanticity	96
VIII.	Vaccine Specificities?	96
IX.	Concluding Remarks	97
	References	98

Chapter 6 Inhibition of Viral Neuraminidase As a New Approach to the Prevention of Influenza

Edwin D. Kilbourne, Peter Palese, and Jerome L. Schulman

I.	Introduction	99
II.	Comparison of Antihemagglutinin and Antineuraminidase Immunization	101

III. Mechanisms of Influenza Virus Suppression 104
References 112
Discussion 112

Chapter 7 **Cellular Immune Response in Viral Infections**

Abner Louis Notkins and Donald Louis Lodmell

I. Introduction 115
II. Different Routes of Spread of Virus Infection 116
III. Immune Mechanisms Preventing Type II Spread 116
IV. Conclusion 128
References 130
Discussion 130

Chapter 8 **Transfer Factor and Cellular Immunity to Viral Infection**

H. Sherwood Lawrence

I. Introduction 135
II. Transfer Factor in the Initiation and Augmentation of CMI 136
III. Immunotherapy with Transfer Factor (TF_D) 139
References 149
Discussion 151

Chapter 9 **Immunological Evidence for a Viral Etiology of Certain Human Malignant Tumors**

H. H. Fudenberg, V. Byers, and A. S. Levin

I. Introduction 153
II. Role of Lymphocytes in Immunology 153
III. Relation of T-Cell Deficiency to Cancer and Autoimmunity 155
IV. Clinical Tests of Transfer Factor 156
V. Combined Effect of Immunodeficiency and Oncogenic Viruses in Cancer 157
References 159
Discussion 160

Chapter 10 **Genetic Transmission of RNA Tumor Viruses**

Robin A. Weiss

I. Introduction 165
II. Replication of RNA Tumor Viruses 167
III. Congenital Infection and Genetic Transmission 169
IV. Evidence for Genetic Transmission 171
V. Vertical Transmission and Host Antiviral Mechanisms 175
VI. Genetic Control of Endogenous Virus Expression 180
VII. Xenotropic Viruses 190
VIII. Inheritance of Acquired Genes 192

IX. Epilogue 199
References 200
Discussion 203

Chapter 11 **Type C Virus Inhibitor Associated with Cells Cultivated from New Zealand Black Mice**

Jay A. Levy

I. Introduction 207
II. Biological and Physico-chemical Properties of the Inhibitor 208
References 214
Discussion 214

Chapter 12 **Inhibition of Rous Sarcoma Virus Replication by Inhibitors of Cell Replication**

Eric H. Humphries

I. Introduction 215
II. Absence of Viral Antigens in Rous Sarcoma Virus-Infected Stationary Chicken Embryo Fibroblasts 217
III. Experimental Systems 219
IV. Appearance of Virus RNA in Rous Sarcoma Virus-Infected Stationary Chicken Embryo Fibroblasts After Stimulation with Serum 221
V. Rous Sarcoma Virus Production in Chicken Embryo Fibroblasts Made Stationary After Activation 227
VI. A Model for the Replication of Rous Sarcoma Virus 228
VII. Summary 229
References 230
Discussion 230

Chapter 13 **Mechanism of Interferon-Induced Block in mRNA Translation**

Michel Revel, J. Content, A. Zilberstein, B. Lebleu, E. Falcoff, and R. Falcoff

I. Introduction 233
II. Materials and Methods 234
III. Results 235
IV. Conclusions 245
References 246
Discussion 247

Chapter 14 **Clinical Utilization of Exogenous Human Interferon**

Thomas C. Merigan, George W. Jordan, and Richard P. Fried

I. Introduction 249
II. Prior Studies with Exogenous Interferon in Man 251

III. Present Studies of Varicella Zoster Infections Treated with Interferon 257
IV. Perspectives for Exogenous Interferon 262
References 264
Discussion 265

Chapter 15 **Interferon Cell Receptor Interactions**

C. Chany, H. Ankel, and M.-F. Bourgeade

I. Cell Species Specificity of Interferon 269
II. Induction of Antiviral State by Interferon 272
III. Conclusion 275
References 277

Chapter 16 **Inhibition of Enveloped Viruses in a Model of Virus-Induced Hyperlipemia**

Sidney E. Grossberg, Frank E. Frerman, and Ernest E. Lengle

I. Introduction 279
II. The Phenomenon of Viral Lipemia 280
III. Specificity of Lipemia 280
IV. Mechanism of Lipemia 282
V. Antiviral Effects of Antilipemic Agents 288
VI. Implications and Discussion 291
VII. Summary 291
References 292
Discussion 293

Chapter 17 **Photodynamic Inactivation of Herpesvirus**

Joseph L. Melnick and Craig Wallis

I. Introduction 297
II. Photosensitization of Virus with Dye 297
III. Dye-Light Therapy of Herpesvirus Infections 302
IV. References 312
Discussion 312

Chapter 18 **Studies on Adenine Arabinoside— A Model for Antiviral Chemotherapeutics**

Richard J. Whitley, Lawrence T. Ch'ien, Robert A. Buchanan, and Charles A. Alford, Jr.

I. Introduction 315
II. Early Investigations of Ara-A and Analogs 316
III. Treatment of Herpetic Infections with Ara-A 323
IV. Treatment of Cytomegalovirus with Ara-A 328
V. Therapeutic Potential of Ara-A 330
References 332
Discussion 332

Chapter 19 **Epilogue**

Roy Stern

Text 337

SUBJECT INDEX 339

PROGRAM COMMITTEE

JORDI CASALS
ROBERT M. CHANOCK
FRIEDRICH DEINHARDT
EDWIN H. LENNETTE
SALVADOR LURIA
KARL MARAMOROSCH
MORRIS POLLARD
FRED RAPP
IRENE STERN
E. L. TATUM
THOMAS H. WELLER

Sponsored by the Gustav Stern Foundation

List of Invited Participants

DR. S. A. AARONSON
National Institutes of Health
Bethesda, Maryland

DR. CHARLES A. ALFORD, JR.
University of Alabama
Birmingham, Alabama

SIR CHRISTOPHER H. ANDREWES
Salisbury, England

MR. STEVEN ANTOPOL
New York, New York

DR. HOWARD L. BACHRACH
Plum Island Animal Disease Station
Greenport, L.I., New York

DR. DAVID BALTIMORE
Massachusetts Institute of Technology
Cambridge, Massachusetts

DR. FREDERIK B. BANG
The Johns Hopkins University
Baltimore, Maryland

DR. LUIZ H. BARBOSA
National Institutes of Health
Bethesda, Maryland

DR. SAMUEL BARON
National Institutes of Health
Bethesda, Maryland

DR. GEORGES BARSKI
Institut Gustave Roussy
Villejuif, France

DR. STEPHEN G. BAUM
Albert Einstein College of Medicine
Bronx, New York

DR. JOHN BEALE
Wellcome Research Laboratories
Beckenham, Kent, England

DR. JOSEPH W. BEARD
Duke University
Durham, North Carolina

Mrs. Dorothy Beard
Duke University
Durham, North Carolina

Dr. Pravin N. Bhatt
Yale University School of Medicine
New Haven, Connecticut

Dr. David H. L. Bishop
Rutgers Medical School
Piscataway, New Jersey

Dr. Francis L. Black
Yale University School of Medicine
New Haven, Connecticut

Dr. Paul H. Black
Massachusetts General Hospital
Boston, Massachusetts

Dr. Neil R. Blacklow
Boston University
Boston, Massachusetts

Dr. Barry Bloom
Albert Einstein College of Medicine
Bronx, New York

Dr. Baruch S. Blumberg
The Institute for Cancer Research
Philadelphia, Pennsylvania

Dr. Howard E. Bond
Electronucleonics Laboratories, Inc.
Bethesda, Maryland

Dr. Ellen Borenfreund
Sloan-Kettering Institute
New York, New York

Dr. F. Brown
The Animal Virus Research Institute
Perbright, Woking, Surrey, England

Dr. Philip A. Brunell
New York University Medical Center
New York, New York

Dr. Gary R. Burleson
University of Notre Dame
Notre Dame, Indiana

Dr. B. R. Burmester
U.S.D.A. Regional Poultry Research Laboratory
East Lansing, Michigan

Dr. T. Burnstein
Purdue University
Lafayette, Indiana

Dr. Robert H. Bussell
The University of Kansas
Lawrence, Kansas

Dr. Victor J. Cabasso
Cutter Laboratories
Berkeley, California

Dr. J. J. Callis
Plum Island Animal Disease Station
Greenport, L.I., New York

Dr. Richard I. Carp
Institute for Basic Research in Mental Retardation
Staten Island, New York

Dr. David H. Carver
The Johns Hopkins University
Baltimore, Maryland

Dr. Jordi Casals
Yale University School of Medicine
New Haven, Connecticut

Dr. R. M. Chanock
National Institutes of Health
Bethesda, Maryland

Dr. Charles Chany
Hospital Saint-Vincent-de-Paul
Paris, France

Dr. Michael A. Chirigos
National Institutes of Health
Bethesda, Maryland

Dr. Purnell W. Choppin
The Rockefeller University
New York, New York

Dr. Gerald A. Cole
The Johns Hopkins University
Baltimore, Maryland

Dr. Richard W. Compans
The Rockefeller University
New York, New York

DR. RICHARD L. CROWELL
The Hahnemann Medical College
Philadelphia, Pennsylvania

DR. EDWARD CURNEN
Columbia University
New York, New York

DR. MARY G. MCCREA CURNEN
Columbia University
New York, New York

DR. SAMUEL DALES
Public Health Research Institute
New York, New York

DR. ALBERT J. DALTON
National Institutes of Health
Bethesda, Maryland

DR. FRED M. DAVENPORT
University of Michigan
Ann Arbor, Michigan

DR. DORLAND J. DAVIS
National Institutes of Health
Bethesda, Maryland

DR. VITTORIO DEFENDI
The Wistar Institute
Philadelphia, Pennsylvania

DR. FRIEDRICH DEINHARDT
Rush-Presbyterian-St. Luke's Hospital
Chicago, Illinois

DR. JEAN DEINHARDT
Rush-Presbyterian-St. Luke's Hospital
Chicago, Illinois

DR. IRVING P. DE LAPPE
National Institutes of Health
Bethesda, Maryland

DR. DONALD C. DELONG
Eli Lilly and Company
Indianapolis, Indiana

DR. PETER B. DENT
McMaster University
Hamilton, Ontario, Canada

DR. WILBUR G. DOWNS
Yale University
New Haven, Connecticut

DR. J. A. DUDGEON
The Hospital for Sick Children
London, England

DR. JAMES T. DUFF
National Institutes of Health
Bethesda, Maryland

DR. RONALD DUFF
Pennsylvania State University
Hershey, Pennsylvania

DR. HARRY EAGLE
Albert Einstein College of Medicine
Bronx, New York

DR. BERNICE E. EDDY
Food and Drug Administration
Department of Health, Education & Welfare
Rockville, Maryland

LTC. GERALD A. EDDY
U. S. Army Medical Research Institute
Fort Detrick, Frederick, Maryland

DR. JOHN F. ENDERS
The Children's Hospital Medical Center
Boston, Massachusetts

DR. M. ESSEX
Harvard University
Boston, Massachusetts

DR. REBECA FALCOFF
New York University
New York, New York

DR. LAWRENCE A. FELDMAN
New Jersey Medical School
Newark, New Jersey

DR. WILLIAM FELLER
Georgetown University
Washington, D.C.

DR. BERNARD N. FIELDS
Albert Einstein College of Medicine
Bronx, New York

DR. AUDREY FJELDE
Roswell Park Memorial Institute
Buffalo, New York

Dr. Jorgen E. Fogh
Sloan-Kettering Institute
Rye, New York

Dr. Caroline Fong
Veterans Administration Hospital
West Haven, Connecticut

Dr. Richard M. Franklin
Public Health Research Institute
New York, New York

Dr. Charlotte Friend
Mount Sinai School of Medicine
New York, New York

Dr. David A. Fuccillo
National Institutes of Health
Bethesda, Maryland

Dr. H. Hugh Fudenberg
University of California Medical Center
San Francisco, California

Dr. George J. Galasso
National Institutes of Health
Bethesda, Maryland

Dr. Robert C. Gallo
National Institutes of Health
Bethesda, Maryland

Dr. John Gavin
Miles Laboratories, Inc.
Elkhart, Indiana

Dr. Anne A. Gershon
New York University
New York, New York

Dr. C. Joseph Gibbs, Jr.
National Institutes of Health
Bethesda, Maryland

Dr. Donald H. Gilden
The Wistar Institute
Philadelphia, Pennsylvania

Dr. Harold S. Ginsberg
Columbia University
New York, New York

Dr. Anthony J. Girardi
The Wistar Institute
Philadelphia, Pennsylvania

Dr. Ronald Glaser
Hershey Medical Center
Hershey, Pennsylvania

Dr. David J. Gocke
Rutgers Medical School
Piscataway, New Jersey

Dr. Eli Gold
Case-Western Reserve University
Cleveland, Ohio

Dr. Allan R. Goldberg
The Rockefeller University
New York, New York

Dr. Irving Gordon
University of Southern California
Los Angeles, California

Dr. A. F. Graham
McGill University
Montreal, Quebec, Canada

Dr. Allan Granoff
St. Jude Children's Research Hospital
Memphis, Tennessee

Dr. John H. Graves
Plum Island Animal Disease Station
Greenport, L. I., New York

Dr. Maurice Green
St. Louis University
St. Louis, Missouri

Dr. Sheldon Greer
University of Miami
Miami, Florida

Dr. Ludwik Gross
Veterans Administration Hospital
Bronx, New York

Dr. Sidney E. Grossberg
The Medical College of Wisconsin
Milwaukee, Wisconsin

Dr. Jack Gruber
National Institutes of Health
Bethesda, Maryland

Dr. William T. Hall
Electronucleonics Laboratories, Inc.
Bethesda, Maryland

Dr. Vincent V. Hamparian
Children's Hospital
Columbus, Ohio

Dr. Hidesaburo Hanafusa
The Rockefeller University
New York, New York

Dr. Teruko Hanafusa
The Rockefeller University
New York, New York

Dr. William D. Hardy, Jr.
Sloan-Kettering Institute
New York, New York

Dr. Janet W. Hartley
National Institutes of Health
Bethesda, Maryland

Dr. Etienne de Harven
Sloan-Kettering Institute
New York, New York

Dr. Edward A. Havell
New York University
New York, New York

Dr. Leonard Hayflick
Stanford University
Stanford, California

Dr. James B. Henson
Washington State University
Pullman, Washington

Mr. James O. Herlands
New York, New York

Dr. Paul K. Hildebrandt
Walter Reed Army Institute of Research
Washington, D.C.

Dr. Maurice R. Hilleman
Merck Institute for Therapeutic Research
West Point, Pennsylvania

Dr. Monto Ho
University of Pittsburgh
Pittsburgh, Pennsylvania

Dr. Horace L. Hodes
Mount Sinai School of Medicine
New York, New York

Dr. Ariel C. Hollinshead
The George Washington University
Washington, D.C.

Dr. Dorothy M. Horstmann
Yale University School of Medicine
New Haven, Connecticut

Dr. John Hotchin
New York State Department of Health
Albany, New York

Dr. G. D. Hsiung
Veterans Administration Hospital
West Haven, Connecticut

Dr. A. S. Huang
Harvard Medical School
Boston, Massachusetts

Dr. Eric Humphries
University of Wisconsin
Madison, Wisconsin

Lt. Col. David L. Huxsoll
Walter Reed Army Institute of Research
Washington, D.C.

Dr. Harry L. Ioachim
Lenox Hill Hospital
New York, New York

Dr. George Gee Jackson
University of Illinois
Chicago, Illinois

Dr. Keith E. Jensen
Pfizer, Inc.
Groton, Connecticut

Dr. Richard Johannes
National Institutes of Health
Bethesda, Maryland

Dr. Kenneth P. Johnson
Cleveland Metropolitan General Hospital
Cleveland, Ohio

Dr. W. K. Joklik
Duke University
Durham, North Carolina

Dr. S. S. Kalter
Southwest Foundation
San Antonio, Texas

Dr. Albert Z. Kapikian
National Institutes of Health
Bethesda, Maryland

Dr. Albert S. Kaplan
Vanderbilt University
Nashville, Tennessee

Dr. David T. Karzon
Vanderbilt University
Nashville, Tennessee

Dr. Joseph R. Kates
State University of New York
Stony Brook, New York

Dr. Michael Katz
Columbia University
New York, New York

Dr. Samuel L. Katz
Duke University
Durham, North Carolina

Dr. Herbert Kaufman
University of Florida
Gainesville, Florida

Dr. Sidney Kibrick
Boston University
Boston, Massachusetts

Dr. E. D. Kilbourne
Mount Sinai School of Medicine
New York, New York

Dr. K. S. Kim
Institute for Basic Research in Mental Retardation
Staten Island, New York

Dr. Morton Klein
Temple University
Philadelphia, Pennsylvania

Dr. Gebhard Koch
Roche Institute of Molecular Biology
Nutley, New Jersey

Dr. Hilary Koprowski
The Wistar Institute
Philadelphia, Pennsylvania

Dr. Bruce D. Korant
E. I. Du Pont de Nemours & Co.
Wilmington, Delaware

Dr. Saul Krugman
New York University
New York, New York

Dr. Jerome Kurent
National Institutes of Health
Bethesda, Maryland

Dr. David J. Lang
Duke University
Durham, North Carolina

Dr. Lloyd W. Law
National Institutes of Health
Bethesda, Maryland

Dr. H. Sherwood Lawrence
New York University
New York, New York

Dr. Robert W. Leader
University of Connecticut
Storrs, Connecticut

Dr. F. Lehmann-Grube
Universitat Hamburg
Hamburg, Germany

Dr. Edwin H. Lennette
California State Department of Public Health
Berkeley, California

Dr. Myron J. Levin
The Children's Hospital Medical Center
Boston, Massachusetts

Dr. Arthur S. Levine
National Institutes of Health
Bethesda, Maryland

Dr. A. S. Levine
Indiana State University
Terre Haute, Indiana

Dr. Hilton B. Levy
National Institutes of Health
Bethesda, Maryland

Dr. Jay A. Levy
University of California
San Francisco, California

Dr. Royce Z. Lockart, Jr.
E. I. Du Pont de Nemours & Co.
Wilmington, Delaware

Dr. Donald Lodmell
Rocky Mountain Laboratory
Hamilton, Montana

Dr. William T. London
National Institutes of Health
Bethesda, Maryland

Dr. Elizabeth H. Macintyre
National Jewish Hospital
Denver, Colorado

Dr. Wayne E. Magee
Indiana State University
Terre Haute, Indiana

Dr. B. W. J. Mahy
Cambridge University
Cambridge, England

Dr. Robert A. Manaker
National Institutes of Health
Bethesda, Maryland

Dr. Karl Maramorosch
Boyce Thompson Institute
Yonkers, New York

Dr. Philip I. Marcus
The University of Connecticut
Storrs, Connecticut

Dr. Giampiero di Mayorca
University of Illinois
Chicago, Illinois

Dr. Sami A. Mayyasi
Chas. Pfizer & Co., Inc.
Maywood, New Jersey

Dr. Robert W. McCollum
Yale University
New Haven, Connecticut

Mrs. Margaret McElwain
National Institutes of Health
Bethesda, Maryland

Dr. D. M. McLean
University of British Columbia
Vancouver, British Columbia, Canada

Dr. Robert C. Mellors
The Hospital for Special Surgery
New York, New York

Dr. Joseph L. Melnick
Baylor Medical College
Houston, Texas

Dr. Thomas C. Merigan
Stanford University
Stanford, California

Dr. G. S. Merz
Institute for Basic Research in Mental Retardation
Staten Island, New York

Dr. Frank J. Michalski
Veterans Administration Hospital
West Haven, Connecticut

Dr. I. George Miller, Jr.
Yale University
New Haven, Connecticut

Dr. Edwin A. Mirand
Roswell Park Memorial Institute
Buffalo, New York

Dr. John B. Moloney
National Institutes of Health
Bethesda, Maryland

Dr. Arnold S. Monto
University of Michigan
Ann Arbor, Michigan

Dr. Councilman Morgan
Columbia University
New York, New York

Dr. Herbert R. Morgan
The University of Rochester
Rochester, New York

Dr. Bernard Moss
National Institutes of Health
Bethesda, Maryland

Dr. Daniel I. Mullaly
National Institutes of Health
Bethesda, Maryland

Dr. Andre J. Nahmias
Emory University
Atlanta, Georgia

Dr. Neal Nathanson
The Johns Hopkins University
Baltimore, Maryland

Dr. Franklin A. Neva
National Institutes of Health
Bethesda, Maryland

Dr. Robert L. Northrup
University of Illinois
Chicago, Illinois

Dr. Abner L. Notkins
National Institutes of Health
Bethesda, Maryland

Dr. Jan van der Noordaa
Laboratorium voor de Gesondheidleer
Amsterdam, The Netherlands

Dr. Timothy E. O'Connor
National Institutes of Health
Bethesda, Maryland

Dr. Lloyd Old
Sloan-Kettering Institute
New York, New York

Dr. Michael N. Oxman
The Children's Hospital Medical Center
Boston, Massachusetts

Dr. Joseph F. Pagano
Smith Kline Diagnostics
Philadelphia, Pennsylvania

Dr. Joseph S. Pagano
University of North Carolina
Chapel Hill, North Carolina

Dr. Amos E. Palmer
National Institutes of Health
Bethesda, Maryland

Dr. John C. Parker
Microbiological Associates, Inc.
Washington, D.C.

Dr. Robert H. Parrott
Children's Hospital
Washington, D.C.

Dr. Cephas T. Patch
National Institutes of Health
Bethesda, Maryland

Dr. Francis E. Payne
University of Michigan
Ann Arbor, Michigan

Dr. Charles J. Pfau
Rensselaer Polytechnic Institute
Troy, New York

Dr. Elmer R. Pfefferkorn
Dartmouth Medical School
Hanover, New Hampshire

Dr. Frank F. Pindak
Brooks Air Force Base
San Antonio, Texas

Dr. Terry Platt
Stanford University
Stanford, California

Dr. Otto J. Plescia
Rutgers University
New Brunswick, New Jersey

Dr. Stanley A. Plotkin
The Wistar Institute
Philadelphia, Pennsylvania

Dr. Morris Pollard
University of Notre Dame
Notre Dame, Indiana

Dr. David D. Porter
U.C.L.A. School of Medicine
Los Angeles, California

Dr. Louis Potash
Flow Laboratories, Inc.
Rockville, Maryland

Dr. Alfred M. Prince
The New York Blood Center
New York, New York

Dr. Robert H. Purcell
National Institutes of Health
Bethesda, Maryland

Dr. H. Gordon Purchase
U.S. Department of Agriculture
Regional Poultry Research Center
East Lansing, Michigan

Dr. Fred Rapp
Pennsylvania State University
Hershey, Pennsylvania

DR. FRANCISCO SALIDO RENGELL
Instituto Nacional de Virologia
Mexico, D.C.

DR. HAROLD E. RENIS
The Upjohn Co.
Kalamazoo, Michigan

DR. MICHEL REVEL
The Weizmann Institute
Rehovot, Israel

DR. LIONEL E. RHULAND
The Upjohn Co.
Kalamazoo, Michigan

DR. FREDERICK C. ROBBINS
Case-Western Reserve University
Cleveland, Ohio

DR. BERNARD ROIZMAN
The University of Chicago
Chicago, Illinois

DR. EUGENE I. ROSANOFF
Wyeth Laboratories, Inc.
Philadelphia, Pennsylvania

DR. HERBERT S. ROSENKRANZ
Columbia University
New York, New York

DR. BENJAMIN A. RUBIN
Wyeth Laboratories, Inc.
Philadelphia, Pennsylvania

DR. DANIEL J. RUBIN
National Institutes of Health
Bethesda, Maryland

DR. ROBERT RUSTIGIAN
Veterans Administration Hospital
Brockton, Massachusetts

DR. ALBERT B. SABIN
Washington, D.C.

DR. F. KINGSLEY SANDERS
Sloan-Kettering Institute
New York, New York

DR. NURUL H. SARKAR
Institute for Medical Research
Camden, New Jersey

DR. PADMAN S. SARMA
National Institutes of Health
Bethesda, Maryland

DR. MORRIS SHAEFFER
National Institutes of Health
Bethesda, Maryland

DR. ANDREAS SCHEID
The Rockefeller University
New York, New York

DR. KLAUS SCHELL
Dow Chemical Co.
Zionsville, Indiana

DR. WILLIAM SCHER
Mount Sinai School of Medicine
New York, New York

DR. WILLIAM F. SCHERER
Cornell University
New York, New York

DR. R. WALTER SCHLESINGER
Rutgers Medical School
Piscataway, New Jersey

DR. MICHAEL SELA
The Weizmann Institute
Rehovot, Israel

DR. DEXTER S. Y. SETO
The Johns Hopkins University
Baltimore, Maryland

DR. JOHN SEVER
National Institutes of Health
Bethesda, Maryland

DR. NEHAMA SHARON
Evanston Hospital
Evanston, Illinois

DR. ALEXIS SHELOKOV
The University of Texas
San Antonio, Texas

DR. ROBERT E. SHOPE
Yale University
New Haven, Connecticut

DR. LOUIS R. SIBAL
National Institutes of Health
Bethesda, Maryland

DR. M. MICHAEL SIGEL
University of Miami
Miami, Florida

DR. GERALD H. SILVER
Schering Corporation
Bloomfield, New Jersey

DR. EDWARD H. SIMON
Purdue University
Lafayette, Indiana

DR. LIONEL N. SIMON
ICN Research Institute
Irvine, California

DR. RICHARD L. SOEHNER
Miles Laboratories, Inc.
Elkhart, Indiana

DR. RUY SOEIRO
Albert Einstein College of Medicine
Bronx, New York

DR. CLARENCE A. SOOTER
National Institutes of Health
Bethesda, Maryland

DR. MARCO SORIA
Harvard Medical School
Boston, Massachusetts

DR. CHESTER M. SOUTHAM
Jefferson Medical College
Philadelphia, Pennsylvania

DR. SOL SPIEGELMAN
Columbia University
New York, New York

DR. EDITH D. STANLEY
The Abraham Lincoln School of Medicine
Chicago, Illinois

DR. ALEX J. STEIGMAN
Mount Sinai School of Medicine
New York, New York

MRS. GUSTAV STERN
New York, New York

DR. ROY STERN
New York, New York

MR. STEVEN STERN
New York, New York

DR. JACK STEVENS
U.C.L.A. School of Medicine
Los Angeles, California

DR. SARAH STEWART
Georgetown University
Washington, D.C.

DR. IGOR TAMM
The Rockefeller University
New York, New York

DR. GIULIO TARRO
Istituto di Clinica Medica Generale
Napoli, Itali

DR. E. L. TATUM
The Rockefeller University
New York, New York

DR. RAYMOND W. TENNANT
Oak Ridge National Laboratory
Oak Ridge, Tennessee

DR. S. S. TEVETHIA
Tufts Medical School
Boston, Massachusetts

DR. HALLDOR THORMAR
Institute for Basic Research in Mental Retardation
Staten Island, New York

DR. GEORGE J. TODARO
National Institutes of Health
Bethesda, Maryland

DR. TADASU TOKUMARU
Columbia University
New York, New York

DR. ROBERT L. TRUITT
University of Notre Dame
Notre Dame, Indiana

DR. DAVID A. J. TYRRELL
Clinical Research Centre
Harrow, Middlesex, England

DR. G. E. UNDERWOOD
The Upjohn Company
Kalamazoo, Michigan

DR. ARTHUR C. UPTON
State University of New York
Stony Brook, New York

DR. JAN VILCEK
New York University
New York, New York

DR. MOGENS VOLKERT
University of Copenhagen
Copenhagen, Denmark

DR. DUARD L. WALKER
The University of Wisconsin
Madison, Wisconsin

DR. JOSEPH L. WANER
Harvard University
Boston, Massachusetts

DR. JOEL WARREN
Nova University
Fort Lauderdale, Florida

DR. ROBIN A. WEISS
Imperial Cancer Research Fund Laboratories
London, England

DR. ARTHUR WEISSBACH
Roche Institute of Molecular Biology
Nutley, New Jersey

DR. THOMAS H. WELLER
Harvard University
Boston, Massachusetts

DR. H. WESTPHAL
Cold Spring Harbor Laboratory
Cold Spring Harbor, L.I., New York

DR. E. FREDERICK WHEELOCK
Jefferson Medical College
Philadelphia, Pennsylvania

DR. RICHARD J. WHITLEY
University of Alabama
Birmingham, Alabama

DR. ROGER E. WILSNACK
Huntingdon Research Center
Baltimore, Maryland

DR. JULIUS S. YOUNGNER
University of Pittsburgh
Pittsburgh, Pennsylvania

DR. BARBARA A. ZAJAC
The Medical College of Pennsylvania
Philadelphia, Pennsylvania

DR. IHOR ZAJAC
Smith Kline & French Laboratories
Philadelphia, Pennsylvania

DR. CHRIS ZARAFONETIS
University of Michigan
Ann Arbor, Michigan

Foreword

In Honor of C.H.A. or Sir Christopher Howard Andrewes

D. A. J. TYRRELL

It has given me great pleasure to prepare at your request a talk about the internationally famous scientist whom we are gathered here to honor. I knew of him by repute almost as soon as I knew anything about virology, and I learned through my mentors, Professor Wilson Smith and Professor Stuart-Harris, of his key part in the discovery of the influenza A viruses and their behavior, both inside and outside the laboratory. I also heard of his ability to be a devastating critic, particularly of the pompous and the obscure, and that he was capable of sustaining an almost polemical opposition to those with whom he disagreed; with this reputation I can remember that as a young man I was of two minds as to whether or not I wanted to meet him. However, when I did so, I was, like so many others, delighted by his charm and enthusiasm and I have enjoyed our meetings ever since.

But I must explain at this point that I am in a very difficult position; I wish to fulfill my commission by giving you a complimentary appreciation of Sir Christopher; on the other hand, I have to return to England and run a Unit to which he is an honorary advisor; there, if I have unfortunately aroused his ire by being too "sugary," I shall be within easy reach of that sustained opposition and devastating criticism to which I have already referred. However I can but do my best.

Like many interesting and stimulating men Christopher Andrewes grew up in an interesting and stimulating family. His father was an eminent medical man, a pathologist who helped to document the effects of the 1918–1919 pandemic of influenza in Britain. His sister was stimulating company, and repartee, jokes, home renderings of Gilbert and Sullivan, and visits to the country gave him much pleasure. He was educated in the English public school system, and began to specialize in the classical languages—his sister became a classicist—but he says he was "fortunate to escape the worst effects of a classical education." He achieved this first of all by having a prolonged illness, during which he watched the robins in the garden winning and preserving their territories which he then mapped out. He was seized, too, with an enthusiasm for butterflies and built up a fine collection. Then, toward the end of his school career he wrote what must have been a remarkable essay on the feet of certain British water birds, how they are constructed and adapted to their function in the life of the bird. He therefore changed over to the science Sixth with an interest in biology, and in those days the best way to pursue this at the professional level was by going into medicine; so he found himself a medical student at St. Bartholomew's Hospital (Bart's) in Central London, where his father had been a medical student and where two of his sons were subsequently to follow him. I know he talks about the disadvantages of his classical upbringing, but to be fair I think it has stood him in good stead. First, it helped his wide-ranging and retentive mind to grow up orderly and systematically, when it might so easily have had as much discipline as the schoolboy butterfly hunter, following and collecting indiscriminately anything which attracted his attention; in addition, he has been able in his two scientific careers to cope effortlessly with elaborate taxonomic systems, the nomenclature of which is rooted in classical languages.

After qualifying, he became more and more interested in the laboratory rather than the wards and clinics and he began to experiment on infectious diseases. Issues of *Bart's Medical Journal* at that time contained descriptions of the way he successfully repeated earlier work in which common colds were transmitted to volunteers—Bart's medical students—by the intranasal inoculation of bacteria-free material. He did not, however, confirm the claim of Dochez to have propagated the virus in chick embryo tissue. Other contributions included the title, "Is sex infectious?"—I can't remember the answer.

He then joined the staff of the Medical Research Council and worked in the old Hampstead Laboratories set up in a converted sanatorium. Great men were there; for example, Dale who subsequently won the Nobel Prize; Burnet was there too for a while as an unknown young man and C.H.A. also spent valuable periods at the Rockefeller Institute in New York. There he was influenced by Tillet and Rivers and made friends with Peyton Rous. He later met Richard Shope, and he and Shope kept in touch as frequently as the postal system allowed. Shope was working on the influenza virus of pigs, while Andrewes back in Britain had succeeded with Wilson Smith, and Laidlaw, the director of the laboratories, in propagating the human influenza virus, and was working on this first in ferrets and then in laboratory mice. However, he was interested in viruses not only as pathogens but also as interesting organisms. For example, he studied quantitatively the neutralization of animal and bacterial viruses with immune sera and, with the biophysicist Elford, enunciated the "percentage law." With Elford, he also made the first systematic study of ultrafiltration of viruses and showed that they must be of different sizes. He also studied a latent virus of rabbits—virus III—and established that viruses would grow freely in tissue cultures made from immune animals. He was working also with viruses that cause tumors in experimental animals—the Rous sarcoma and Shope papilloma—and was chosen to give the Cantor lecture on this subject in 1933. He showed with Ahlström that tar treatment enhanced the response of skin to a tumor virus.

During this time his other interests continued. He had married another scientist who was teaching physics. They had three sons, with whom he spent as much time as possible, introducing them to the natural life of the countryside and trying to tire them out by marching across hills and commons, singing light-hearted verses he had invented to the main themes of great orchestral works, Beethoven symphonies, for example.

All this was interrupted by the war, during which he worked in a laboratory of the Emergency Public Health Laboratory Service near Dartmoor; but he had to return to London for more important work, first on Air Hygiene, then on typhus vaccines and chemoprophylaxis, during which time he had a nasty attack of the disease.

When the war was over he conceived the idea of starting up research on common colds in volunteers who were rigorously isolated and observed. He has described how he came to set up specialized volunteer research facilities in the buildings vacated by the American Red Cross Harvard Hospital—and the Unit has operated continuously as the Common Cold Research Unit since 1946. The Unit was set up as a short-term project at first, but once the methods of clinical observation were worked out, it became clear that although the common cold virus could to some extent be characterized and its spread from person to person could be studied, it was not possible to grow it in the way various authors had claimed. He regarded the growth of the virus in the laboratory as the key to the whole field but the results came slowly and were depressing, as negative experiments usually are, but they laid a valuable foundation for future work, particularly for H.G. Pereira and subsequently for myself, who worked at the Unit in succession until 1960. Just before the retirement of Sir Christopher, as he was by then, the first method for growing rhinoviruses with reproducibility was developed.

A key to this success story was the enthusiasm and unshakeable optimism of C.H.A., who pleaded with successive administrative bodies to believe that success *would* come, that this *was* the right approach and that the right tricks would finally be learned. He was clearly a brilliant and successful advocate—but there was probably more than one occasion on which the Unit was almost closed, and had this been so, he would have been classified instead as stubborn and misguided—but fortunately that did not happen.

In parallel with this he was engaged in a variety of laboratory studies, one of which in retrospect was very significant. Working with a visiting American, Dr. Dorothy Horstmann, he showed that viruses could be classified quite unequivocally into those that were inactivated by ether and those that were not. Like most of his bench work, the experiments were uncomplicated and the results unambiguous and suggested that viruses might differ profoundly, not only in size, but also in the fundamental methods of their construction.

His interests were always wide, and were further widened by acting as a consultant to the Council laboratories in West Africa, and also by his position on the governing body of the Animal Virus Research Institute at Pirbright, south of London, but the widest of all was probably his pioneering of the International Committee for the Nomenclature of Viruses. There was ample opportunity to argue with those who wanted to classify viruses by trivial characters such as tissue tropism, or those who wanted to be overly rigid and comprehensive— but I doubt whether virus taxonomy would be in its present evolving and creative state without his influence. In the last years before retirement, he personally compiled a comprehensive index of papers, all of which he had himself critically evaluated and from which he extracted the best data on the properties and taxonomic position of the animal viruses. He then sat down and wrote the only one-man encyclopedic taxonomy of the "Viruses of Vertebrates," a book which has become indispensible to most virologists, and one which he has, with Pereira, twice rewritten during his official retirement. He tells me that he is not going to undertake the fourth edition as it is not right that a scientific treatise should be produced by a man aged over 80! In his retirement he has written a series of books on virology for the general public that are models of clear, interesting writing and which have enhanced his reputation as one with an uncommon skill in the use of English.

Another imaginative "first" was his idea of establishing a center in his laboratory to collect and study the influenza viruses from epidemics which occurred in all quarters of the globe. Under the aegis of the World Health Organization, a world-wide network of influenza centers was set up. It was the antigenic analysis of viruses gathered in this way, that first showed for certain, that new serotypes of virus appear in one country and then spread across the world causing epidemics in the northern or southern hemispheres according to the season. The network also served as a model for other WHO reference centers. He had a sustained argument with the Ann Arbor group who believed in the cycling of influenza virus serotypes and building up a basic immunity by vaccination with a group of influenza A viruses. He was quite sure that it was the new antigen that determined the ability of viruses to spread and which should therefore be included in vaccines. Francis may have had more right on his side than was thought, but certainly, in practice, the best vaccines have the latest strains.

I think, however, you should also know that this Dr. Jekyll is often transformed into a Mr. Hyde who scrambles in the New Forest, round noisome bogs, and sometimes into them, catching small creatures and killing them off. Seriously, in addition to the virologist we all know, there has always been for other entomologists the Dr. Andrewes "who does something with viruses." While in New York, he made a collection of American bees and wasps which is now part of the collection of the British Natural History Museum. He has written a book on "Bees and Wasps of Europe and North America," and is an authority on diptera; he has recognized new British species of these and in recent years has made entomological surveys of reserves for the Nature Conservancy, living the while with his wife in a hut on the side of a remote mountain.

You will realize by now that you have set me an impossible task—I have been unable to compress into a few thousand words the character, the achievements, and the career of such a varied and outstanding man. But there has been a sort of theme, a pattern of all these aspects. In a period during which virologists have become more and more specialized, he has refused to limit his interests, and because of his retentive memory and grasp of detail he has been able to understand the fundamentals of the epidemiology and the life of organisms in relation to all the complexities of their environment—this viewpoint has been set out in his book "The Natural History of Viruses"; he has also been at home with studies on the fundamental nature of viruses and the mysteries of their interactions with host cells. He has probably had at least as much influence on science by the impact he has had on all those who worked with him. (Alick Isaacs was perhaps the most brilliant of these and made important contributions to the epidemiology of influenza and not only discovered interferon, but set out many of the key ideas by which the subject has subsequently been developed.) How did C.H.A. have this effect? He could have overpowered people by his intellect and has made me at times realize that I was stupid or ill-informed. Instead, he has communicated his great sense of enthusiasm, the idea that everything he is interested in is important and exciting and that the whole of life can be fun—one of his famous lectures was entitled "Adventures with Viruses." Since this has usually been linked to a careful and critical review of the facts and a judicious caution in deciding what is actually known, his colleagues have absorbed most of the secrets of the conduct of successful research. He still has a certain boyishness, which I greatly appreciate; under the white hair, there is still the schoolboy fascinated with every aspect of nature, and full of ideas on how to classify its creatures and understand how it all works. Sir Christopher, you may not have the secret of eternal youth, but you seem to have got as near to it as mortal man is likely to do. You have inspired, informed, aggravated, and amused us all; long may you continue to do so—we are all in your debt.

Preface

The prevention and treatment of virus infections represent a significant frontier in biological science. Much has been accomplished in the area of immunization against virus infections. However, there are limitations to the control of some viral diseases because (a) the virus is antigenically labile, or (b) the host is tolerant of the virus because of congenital passage, or (c) the virus cannot be propagated in the laboratory, or (d) because of multiple types it is not economically feasible to produce a vaccine, or (e) the virus is sequestered in an "immunologically privileged" site. This symposium reports on new improved vaccines, on enhancement of host responses to infection, and on procedures for direct therapeutic assaults upon viruses by chemical agents.

The symposium honors Sir Christopher Andrewes who is responsible for the great advancement in the understanding and control of influenza, one of the important life-limiting plagues of contemporary society. This volume recognizes his very significant contributions to the field of virology in the roles of investigator, author, and administrator.

We express our great thanks for the continued support to virology by Mrs. Irene Stern, and the Gustav Stern Foundation, which she represents. Virology is indeed the beneficiary of her interest for which we express our grateful thanks.

MORRIS POLLARD

Notre Dame, Indiana
February, 1974

Welcome

The title of this year's Gustav Stern Symposium is "Antiviral Mechanisms," which indicates a somewhat different philosophy adopted for "Perspectives in Virology," Volume IX. This year we are attempting to concentrate on the clinical application of research in virology, and, therefore, many of the papers concern themselves with the different ways in which the human body attempts to defend itself against viral attack. Others deal with artificial means of interfering with the life cycle of viruses. All these works seek the ultimate goal of all research in virology, which is, a means to prevent or to neutralize viral infections. As this year's symposium is philosophically unique, it is also unique in that both the Guest of Honor and the recipient of the Gustav Stern Award are from England. We are proud to honor Sir Christopher Andrewes, a long-time participant of these symposia, whose multiple past accomplishments include discovery of the influenza virus and wide-ranging research in respiratory diseases. It is also a pleasure to honor Dr. Robin A. Weiss, of the Imperial Cancer Research Fund Laboratories. Their accomplishments have matched the philosophy of this meeting.

It is, therefore, with the same hope and faith expressed so many times before—by my late husband and by myself—a hope, that this year's symposium will add another page to the volume of human knowledge; and a faith, that those here tonight have the capacity and willingness to sacrifice themselves toward that goal, that I welcome you to the Ninth Gustav Stern Symposium on "Perspectives in Virology."

MRS. GUSTAV STERN

CHAPTER 1

Perspectives in Virology: Vaticinations

Edwin H. Lennette

In 1970, at the seventh symposium on "Perspectives in Virology," I had the pleasure of recounting my personal view of the history and development of virology over the preceding four decades. In 1972, at the eighth symposium, Frank Fenner gave a brief and concise summation of what he considered important developments and contributions to virology over the interval 1967–1972. During this same period, there appeared the thoughtful and provocative book by Gunther Stent (*12*) on "The Coming of the Golden Age: A View of the End of Progress," depicting the rise and fall of molecular genetics in particular and molecular biology in general and presenting the argument that progress in science (and in other creative activities) is a self-limiting process that culminates in an age of decline, paradoxically labeled The Golden Age, and one now upon us.* Perhaps so; I do not profess to know where we stand. Perhaps we have entered The Golden Age, but when we list, as an offset against the manifold problems that adversely affect the well-being of mankind, the impressive advances that are being made, it seems to me that biomedicine is not only alive and well but still highly imaginative and innovative and has yet to reach apogee.

* Konrad Lorenz seems of the same mind when he points to the imminent downfall of civilization. In an address given before the Klub für Bildungs- und Wissenschafts-journalisten in Vienna, March 20, 1973, he stated that "there can be little doubt that humanity today is in the process of rapid involution . . . our science is less affected, for the moment . . . than our morals and our art. . . . The mind of man has abolished creative selection, thereby dooming itself to total involution" [*Intellectual Digest,* 4, 62 (1974); translated and abridged by Tatyana A. Schmidt.]

I remember reading some time at the end of the 1940s a report from one of our philanthropic foundations stating that in the field of microbiology, the first half of the twentieth century belonged to the bacteriologist but that the second half would belong to the virologist. It seems reasonable to conclude that on the whole this prediction has come true and that over the remaining years of this century virology will continue as a dominant force in microbiology. Current knowledge and developments suggest that we will witness notable progress in the elucidation of the etiology of chronic and degenerative diseases and in the prevention and therapy of viral infections.

First off, it seems reasonable that in the future we will continue to uncover new viruses and new viral infections just as we have in the past. With the advent of cell culture, the suckling mouse, and the sentinel monkey as host systems, hundreds of new viruses have been isolated, and the addition of ancillary methods, such as immune electron microscopy and radioimmunoassay, should expand our ability to detect additional new viruses and eventually to propagate and to characterize them. A notable proportion of these hundreds of viruses I have mentioned has been isolated from arthropods, but only relatively few have been associated with pathogenicity for man or animals. The natural history of most of the arthropod-borne agents is unknown, but as man intrudes upon virgin lands and disrupts the existing ecological balance through settlement or agricultural pursuits, he will quite likely find himself confronted with new viruses and diseases or he may experience disease produced by viruses already well recognized but currently of unknown importance. New diseases have emerged from time to time in recent history, and we should expect nature to have similar surprises for us in the future.

In 1967, there was an outbreak of severe hemorrhagic fever among monkey handlers and laboratory workers in Marburg and Frankfurt, Germany, and in Belgrade, Yugoslavia. The disease, now known as Marburg disease (8), was recognized as acquired from direct contact with the tissues of African green monkeys, imported from Uganda. However, subsequent studies showed that the green monkey is not the reservoir of the virus, and it is unknown how the involved animals acquired the infection between their departure from Africa and their arrival at their European destinations. The basic natural cycle of the virus is still unknown.

In 1969, a new viral disease, now known as Lassa fever, was reported from northeastern Nigeria. In 1970, the disease was reported from Sierra Leone and in 1972 from Liberia. The causal agent belongs to the arenavirus group, which includes Bolivian hemorrhagic fever (Machupo virus); Argentinian hemorrhagic fever (Junin virus); and lymphocytic choriomeningitis virus, all of which are rodent-associated. The animal reservoir of the Lassa virus is thus presumed to be a rodent and an active search for the actual reservoir is currently under way in Nigeria.

Tanapox is a newly described disease in the Tana River basin (Kenya). The area is being brought under irrigation for agricultural development and the disease, which is probably transmitted mechanically by mosquitoes from a monkey reservoir, may assume some importance when the largely nomadic population is replaced by settlements and the land is cultivated.

Of special interest is acute hemorrhagic conjunctivitis because, with the exception of influenza, it is the only example in recent history of a newly emerging virus giving rise to a pandemic. This new disease broke out in Ghana in 1969, spread along the west coast of Africa in 1970 and the Mediterranean coast in 1971 and then over nearly all of Asia and the Indian subcontinent. The disease is caused by a virus that has the characteristics of an enterovirus, is antigenically distinct from other enteroviral immunotypes, and is present in the ocular conjunctiva rather than the intestine. The virus in the future may well have to be considered as an important cause of ocular disease and its role in acute neurological disorders assessed, since such disorders have been found to follow the conjunctival disease.

Recognition of the role of so-called slow viruses in disease of the central nervous and lymphoreticular systems marks an important step and one that has opened up a fascinating area of investigation for the future. Such persistent nonlytic infections may eventually induce cellular changes that may impair organ function, may enhance or depress immunological responses, or may result in neoplasia. Because cellular dysfunction may result from erosion of the cell by a direct attack by the virus or through immunopathological processes, and since the disease process is a long continued one, the implications for a viral role in chronic disease are obvious. The cumulative evidence in the medical literature suggests strongly that viruses may in some cases underlie chronic disease of the heart, the kidney, and the endocrine system. We need not necessarily expect that new and unknown viruses will turn up but rather that some of our long-known, garden variety viruses may play a hitherto unsuspected role. Thus, several viruses, e.g., encephalomycarditis virus in the mongoose and especially Coxsackie viruses n the mouse, have been shown to induce cardiac lesions such as endocarditis, valvulitis, and myocarditis and, in addition, the Coxsackie viruses have been shown to produce a fatal myocarditis in the human newborn infant. There is increasing, even if indirect, evidence (such as that obtained by immunofluorescence studies) that the Coxsackie viruses play a role in chronic heart disease. Epidemiologic evidence also tends to incriminate certain enteroviral immunotypes in cardiomyopathies. Experimentally and clinically, much the same obtains with repect to kidney disease; glomerular lesions have been produced in mice by injection of an echovirus, and fluorescent antibody studies in man suggest the possibility of Coxsackie viruses as incitants. Similarly, viral infections may involve the endocrine system, e.g. the pancreas. Seroepidemiologic surveys have pointed to the possibility that a Coxsackie virus (Type B4) may be a causative agent in the juvenile form of diabetes mellitus, and experimentally the condition can be reproduced in mice by inoculation with encephalomyocarditis virus. None of these conditions—cardiopathies, nephropathies, and endocrinopathies—has been adequately studied from the virologic standpoint, and comprehensive studies combining the clinical, laboratory, and epidemiologic aspects should prove rewarding. In the experimental laboratory, a search should be made for an animal model for the disease under study in man. Some current examples are diabetes mellitus syndrome produced in mice by encephalomyocarditis virus; so-called old dog encephalitis induced by distemper virus as a model for multiple sclerosis; and the disease in the New Zealand black mouse which is a counterpart of systemic lupus erythematosus in man.

Present-day corrections of some of these chronic conditions such as dialysis or renal transplant in kidney disease or substitution of mechanical parts for a malfunctioning valve or transplantation of the whole organ in cardiac disease, represent what Lewis Thomas has referred to as "half-way technology," stop-gap measures to which we resort because we do not as yet possess the means of preventing, from the very beginning, the development of the offending pathological changes. The most obvious approach to this objective is through vaccines, which have been in the forefront of the armamentarium for prophylaxis of disease in man and animals. Once a virus has been isolated and may be cultivated in the laboratory, a vaccine becomes an attainable objective and usually one promptly sought by public health officials. Because of the considerable morbidity associated with hepatitis A and hepatitis B viruses, vaccines for the control of these diseases will probably be given a priority. Vaccines will probably also be sought for chronic diseases that are high on the list of leading causes of death, and it has been suggested that such a vaccine will be developed for cancer if and when a viral etiology is demonstrated.

These comments lead to a consideration of the whole problem of vaccines; for instance, how many vaccines can be administered, how many will the public accept, the matter of a reasonable antigenic mass when polyvalent vaccines are used; and reactogenicity.

A good example is the 100 or more rhinovirus immunotypes responsible for so much common upper respiratory disease and the economic losses that are incurred through such illnesses. [Speaking of cost–benefit ratios, a recent study (5) on the efficacy of adenovirus vaccines in reducing acute respiratory disease in military recruits revealed a savings of several million dollars in hospitalization costs over a 5-year period.] Vitamin C or any other chemoprophylactic product, if proved effective against such a multiplicity of disease incitants, might provide the answer, but failing that, vaccines might best be developed against those immunotypes most commonly involved. Even so, one is still confronted with the question of the efficacy of such polyvalent vaccines as well as their reactogenicity.

A more logical approach is to ascertain which viral antigens are necessary for protection against a given viral disease, so that vaccines can be prepared from isolated antigenic components rather than from the whole virion. Current studies on inactivated influenza virus vaccines provide a case in point. These vaccines contain the envelope antigens, hemagglutinin and neuraminidase, and are relatively free of such unnecessary components as the internal proteins and nucleic acid of the virus. Similar subunit vaccines against rabies and adenoviral infections are also under study. After the chemical nature of the viral subunits mandatory for induction of a protective immune response is known, the next step will be to attempt their synthesis.

In the estimate of some workers, prevention of disease is best achieved through live virus vaccines since they evoke immune responses above and beyond those elicited by killed virus vaccines. However, in the judgment of others, live virus vaccines are unacceptable because of the potential hazards associated with injection of foreign genetic material, e.g., the hazard that subacute sclerosing panencephalitis might follow the administration of live measles virus vaccines.

But again, need such introduction of foreign genetic material necessarily be always detrimental?

In nature's scheme of things, some useful function can usually be discerned for each creature within its biome, and hence one may doubt that the sole function of viruses is the induction of disease and death—these appear to be aberrant or accidental end results of infection. One such function and one that might redound to the advantage of the infected individual and, through him, to his species, would lie in carriage, via viral transduction, of genetic material across species lines and between phyla (and between kingdoms?).*

As an example, recent evidence indicates that a human strain of influenza A virus can cross species barriers and infect a wide range of potential hosts and that recombination between influenza viruses occurs under experimental conditions *in vivo* and may be accompanied by reassortment of genetic information coding for antigenic character and virulence (*15*). Such exchange of genetic information could play a major, and perhaps key role in evolution. Cellular mutations brought about by the new genetic material will form part of the future heritage provided that the altered cell survives and that the newly acquired genetic information can be transmitted to its descendants.

Mourant (*9*) suggests that the parallel and rapid evolution from soft-bodied ancestral forms in many different phyla of animals with mineral skeletons cannot be attributed to dissemination of the character via ordinary sexual reproduction and natural selection, but broad dissemination of the responsible gene could be explained by the transduction mechanism. In an examination of a number of skeletal variants encountered in several substrains of the C-57B1 mouse derived from a mating initiated several decades back, Gruneberg (*6*) concluded that certain common differences might best be explained not as originating on the basis of changes in the nuclear genes but as due to some vertically transmitted, self-reproducing entity in the cytoplasm, probably a latent virus. Viroids, too, that lack the capacity of horizontal transmission might serve as carriers of epigenetic inheritance. What role viruses or viroids may play in evolution and in the genetic background is an interesting and essentially unexplored area. Studies in animals, for example, on the role of a vertically transmitted virus in the production of observable or measurable (biochemical) differences in an experimentally infected inbred line of animals, should yield information unobtainable by other means, for

* Abraham (*1*) cites Vendeg (e.g., *13, 14*) as having reported many years ago that some animal viruses could give rise to disease in plants and that some plant viruses could initiate infection in man and animals. Abraham's own experiments suggested that a coxsackievirus could replicate in plant tissue, and, more recently, Banatvala and Payne (*3*), and also Sänger (*11*), reported induction of leaf necrosis in tobacco plants by inoculation with HBAg-positive sera [but see Beasley and Su (*4*) and also Payne and Banatvala (*10*)]. The use of viruses for the control of insect pests and disease vectors, projected for the future as a possible biological weapon to replace chemical pesticides, may increase the opportunity for transmission of these viruses to man and animals. Longworth *et al.* (*7*) recently reported naturally occurring antibodies to *Gonometa* virus (a virus of the moth, *Gonometa podocarpi*) in several mammalian species, which suggests exposure either to this picornavirus or to an antigenically related one.

example, by comparative studies of amino acid sequences to detect and systematize evolutionary relationships. As Norman Anderson (2) points out, such chemical comparisons concern homologous proteins that are already well established and do not tell us much how new proteins may arise to replace older, quite different ones.

The phenomena of viral latency and viral persistence have long been known, and their role in immunological processes and the pathogenesis of disease are receiving the attention their clinical importance merits. Twenty years ago susceptibility to viral leukemogenesis in the mouse was reported to have a genetic basis, and was linked to the histocompatibility-2 (*H-2*) locus. Recently, evidence has been adduced that susceptibility to lymphocytic choriomeningitis, the classic model of a vertically transmitted and lifelong infection, is under the control of a dominant gene analogous to the *Ir* (immune response) gene closely linked to the *H-2* locus. The nature of the receptors on T cells of the immune system is unknown, but some immunologists believe these receptors are *Ir* gene products and may actually be histocompatibility antigens. Hence, immunological research into transplantation and transplant rejection, the mechanisms of immune responses, and the role of infection in provoking immune responses or autoimmune reactions should find a common meeting ground, especially because of their importance for medical applications. One example for serious consideration is the accumulating evidence that viral infection may initiate the events leading to rejection of a transplanted organ. In the case of renal allografts, most instances represent infections with herpesviruses, cytomegaloviruses in particular, although other viruses, such as influenza, may be involved. Although it is possible that the process of rejection may lead to concomitant activation of a latent virus, it seems more probable that depression of cellular immunity in these patients activates a latent virus, which, in turn, sets off the rejection mechanism. This finds support in the clinical observation that exogenous infections, such as influenza, may be associated with transplant rejection. Obvious areas for research come to mind, such as the question of whether patients with transplants should receive vaccines, and especially live virus vaccines, as a defense against transplant rejection.

I hope this discussion has served to convey my ideas and impressions on what some of the emphases of virology will be over the years to come.

REFERENCES

1. Abraham, A. S., *Lancet* **2**, 1327 (1973).
2. Anderson, N., *Nature* (*London*) **227**, 1346 (1970).
3. Banatvala, J. E., and Payne, C., *Lancet* **1**, 1359 (1973).

4. Beasley, R. P., and Su, H.-J., *Lancet* **2,** 1203 (1973).
5. Collis, P. B., Dudding, B. A., Winter, P., Russell, P., and Buescher, E. L., *J. Infec. Dis.* **128,** 745 (1973).
6. Gruneberg, H., *Nature* (*London*) **225,** 39 (1970).
7. Longworth, J. F., Robertson, J. S., Tinsley, T. W., Rowlands, D. J., and Brown, F., *Nature* (*London*) **242,** 314 (1973).
8. Martini, G. A., and Siegert, R., eds., "Marburg Virus Disease." Springer-Verlag, Berlin and New York, 1971.
9. Mourant, A. E., *Nature* (*London*) **231,** 466 (1971).
10. Payne, C., and Banatvala, J. E., *Lancet* **2,** 1327 (1973).
11. Sänger, H., cited by Payne and Banatvala (*10*).
12. Stent, G., "The Coming of the Golden Age: A View of the End of Progress." Natural History, New York, 1971.
13. Vendeg, V., *Nature* (*London*) **208,** 598 (1965).
14. Vendeg, V., *Orvoskepzes* **48,** 103 (1973).
15. World Health Organization, Document VIR/73.9, (1974).

CHAPTER 2

Detection and Identification by Immune Electron Microscopy of Fastidious Agents Associated with Respiratory Illness, Acute Nonbacterial Gastroenteritis, and Hepatitis A

Albert Z. Kapikian
Stephen M. Feinstone
Robert H. Purcell
Richard G. Wyatt
Thomas S. Thornhill
Anthony R. Kalica
Robert M. Chanock

I. INTRODUCTION

It was in 1941 that Anderson and Stanley (*10*) in the United States and von Ardenne, Friedrich-Freksa, and Schramm (*11*) in Germany independently reported electron-microscopic studies of the reaction between tobacco mosaic virus (TMV) and its antiserum. The latter group of investigators observed that a mixture of TMV and anti-TMV rabbit

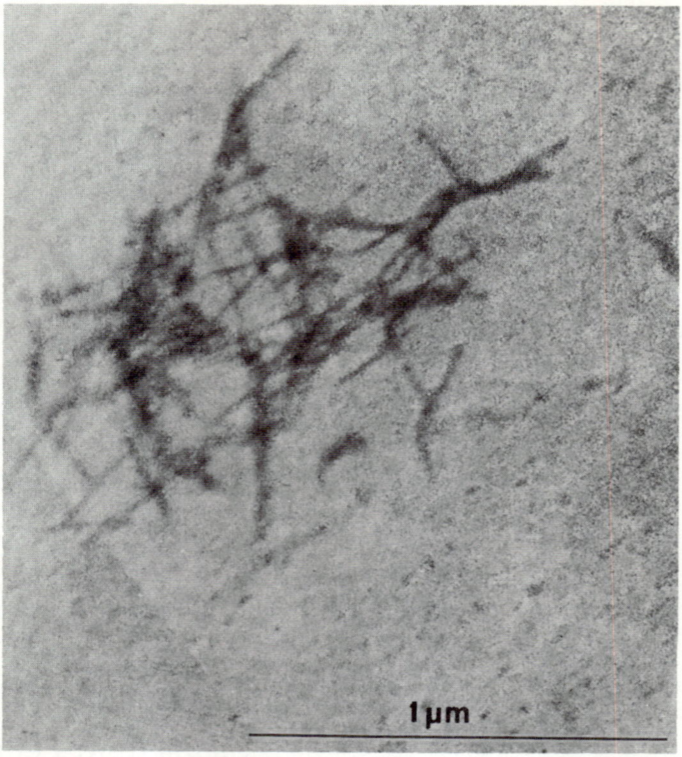

Fig. 1. Tobacco mosaic virus (TMV) particles aggregated following reaction with a 1:1000 dilution of TMV rabbit antiserum. Preparation treated with osmium tetroxide. [After von Ardenne *et al.* (*11*).]

serum resulted in aggregation of the virus particles (Fig. 1), whereas by contrast such aggregation did not occur when the virus was reacted with a control normal rabbit serum (Fig. 2). Anderson and Stanley made similar observations and, in addition, demonstrated the electron-microscopic specificity of the reaction by observing the lack of such reactivity between TMV and rabbit antiserum to tomato bushy stunt virus (Fig. 3), and between tomato bushy stunt virus, and anti-TMV rabbit serum (Fig. 4). These studies were, indeed, the first demonstration of the method that was later termed immune or immuno-electron microscopy (IEM)—a technique which may be defined as the direct observation by electron microscopy of antigen–antibody interaction (*8, 53a*).

In ensuing years, in spite of the development of electron microscopes with greater resolving power and of methods, such as negative staining

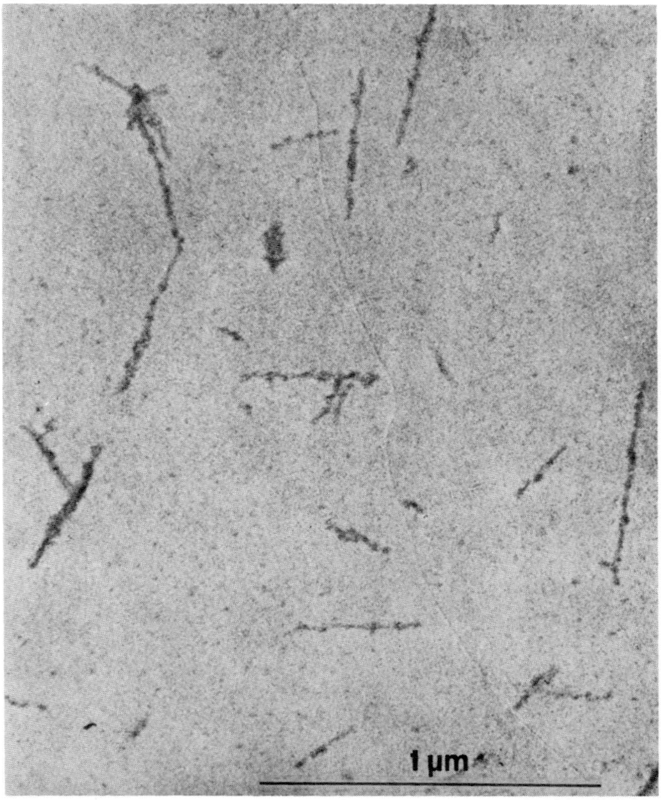

Fig. 2. Tobacco mosaic virus particles were not aggregated following reaction with a 1:1000 dilution of normal rabbit serum. Preparation treated with osmium tetroxide. [After von Ardenne *et al.* (*11*).]

(*16*), that greatly increased contrast, this technique was relatively underutilized (*8*).

II. DETECTION OF RHINOVIRUSES

My (A.Z.K.) exposure to the technique came in 1970 while a guest worker for about 6 months in the laboratory of June Almeida, a pioneer in the application of this technique (*4–8*), in Anthony Waterson's Department of Virology in the Royal Postgraduate Medical School in London. We were working originally on antigenic and serological studies of coronaviruses using the technique of IEM. We also applied the technique, in collaboration with E. J. Stott, to the study of rhinoviruses (*46*) since these agents, although visible in thin sections of infected cells (*13, 36, 52, 54*), and in purified preparations, (*20, 31, 58–60*), had been

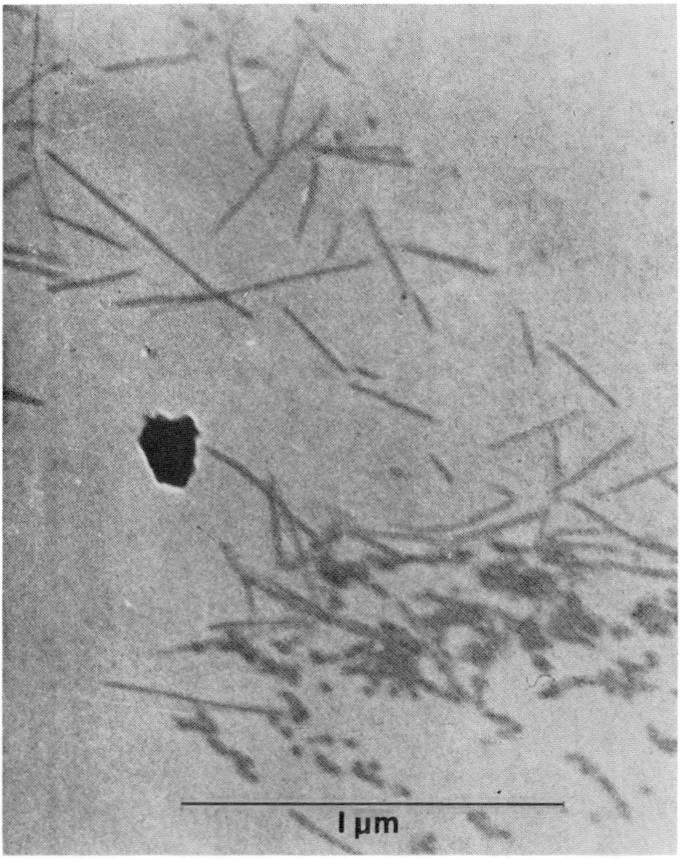

Fig. 3. Tobacco mosaic virus (TMV) and tomato bushy stunt virus after reaction with tomato bushy stunt virus rabbit antiserum. The TMV was not aggregated by this serum, whereas the tomato bushy stunt virus particles were aggregated by it. [After Mudd (*67*) in which the figure was reproduced after Morton (*66*), from the study of Anderson and Stanley (*10*).]

difficult to visualize in partially purified suspensions by conventional negative staining techniques, because they do not normally grow to sufficiently high titer, do not display a very distinctive morphological appearance by negative staining, and are among the smallest viruses known.

In an experiment in which rhinoviruses 1A and 13 were incubated with either phosphate-buffered saline (PBS) or a homologous goat antiserum before centrifugation and further preparation for electron microscopy, virus particles were seen in all specimens but with greatly differing ease and certainty of recognition (*46*). Figure 5 shows a single particle

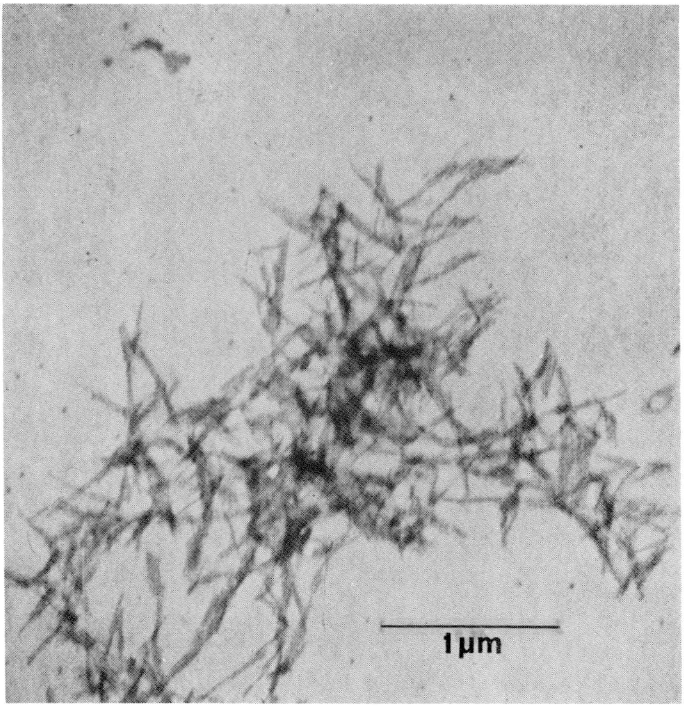

Fig. 4. Tobacco mosaic virus (TMV) and tomato bushy stunt virus after treatment with TMV rabbit antiserum. The TMV particles were aggregated by this serum, whereas the tomato bushy stunt virus particles were not aggregated by it. [After Mudd (*67*) in which the figure was reproduced after Morton (*66*), from the study of Anderson and Stanley (*10*).]

of rhinovirus 1A from a control preparation in which the virus, which had been grown in monolayer cell cultures and had a titer of 10^5 $TCID_{50}/0.2$ ml, was reacted with PBS. The particles shown in this figure is probably the virus, but it was often difficult to determine with certainty whether or not a particular structure was a virus particle especially since even such probable virus particles were seen only infrequently. Figure 6 shows several rhinovirus 13 particles from another control preparation. The particles were present in greater number since this suspension had been prepared in spinner culture (*77*) and was higher titered ($10^{7.5}$ $TCID_{50}/0.2$ ml) (*46*). In spite of this, it was still difficult to be certain whether a particular structure was or was not a virus particle. In both of these preparations the scattered particles were distributed randomly over the grid squares. By contrast, in the virus and homologous goat antiserum mixtures, both large and small aggregates were observed,

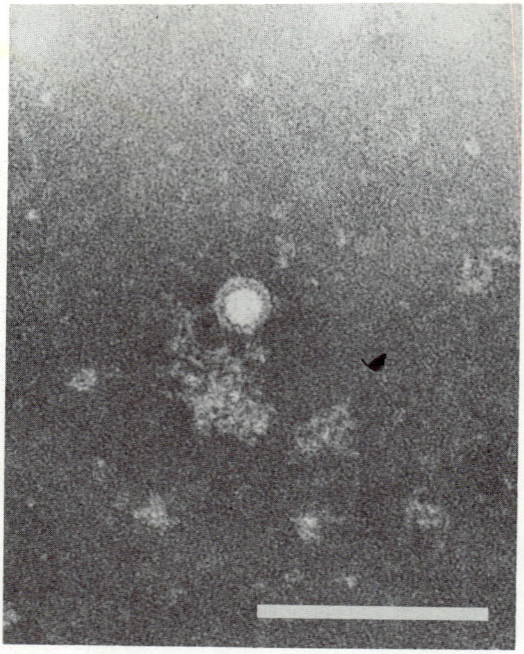

Fig. 5. A single particle of rhinovirus 1A from a control preparation in which rhinovirus 1A was incubated with phosphate-buffered saline prior to further preparation for electron microscopy. Particles were randomly distributed, and it was difficult to determine whether certain objects were virus particles. The bar = 100 nm. [After Kapikian *et al.* (*46*).]

as shown for rhinovirus 13 in Fig. 7. The particles were no longer randomly distributed but were present as aggregates that stood out clearly from the background matter. There was no longer any doubt about the appearance of the virus since the virus–antibody reaction had resulted in the formation of distinctive aggregates that could be easily visualized, as shown also for rhinovirus 1A in Fig. 8. Optimal antibody concentration resulted in maximal clumping of particles, whereas extreme antibody excess not only tended to obscure the particles but also often hindered the formation of aggregates.

III. DETECTION OF CORONAVIRUS STRAIN 692

During these studies it occurred to me that the technique of IEM, which has the distinctive feature of increasing the sensitivity for the detection, and facilitating or enabling the recognition of a specific low-titered agent by virtue of the immunological specificity of antibody

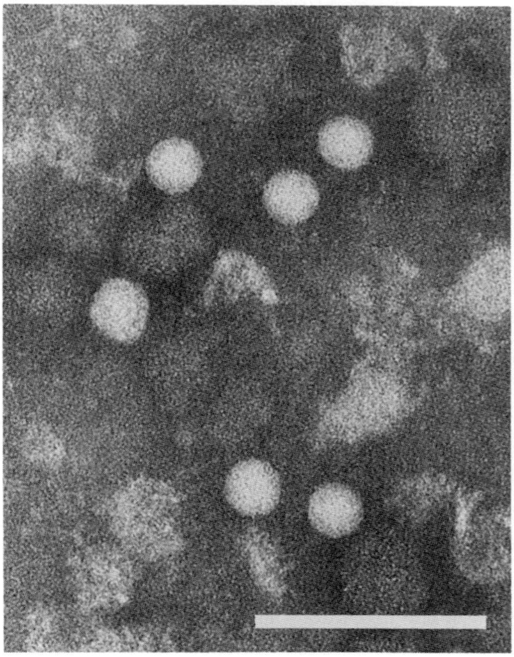

Fig. 6. Several rhinovirus 13 particles from another control preparation in which rhinovirus 13 was incubated with phosphate-buffered saline prior to further preparation for electron microscopy. Particles were present in greater number than in Fig. 5 since the rhinovirus 13 suspension was higher titered, but it was still difficult to determine whether certain objects were virus particles. Particles were randomly distributed. The bar = 100 nm. [After Kapikian *et al.* (*46*).]

for that very agent, might be successfully applied in a very practical way to the detection and identification of fastidious agents of human disease if a patient's convalescent serum could be utilized as the source of specific antibody in the hope that the latter would enable the detection and identification of an otherwise unidentifiable low-titered fastidious agent from that same patient. Therefore, following my return to the National Institutes of Health, we began to look for one of these fastidious agents of human disease, the coronaviruses, using IEM.

Coronaviruses have been shown in several seroepidemiologic studies to be not uncommon in both pediatric and adult populations (*18, 19, 38, 40, 48, 53, 61, 62, 75*) but, in spite of this, recovery of these agents has been quite rare, as only thirty-two isolations have been reported since 1965 when the first strain, B814, was recovered in human embryonic tracheal organ culture (HETOC) by Tyrrell and Bynoe (*15, 38, 45,*

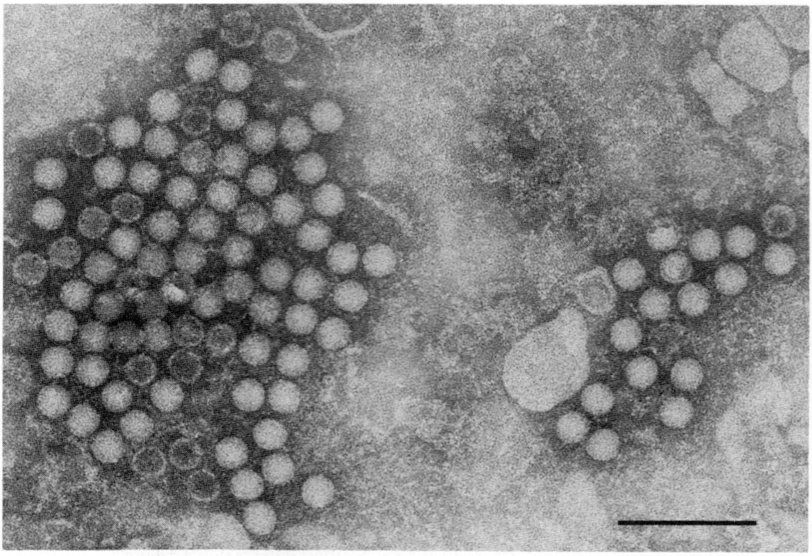

Fig. 7. Micrograph shows particles of rhinovirus 13. The rhinovirus 13 suspension was incubated with a 1:180 dilution of rhinovirus 13 goat antiserum prior to further preparation for electron microscopy. There is now no doubt about the appearance of the virus as it has been aggregated by the specific antibody (compare with Fig. 6). The bar = 100 nm. [After Kapikian *et al.* (*46*).]

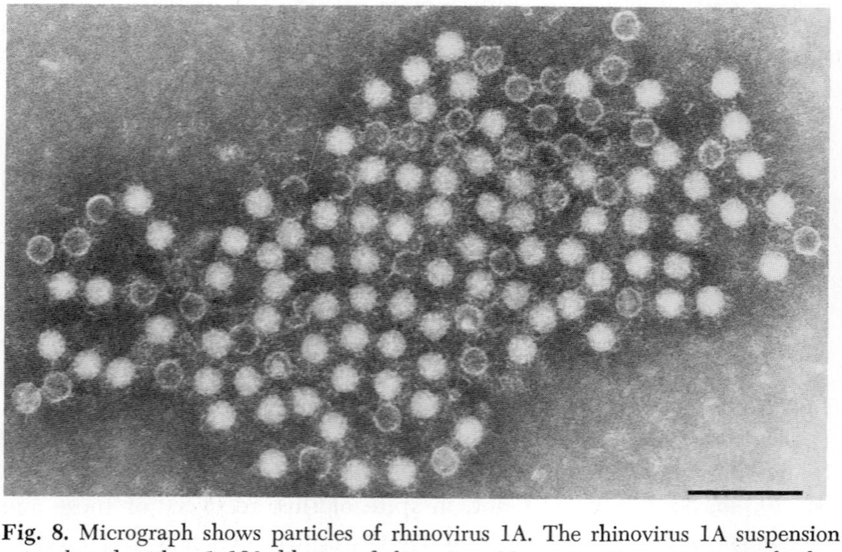

Fig. 8. Micrograph shows particles of rhinovirus 1A. The rhinovirus 1A suspension was incubated with a 1:180 dilution of rhinovirus 1A goat antiserum prior to further preparation for electron microscopy. The micrograph illustrates a large distinctive aggregate that may be observed utilizing the technique of immune electron microscopy. The bar = 100 nm. [After Kapikian *et al.* (*46*).]

68, 80, 81). Ten of the thirty-two were originally isolated in HETOC and identified by conventional electron microscopy, whereas the remainder, all of which were similar to a prototype strain designated 229E (*39*), were recovered with difficulty in monolayer cell cultures (*15, 38, 45, 50, 68, 75a, 80*). In attempting to find new methods for the detection of these fastidious agents, we adapted the technique of IEM to the study of HETOC harvests derived from washings from adults with upper respiratory tract illnesses, and by these methods recovered one of the thirty-two coronavirus strains (*50*). These studies (*50*) are described below.

A diluted nasal-nasopharyngeal washing from an NIH employee (study #692), enrolled in our "common cold study," was inoculated into roller tube cell cultures of primary rhesus monkey kidney (MK), primary human embryonic kidney, Hep-2, semicontinuous human embryonic intestine, and cultures of human diploid cell strain WI38. Since neither cytopathic effect (CPE) nor hemadsorption (in MK cell cultures) were observed, the diluted washing was carried through three passages in HETOC. Harvests from various days of the third passage were pooled and inoculated into the roller tube cell cultures previously mentioned, except for Hep-2. Since again neither CPE nor hemadsorption were observed, the pooled harvests of the third HETOC passage were examined for the presence of virus particles by IEM utilizing the patient's convalescent serum as the source of specific antibody in the hope that virus particles, if present, would appear in the form of aggregates thereby enabling the visualization of a low-titered coronavirus suspension that might not have been detectable without reaction with the specific antibody in the convalescent serum.

Briefly, the procedure involved incubation of 0.1 ml of the organ culture harvest for 1 hour at room temperature with 0.1 ml of a 1:20 dilution of the patient's uninactivated convalescent serum, and also with 0.1 ml of PBS (the latter as a control to the IEM procedure). Then, 0.8 ml of PBS was added to each tube and the mixture centrifuged at 17,000 rpm for 90 minutes in a Sorvall RC2 centrifuge with an SS34 fixed-angle rotor. The supernatant fluid was discarded, the pellet or residue resuspended with a few drops of distilled water, stained with 3% PTA, placed on a 400-mesh Formvar carbon-coated grid, and the excess fluid withdrawn with filter paper. The grid was examined at a magnification of approximately 44,000, with a Siemens Elmiskop 1A electron microscope.

Reaction of the organ culture harvest with the patient's convalescent serum resulted in the appearance of both large and small aggregates of coronavirus-like particles which appeared to be heavily coated with antibody (Fig. 9A, B, and C). The particles were not randomly distributed but were present almost exclusively as groups that stood out clearly from the surrounding matter. Interestingly, a single particle that ap-

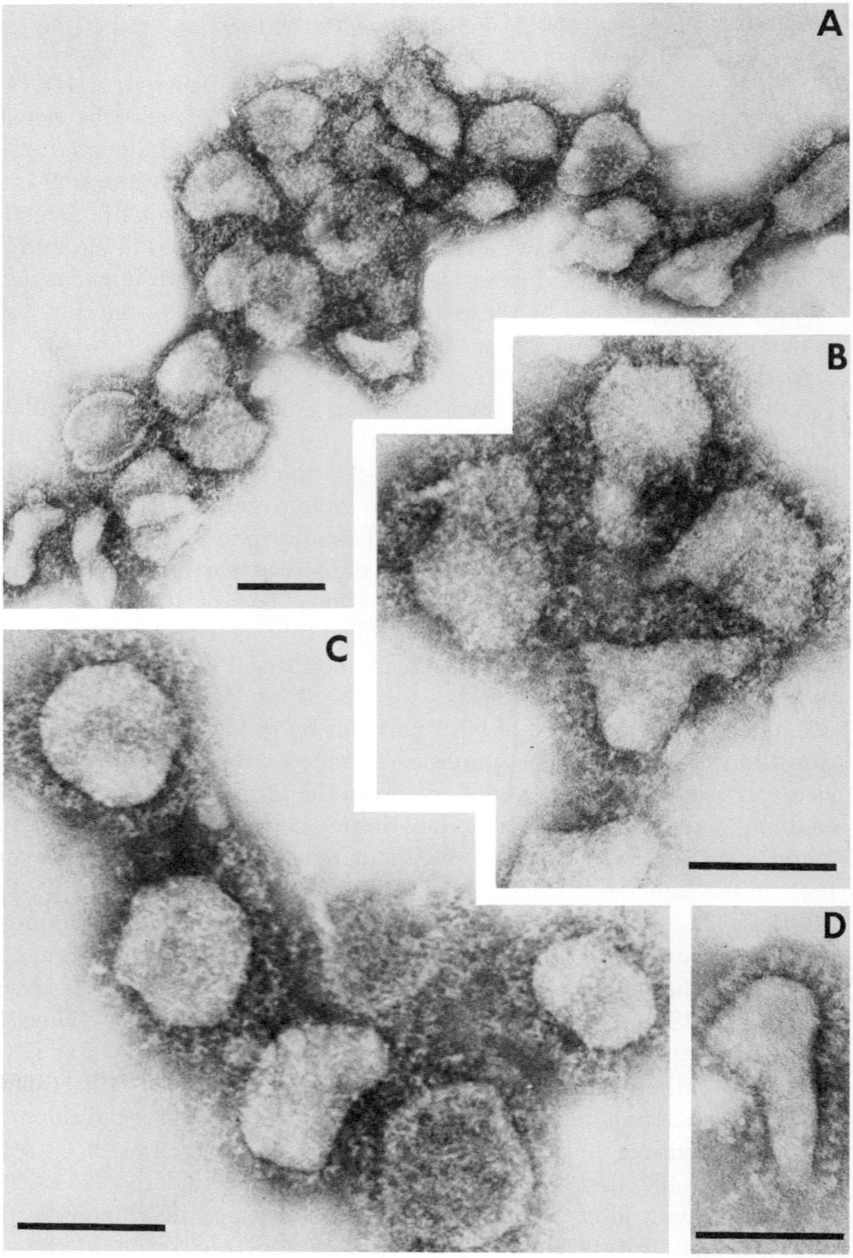

Fig. 9. (A–C) Large and small aggregates of coronavirus strain 692. The human embryonic tracheal organ culture harvest derived from a washing from a patient in the NIH common cold study was incubated with a 1:20 dilution of the patient's convalescent serum prior to further preparation for electron microscopy. Particles heavily coated with antibody were not randomly distributed but were present almost

peared to be covered with little, if any, antibody (Fig. 9D) was also observed. Examination of the HETOC harvest–PBS mixture did not reveal any coronavirus paricles but only predominantly round or oval-shaped structures (Fig. 10A–D). Additional studies were made by IEM to determine if this coronavirus strain, designated as 692, was related to coronaviruses 229E and OC43, the only known distinct coronaviruses of human origin that we have been able to propagate in either cell cultures or an animal system or both in our laboratory (*17, 39, 45, 48, 49*). In serological tests by IEM in which the presence or absence of aggregation with different sera was used as the measure of serological reactivity (*4, 5, 8*), aggregates, although less numerous than in the initial experiment, were observed following incubation of strain 692 with the homologous convalescent serum, but not with specific animal antisera to the other two coronaviruses, indicating that by this test system strain 692 was not related serologically to strains OC43 and 229E (*50*). Thus, the technique of IEM not only enabled the detection of a fastidious coronavirus but also demonstrated its serological distinctness.

IV. IDENTIFICATION OF AN AGENT ASSOCIATED WITH ACUTE NONBACTERIAL GASTROENTERITIS

We also attempted to utilize the IEM technique for the identification of fastidious agents of acute infectious nonbacterial gastroenteritis, since, in spite of intensive efforts, an etiological agent for this syndrome had not been found (*12*). This self-limited disease, which is characterized by a spectrum of clinical symptoms which may include vomiting, diarrhea, and abdominal pain, or a combination thereof, affects a broad segment of the population and was the second most common disease experience in the Cleveland family study (*12, 23–25, 35*). The disease, which has been given various descriptive names, such as winter vomiting disease and epidemic diarrhea and vomiting, was transmitted to volunteers in the 1940s and 1950s, and again more recently in the 1970s, but all attempts to cultivate and characterize definitively an etiological agent *in vitro* have failed (*12, 21, 26, 27, 30, 33–36, 43, 55, 72, 76*). In one of these studies, by Dolin *et al.* (*27*), a filtrate prepared from a rectal swab specimen from an adult who developed a secondary case of acute nonbacterial gastroenteritis during an outbreak in Norwalk,

---

exclusively as groups that stood out clearly from the surrounding matter. (D) A single particle of coronavirus strain 692 was also found on the same grid on which particles in A, B, and C were observed. This particle appears to be covered with little, if any, antibody. The bars = 100 nm. [After Kapikian *et al.* (*50*).]

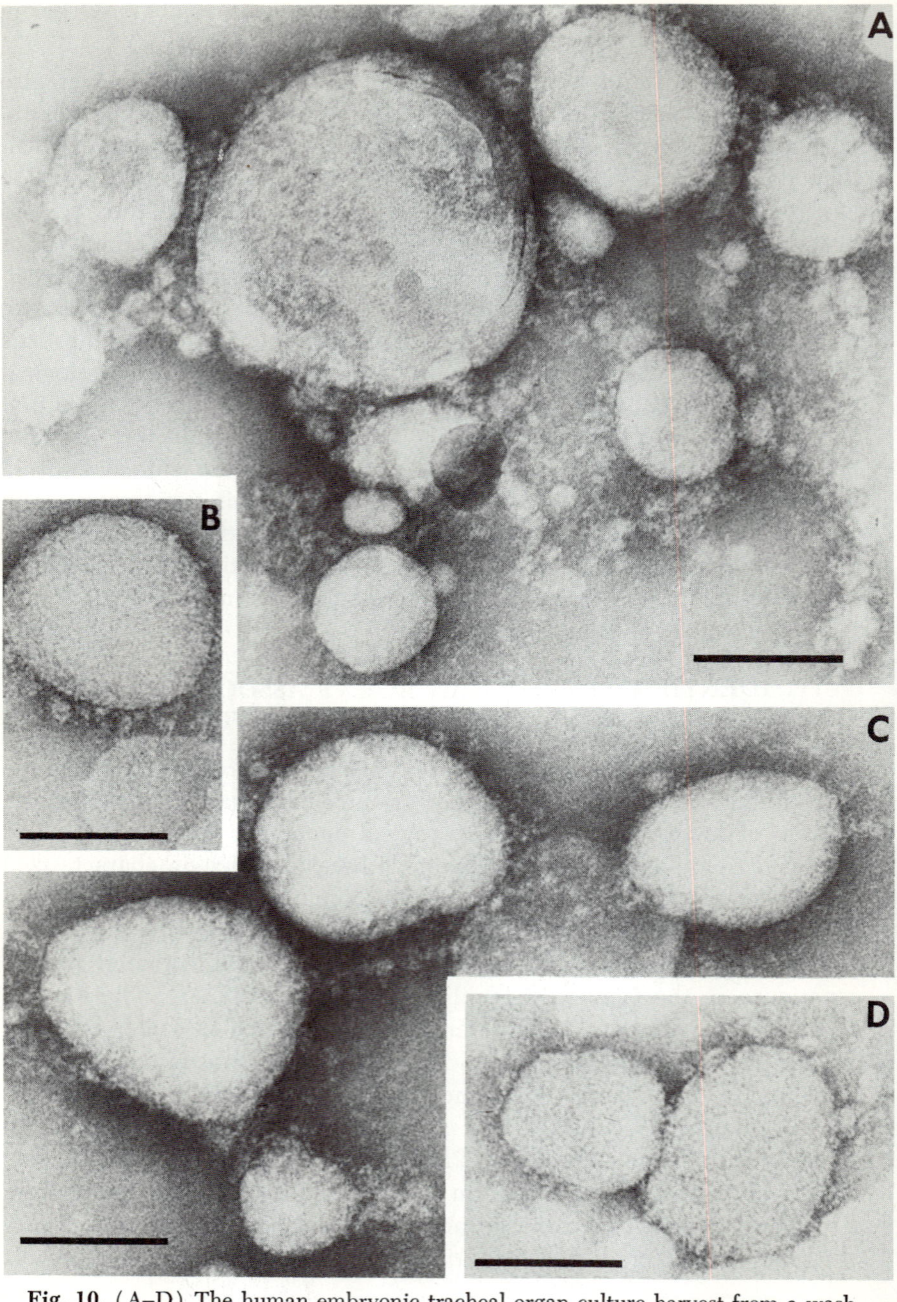

Fig. 10. (A–D) The human embryonic tracheal organ culture harvest from a washing from patient 692 was incubated with phosphate-buffered saline prior to further preparation for electron microscopy in the same experiment described for Fig. 9. No coronavirus particles were observed, only predominantly round or oval-shaped objects. The bar = 100 nm. [After Kapikian *et al.* (*50*).]

Ohio (*3*), induced the typical illness in two of three volunteers following its oral administration (*27*). It was serially passaged two additional times in volunteers by Dolin *et al.* (*26, 27*) and again induced the typical illness. Characterization studies (*26*) in volunteers revealed that the infectious agent in Norwalk outbreak-derived filtrates was less than 66 nm in diameter and probably less than 36 nm, and, in addition, was not inactivated by ether, acid, or heating at 60°C for 30 minutes. However, attempts to cultivate the infectious agent in cell cultures and animals were uniformly unsuccessful, and attempts in human embryonic intestinal organ cultures, inconclusive (*12, 26*).

In an attempt to detect these fastidious, presumably viral gastroeneritis agents, we adapted the technique of IEM to the study of stool filtrates derived from the Norwalk outbreak (*51*). A 2% second human passage Norwalk-derived stool filtrate (designated 8FIIa) was used as antigen in the IEM studies (*51*). It had been filtered through a 1200- and a 450-nm membrane filter and was known to contain an infectious agent since it had induced gastroenteritis in 6 of 10 volunteers, but extensive attempts to recover or detect an etiological agent by conventional techniques were unsuccessful (*83*). We, therefore, examined this filtrate for the presence of virus particles by IEM utilizing, as in the coronavirus study, a volunteer's convalescent serum as the source of specific antibody; the serum was obtained from an experimentally infected volunteer who had developed the typical illness. This approach was taken in the hope that virus particles would appear in the form of aggregates thereby enabling the visualization and the recognition of a viral agent present in low titer. The IEM technique was similar to that outlined previously in the coronavirus study except for minor modifications (*51*).

Reaction of 0.4 ml of the stool filtrate with 0.1 ml of a 1:10 dilution of a convalescent serum resulted in the appearance of aggregates of viruslike particles (Fig. 11). The particles, which were heavily coated with antibody, were not randomly distributed but were present as groups that stood out clearly from the surrounding matter and could be easily differentiated from other round structures or other groups of round structures devoid of antibody in the stool filtrate. Morphologically, the antibody-coated particles resembled the picorna- or parvoviruses. In stool filtrate–PBS control preparations, particles or groups of particles without or with little apparent antibody were seen only rarely in comparison; their significance would have been difficult to evaluate without previous experience from examining similar but heavily coated particles aggregated by antibody. The particles measured approximately 27 nm in their shortest and 32 nm in their longest diameters.

We used this stool filtrate, which had readily detectable particles by

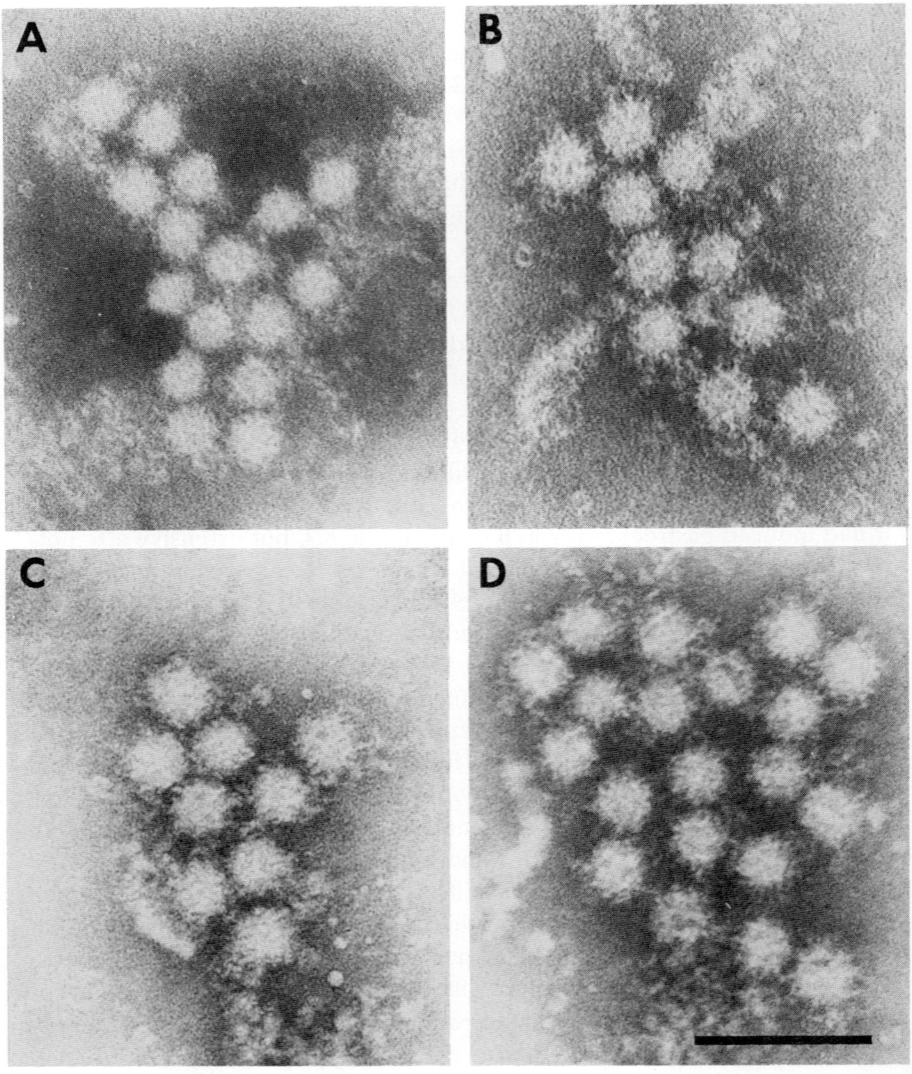

Fig. 11. Aggregates observed following (a) incubation of 0.4 ml of the 8FIIa stool filtrate with 0.1 ml of a 1:10 dilution of convalescent serum from volunteer A who developed gastroenteritis following challenge with the Norwalk filtrate and (b) further preparation for electron microscopy. (A) Aggregate observed in the initial 8FIIa experiment (incubation of serum-stool filtrate 1 hour) and (B–D) aggregates observed in the second 8FIIa experiment (incubation of serum-stool filtrate $1\frac{3}{4}$ hours). The particles are coated with antibody. The bar = 100 nm. [Fig. 11B from Kapikian *et al.* (*51*).]

IEM, to examine several groups of paired sera for antibody to this virus-like antigen in order to determine if the particle was related to gastroenteritis. In a manner similar to that described for the coronavirus studies, 0.2 ml of a 1:5 dilution of uninactivated serum was incubated with 0.8 ml of the stool filtrate and prepared for IEM. Antibody evaluations were made on coded specimens in order to eliminate the possibility of biased interpretation. Routinely, five squares on each grid were examined and the relative concentration of antibody in each serum specimen was estimated by judging the amount of antibody coating the aggregated particles on a 0–4+ scale. Three or more particles in a group were considered to constitute an aggregate. A rating of 0 (i.e., no antibody) was assigned when no aggregates were seen, whereas a 4+ rating indicated that the aggregated particles were so heavily coated with antibody that they were almost obscured. Single and paired particles that were heavily coated with antibody, thereby enabling their easy recognition, were observed along with aggregates in some convalescent sera which had high levels of antibody. Ratings of 1+, 2+, and 3+ indicated the presence of antibody but in lesser amounts than that rated 4+. A 1+ change in antibody rating between paired sera was considered to be significant. Examples of aggregates scored as 1+ and 4+ are shown in Fig. 12. Serum of volunteer A was scored as 1+ since the particles were well-defined and appeared to be covered with very little antibody, whereas that of volunteer B was scored as 4+ since the particles so heavily coated with antibody were almost obscured (*51*). Heavily coated particles were usually found in small aggregates (or as single or paired units), whereas those with less antibody usually formed larger aggregates. An example of what is considered to be a significant seroresponse, as observed by IEM, with a single volunteer's paired sera is shown in Fig. 13.

As seen in the Table 1, all 4 volunteers who developed gastroenteritis following administration of a second human passage Norwalk stool filtrate developed serological evidence of infection to the 27-nm particle as determined by IEM. A fifth volunteer, who did not develop illness, did not demonstrate a significant antibody response to this agent. Representative aggregates observed by IEM with this volunteer's pre- and postinoculation sera (rated 1+ and 1–2+, respectively) are shown in Fig. 14. Aggregates were not observed in PBS–Norwalk stool filtrate mixtures when such controls were utilized in certain of these serological studies (Table 1).

Although all 4 volunteers who developed illness after challenge with the stool filtrate developed serological evidence of infection, it was possible that the observed particles might represent a virus not related to the etiological agent of the Norwalk outbreak, since it was conceivable

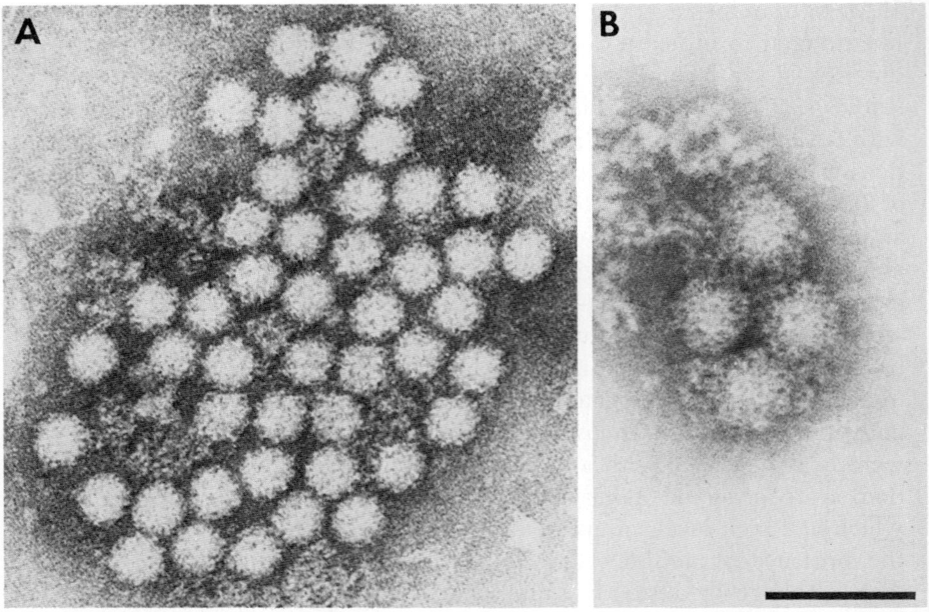

Fig. 12. (A) An aggregate observed after incubation of 0.8 ml of Norwalk (8FIIa) stool filtrate with 0.2 ml of a 1:5 dilution of prechallenge serum of volunteer A and further preparation for electron microscopy. The quantity of antibody on these glistening particles was rated as 1+. (B) An aggregate observed after incubation of 0.8 ml of Norwalk (8FIIa) stool filtrate with a 1:5 dilution of postchallenge serum of volunteer B and further preparation for electron microscopy. The particles were very heavily coated with antibody. Heavily coated particles were usually found to be in small aggregates, whereas those with less antibody were usually in larger aggregates. The quantity of antibody on these particles was rated as 4+. The bar = 100 nm and applies to both A and B. [Fig. 12 A from Kapikian *et al.* (*51*); Fig. 12 B after Kapikian *et al.* (*51*).]

that an adventitious agent could either have been present in the stool of the patient from the original Norwalk outbreak or could have been acquired during passage through volunteers. We examined this possibility by testing paired sera from 6 individuals from the original Norwalk outbreak. Four were primary cases, 1 a secondary case, and 1 a contact who did not become ill. [Serum samples were kindly supplied by Drs. Hatch, Adler and Zickl (3).] As shown in Table 1, 3 of the 5 individuals with naturally acquired Norwalk gastroenteritis developed an increase in antibody to the 27-nm particle as determined in tests of paired sera by IEM, whereas the 2 who did not show such an increase demonstrated a high level of antibody in both acute and convalescent sera. The acute

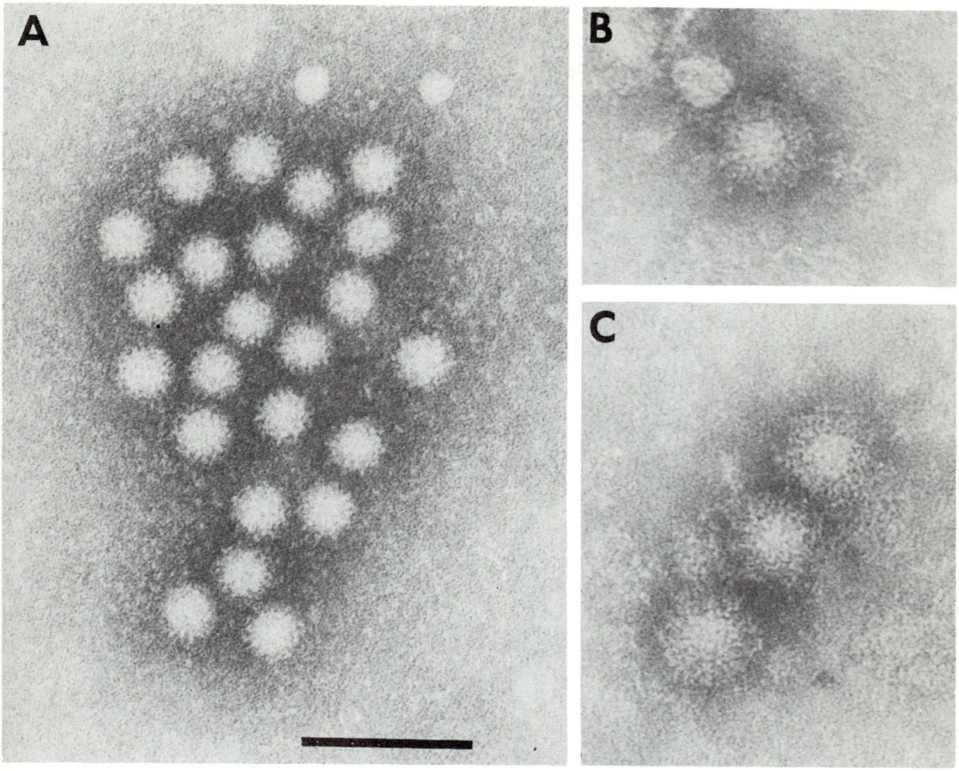

Fig. 13. (A) An aggregate observed after incubation of 0.8 ml of Norwalk (8FIIa) stool filtrate with 0.2 ml of a 1:5 dilution of a volunteer's prechallenge serum and further preparation for electron microscopy. This volunteer developed gastroenteritis following challenge with a second passage Norwalk filtrate which had been heated for 30 minutes at 60°C (*26*). The quantity of antibody on the particles in this aggregate was rated 1–2–2+ and this prechallenge serum was given an overall rating of 1–2+. (B) A single particle and (C) three single particles observed after incubating 0.8 ml of the Norwalk (8FIIa) stool filtrate with 0.2 ml of a dilution of the volunteer's postchallenge convalescent serum and further preparation for EM. These particles are very heavily coated with antibody. The quantity of antibody on these particles was rated 4+ and the serum was given an overall rating of 4+ also. The difference in the quantity of antibody coating the particles in the prechallenge and postchallenge sera of this volunteer is clearly evident. The bar = 100 nm and applies to A, B, and C.

phase sera were not collected until several days after the onset of disease and this could explain our failure to detect a response in 2 of the patients. The contact who did not become ill did not exhibit a serum response. It is of interest that among the seroresponders was donor (J) of the original

TABLE 1. SEROLOGICAL EVIDENCE OF INFECTION DETECTED BY IMMUNE ELECTRON MICROSCOPY: STUDY OF VOLUNTEERS WITH EXPERIMENTALLY INDUCED ILLNESS AND INDIVIDUALS IN ORIGINAL NORWALK OUTBREAK[a]

Category	Inoculum	Individual[b]	Gastrointestinal illness	Serum	Amount of antibody present on particles†	No. of aggregates observed in five squares
Experimentally infected volunteers	Second passage stool filtrate (8FII)[c] derived from stool of individual L	A	Yes	Prechallenge	1	1
				Postchallenge	4	ca. 15
		B	Yes	Prechallenge	2	4
				Postchallenge	4	10
		C	Yes	Prechallenge	1–2	3
				Postchallenge	4	3[d]
		D	Yes	Prechallenge	2–3	4[e]
				Postchallenge	4	4
		E	Yes	Prechallenge	1	3
				Postchallenge	1–2	2
Naturally occurring illness (Norwalk, Ohio, outbreak)		F[f]	Yes	Acute	1–2	4
				Convalescent	3	15
		G[f]	Yes	Acute	4	6
				Convalescent	4	8
		H[f]	Yes	Acute	2	5
				Convalescent	4	6
		I[f]	Yes	Acute	3–4	8
				Convalescent	3–4	6
		J[g]	Yes	Acute	<1	3
				Convalescent	2	9
Contact (Norwalk, Ohio, outbreak)		K	No	Early	1	7
				Late	0	0
Experimentally infected volunteers	Filtrate from rectal swab from individual J	L	Yes	Prechallenge	2–3	7
				Postchallenge	3–4	3
		M	Yes	Prechallenge	1	4
				Postchallenge	0[d]	

† After incubation with individual serum and further preparation for EM.

[a] After Kapikian *et al.* (*51*).

[b] In experiments with sera from individuals A, B, C, D, E, F, and K, filtrate 8FIIa was also incubated with phosphate-buffered saline as control, and no aggregates were observed.

[c] Filtrates 8FII and 8FIIa derived from stools obtained from same individual during same illness.

[d] Six squares counted.

[e] Eight squares counted.

[f] Primary case.

[g] Secondary case.

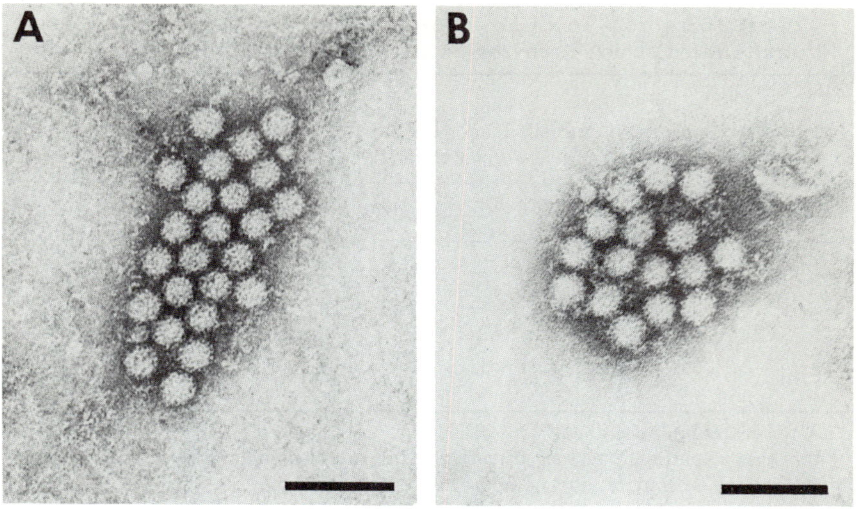

Fig. 14. Aggregates observed after (1) incubation of 0.8 ml of Norwalk 8FIIa stool filtrate with 0.2 ml of a 1:5 dilution of prechallenge serum (A) and a 1:5 dilution of postchallenge serum (B) from volunteer E who did not develop illness following challenge with the Norwalk filtrate and (2) further preparation for electron microscopy. No significant change in antibody was observed between the pre- and postchallenge sera. The prechallenge serum was rated 1+ and the postchallenge, 1–2+. [After Kapikian *et al.* (*51*).]

rectal swab specimen which had been used to induce illness in 2 of the 3 volunteers in the 1971 study (*27*). We examined by IEM the prechallenge and convalescent sera of these 2 volunteers and found that 1 developed an increase in antibody whereas the other did not. The development of a significant increase in antibody by 3 of 5 individuals with naturally induced illness suggested strongly that the 27-nm particle was not an adventitious agent acquired during passage in volunteers and, furthermore, indicated that this viruslike agent had also infected patients in the Norwalk outbreak.

Finally, we investigated the possibility that gastrointestinal disease might stimulate the observed antibody responses by a nonspecific mechanism. In another study, volunteers were first administered either a stool filtrate containing an infectious agent from a secondary case of gastroenteritis that had occurred in Honolulu, Hawaii, or from a secondary case that had occurred in Montgomery County, Maryland (*83*). Those volunteers who became ill after primary challenge were later rechallenged with the same agent and homologous immunity was demon-

TABLE 2. SEROLOGICAL EVIDENCE OF INFECTION DETECTED BY IMMUNE ELECTRON MICROSCOPY: STUDY OF SEQUENTIALLY CHALLENGED VOLUNTEERS[a]

Serum sequence tested [prechallenge (Pre) and days postchallenge (indicated by number)]			Estimated quantity of antibody associated with aggregates of 27-nm particles[b,c]	
1st challenge	2nd challenge	3rd challenge	Volunteer C	Volunteer D
Pre	—	—	2	2[d]
27	—	—	1–2	3
52	4	—	1	2–3[e]
83	35	Pre	1–2[f]	2–3[d]
104	56	21–25	4[g]	4

[a] After Kapikian *et al.* (*51*).

[b] In these experiments, stool filtrate 8FIIa was also incubated with phosphate-buffered saline as control, and no aggregates were observed.

[c] Volunteers C and D developed gastroenteritis after first and third challenges. Volunteer C challenge sequence: H filtrate, H filtrate, 8FIIa filtrate. Volunteer D challenge sequence: MC filtrate, MC filtrate, 8FIIa filtrate [H and MC filtrates were each derived from a secondary case of gastroenteritis in Honolulu (H), Hawaii, and in Montgomery County (MC), Maryland, respectively].

[d] Eight squares counted.

[e] Nine squares counted.

[f] Plasma utilized since serum not available.

[g] Six squares counted.

strated. Subsequently, challenge with a filtrate derived from the Norwalk outbreak produced disease, indicating that the infectious agent in the filtrate derived from the Hawaii and from the Maryland cases differed from that of Norwalk. Sequential sera from 2 volunteers who underwent the sequence of the three challenges just described were studied by IEM with the Norwalk filtrate (8FIIa) again as antigen (Table 2). The volunteer who received the filtrate derived from the Hawaii case failed to develop an increase in antibody following gastroenteritis induced by the primary challenge, but did develop a significant response after illness induced by the Norwalk filtrate, suggesting that the IEM seroresponse was specific. The other volunteer developed an increase in antibody after gastroenteritis induced by primary challenge, and a further increase after illness induced by the Norwalk filtrate, suggesting that the infectious agents in the Maryland and Norwalk filtrates may have been antigenically related (*51*).

The data from our studies suggested that the 27-nm Norwalk particle was the etiological agent of Norwalk gastroenteritis. Additional evidence to support this view has emerged from recent studies in our laboratory, in which the technique of IEM was used to determine shedding patterns of the Norwalk agent (78). These studies revealed by IEM the presence of the 27-nm Norwalk particle in stool filtrates derived from 11 of 23 volunteers who developed illness following challenge with Norwalk outbreak-derived stool filtrates. A total of seventy-seven stool filtrates were studied by IEM using a single convalescent serum known to contain a high level of antibody to the Norwalk particle. It was noteworthy that the 27-nm particle was observed in none of 12 specimens obtained prior to the onset of illness, in 26 of 54 collected during the first 72 hours after the onset of illness, and in only 2 of 11 obtained 72 hours following the onset of illness. Tabulation of the total number of particles observed by IEM in five grid squares in stool filtrates derived from serial stool samples from 8 particle-positive volunteers revealed that the shedding of the 27-nm particle was maximal at the onset of experimental illness and shortly thereafter (Fig. 15). Three of the particle-posi-

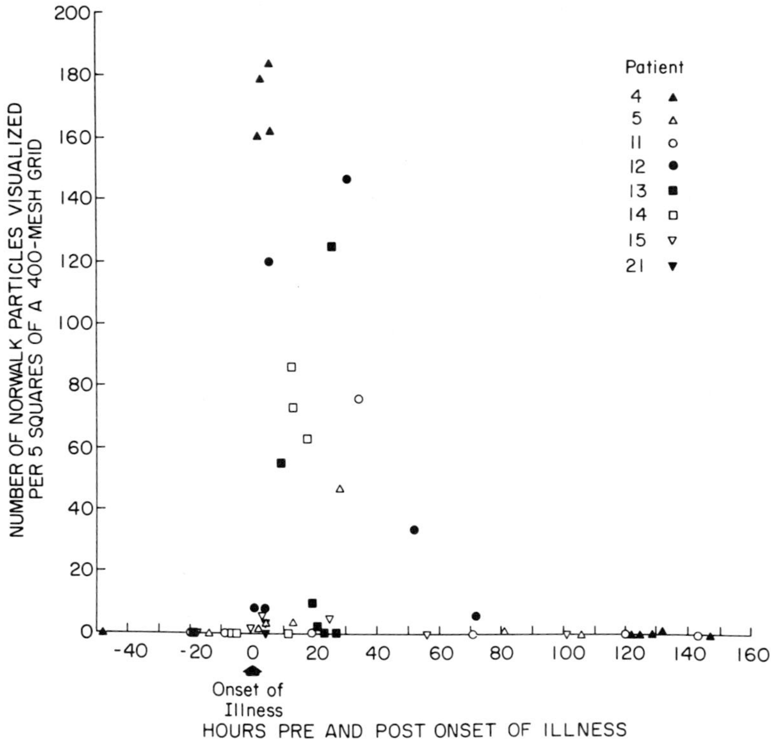

Fig. 15. Shedding pattern of the 27-nm Norwalk particle in stool filtrates derived from stool specimens from 8 volunteers experimentally infected with Norwalk filtrates. [From Thornhill *et al.* (78).]

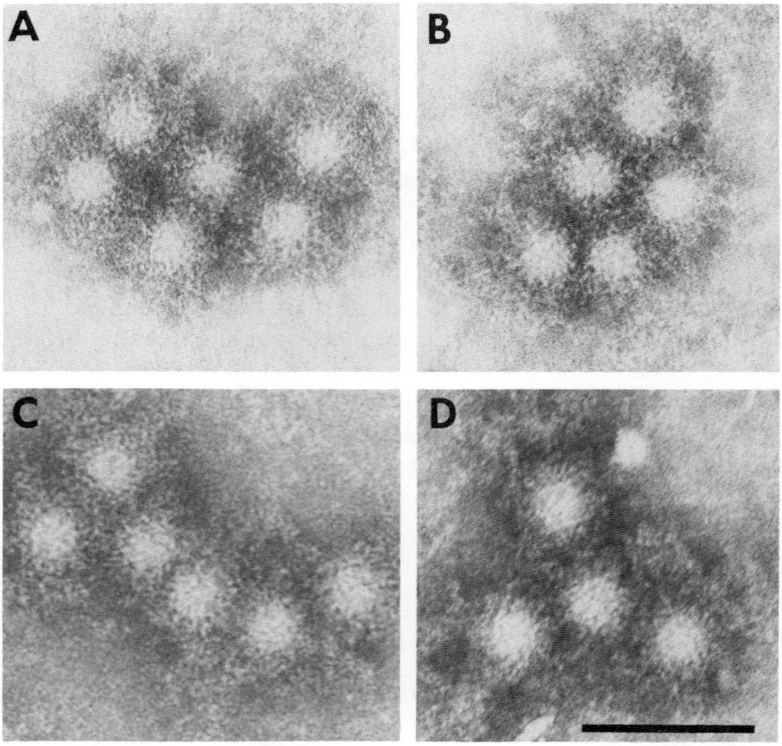

Fig. 16. Aggregates observed after incubation of 0.8 ml of Norwalk (8FIIa) stool filtrate with 0.2 ml of a 1:5 dilution of immune human serum globulin followed by further preparation for electron microscopy. These particles were heavily coated with antibody. The quantity of antibody on these particles was rated as 4+ for A, and 3–4+ for B, C, and D with an overall rating of 3–4+ for the 1:5 dilution of the immune human serum globulin. The bar = 100 nm and applies to A–D.

tive volunteers were excluded from the figure: 2 because their only positive stool contained only one particle and 1 because stools collected during the 48 hours after onset of symptoms were pooled.

It is of interest that in a further study by IEM we found that incubation of the 8FIIa Norwalk filtrate with a 1:5 dilution of an aliquot from a preparation of commercial immune human serum globulin resulted in the appearance of particles very heavily coated with antibody as shown in Fig. 16, indicating that antibody to this or a related agent must occur more than rarely in the general population.

We have also recently shown that 5 volunteers, who developed gastroenteritis following oral administration of the "W" gastroenteritis agent (*69*) of Clarke *et al.* (*21*), did not develop serological evidence of infection by IEM to the 8FIIa Norwalk agent indicating these two agents are probably not related serologically; 2 had developed the illness in volunteer studies by Clarke *et al.* (*21*), and 3 in such studies by our laboratory in collaboration with the University of Maryland (*85*). In addition, in another study from our laboratory in which IEM was utilized, the successful transmission of the Norwalk particle to chimpanzees was demonstrated (*84*). Although illness did not occur, serological evidence of infection with the 27-nm particle was shown by IEM.

In additional studies, in which we collaborated with Dr. John Gerin, we determined the buoyant density of the 27-nm particle in cesium chloride by examining each of sixteen gradient fractions by IEM using an appropriate human serum to aggregate the particles (*47*). The total number of particles in the gradient was determined and the number in each fraction expressed as a percentage of the total number of particles; the highest percentage of particles was seen in Fraction 4 which had a mean density of 1.403 (Fig. 17). Examples of some of the aggregates observed in Fraction 4 and in Fraction 5, which had the second highest percentage of particles, are shown in Fig. 18. In two additional isopycnic banding experiments in cesium chloride, the peak fractions were found by IEM to be at densities of 1.376 and 1.413 gm/cm^3, respectively. On the basis of its buoyant density, it appears that the Norwalk agent is not an enterovirus, the latter having a lower buoyant density in cesium chloride (*74*). Although the Norwalk agent has a density

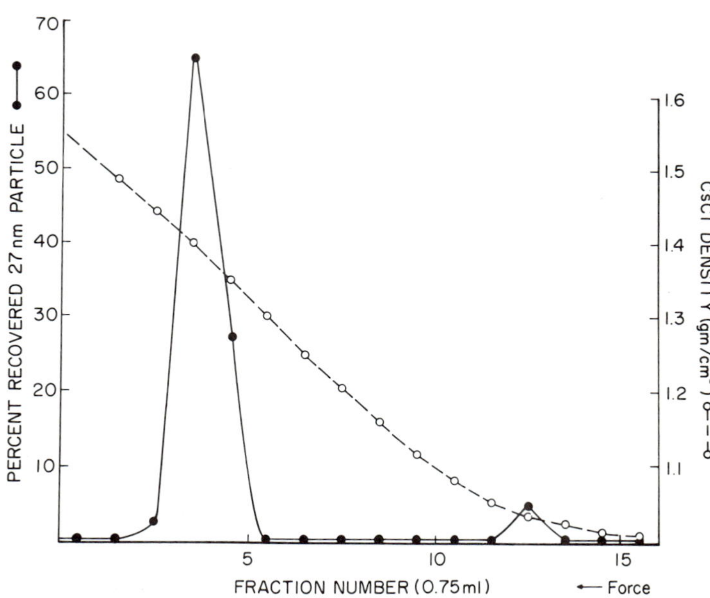

Fig. 17. Isopycnic banding of the 27-nm 8FIIa Norwalk particle in a cesium chloride density gradient. Quantitation of particle recovery was by IEM. [From Kapikian *et al.* (*47*).]

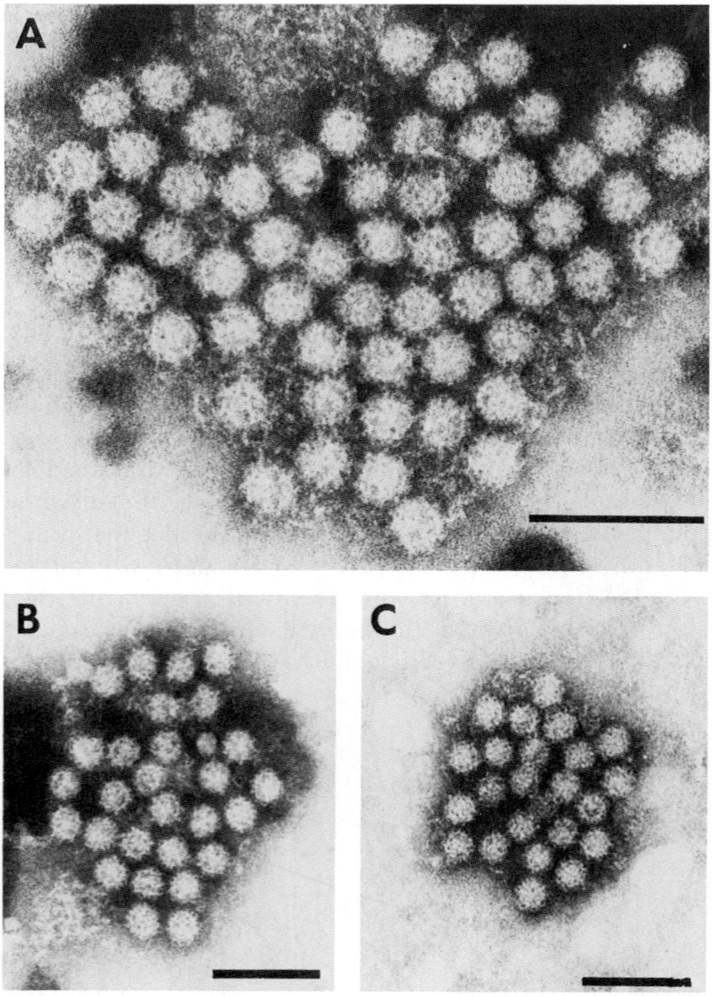

Fig. 18. Aggregates observed after incubation of 0.8 ml of a 1:3.3 dilution of each of Fractions 4 and 5 with 0.2 ml of a 1:5 dilution of prechallenge serum from volunteer E (see Table 1). (A, B) Aggregates observed in Fraction 4; (C) an aggregate in Fraction 5. [After Kapikian *et al.* (*47*).]

similar to that of a rhinovirus in cesium chloride, it can also be excluded from this subgroup of the picornaviruses since rhinoviruses are acid-labile, whereas an infectious stool filtrate derived from the Norwalk outbreak has been shown by Dolin *et al.* to be acid stable (*26, 37, 44, 74, 79*). Although the nucleic acid content of the Norwalk particle has not yet been determined, it appears that this agent, as previously suggested by Dolin *et al.* from volunteer studies (*26*), is parvovirus-like since it not only has a similar buoyant density in cesium chloride but also shares other properties with this group such as morphology, and, as shown in volunteer studies cited above, ether, acid, and relative heat stability (*26, 41*).

V. IDENTIFICATION OF AN AGENT ASSOCIATED WITH HEPATITIS A

At the conclusion of this phase of the gastroenteritis studies, we began to look by IEM for the fastidious agent of hepatitis A in stool filtrates obtained from different sources utilizing techniques similar to those in the gastroenteritis studies. Stool specimens for the study described here (*29*) were supplied under code as normal saline extracts by Dr. Dean Gibson of the Armed Forces Institute of Pathology. These extracts consisted of the supernatant fluids of approximately 20% fecal suspensions in normal saline which had been shaken for homogenization and then centrifuged at low speed (*32*). The original specimens had been obtained from each of 4 adult volunteers, from the Joliet, Illinois studies, who developed hepatitis after receiving either orally or parenterally the MS-1 or MS-1-derived strain of hepatitis A virus (*14, 56*). The extracts, which were derived from stools obtained prior to challenge and during the acute phase of illness from each volunteer, were further diluted to a 2% suspension and passed through a 450-nm membrane filter. All of the stool filtrates were examined by IEM using a volunteer's convalescent phase serum as a source of specific antibody for reasons similar to those in the gastroenteritis studies; commercial immune human serum globulin was also used to examine two pairs of these filtrates.

With our previously described approach, we found small viruslike particles measuring approximately 27 nm in diameter in one of two stool filtrates derived from stools from 2 volunteers in the acute phase of the disease, after the filtrates had been incubated with either a volunteer's convalescent phase serum or with human immune serum globulin (*29*). No significant particles were observed in two stool filtrates derived from prechallenge stools of these two volunteers, after incubation of the filtrates with the same convalescent serum or immune serum globulin as above. The 27-nm particles stood out clearly from the surrounding matter and were easily differentiated from other numerous spherical objects usually found in stool filtrates because the former occurred either as single particles heavily coated with antibody or as particles in groups aggregated by antibody (Fig. 19). Later in the study, we found similar appearing particles, which were much fewer in number and occurred as single particles, in one of two stool filtrates derived from stools of

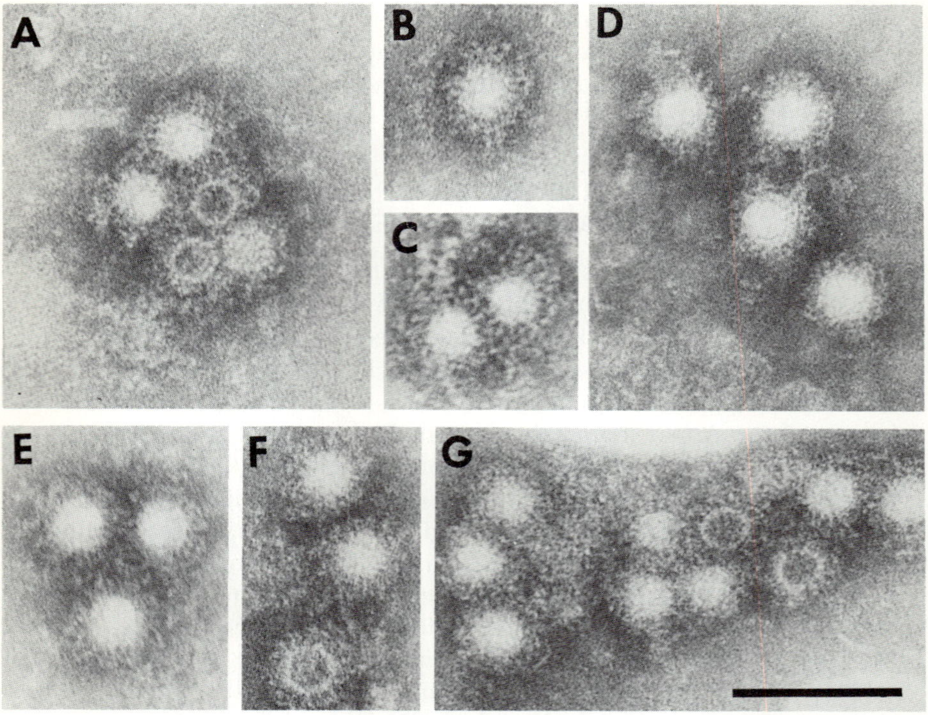

Fig. 19. Hepatitis A viruslike particles observed after 1 ml of the 2% stool filtrate was incubated with 0.1 ml of a 1:10 dilution of a convalescent serum or a 1:20 dilution of immune human serum globulin followed by further preparation for electron microscopy. An aggregate (A), a single particle (B), and a doublet (C) observed following incubation of the stool filtrate with a convalescent serum from a patient from the Joliet, Illinois, study. The particles are heavily coated with antibody. The quantity of antibody in this serum to the hepatitis A particle was rated 3–4+. (D, E, F, and G) Hepatitis A, viruslike particles observed following incubation of 1 ml of the stool filtrate with immune serum globulin. The particles are heavily coated with antibody also. The quantity of antibody in this lot of immune serum globulin to the hepatitis A particle was also scored 3–4+. The bar = 100 nm and applies to A–G. [Micrographs A, B, F, G from Feinstone *et al.* (*29*).]

2 other volunteers in the acute phase of the disease, after the filtrates had been incubated with the same convalescent serum as above. Again, no significant particles were found in filtrates derived from prechallenge stool specimens from these 2 volunteers, after the filtrates had been incubated with the same convalescent serum as above.

Using the initially positive stool filtrate as antigen for serological studies by IEM, we examined several sets of paired sera for antibody to this viruslike agent in order to determine if it was related to hepatitis

A. Usually, 0.1 ml of a 1:10 dilution of an uninactivated serum was mixed with 0.9 ml of the stool filtrate and examined by IEM with minor modifications of the technique used in the gastroenteritis studies. The relative concentration of antibody in each serum specimen was again estimated by scoring the amount of antibody coating the particle on a 0 to 4+ scale. One difference between the 8FIIa gastroenteritis and hepatitis A filtrates was the latter contained fewer particles and, therefore, in the presence of antibody, the hepatitis A antigen occurred frequently as individual particles coated with antibody and less often in the form of antigen–antibody aggregates. However, although the particles were fewer, the rating for antibody was somewhat simplified since antibody-coated single particles were generally more quickly visible, since they appeared to be randomly distributed whereas aggregates generally were not. As before, antibody evaluation of paired sera were made on coded specimens in order to eliminate the possibility of biased interpretation.

Four paired sera from experimentally infected hepatitis A individuals from the Joliet, Illinois study (*14*) and 2 paired sera from experimentally infected hepatitis A individuals from the Willowbrook, New York study (*56*) were examined by this method. As shown in Table 3, all 6 of these experimentally infected hepatitis A patients developed serological evidence of infection with the 27-nm, hepatitis A, viruslike particle as determined by IEM. However, since the hepatitis A containing inoculum used in both experimental studies was derived from the same source (MS-1), it was conceivable the seroresponses might have been due to an unknown adventitious agent in the original MS-1 inoculum and thus may have had no relationship to hepatitis A. We examined this possibility by testing paired sera from 3 individuals who developed hepatitis A in a small naturally occurring hepatitis A outbreak in American Samoa in 1972 (*73*) and, in addition, from 3 individuals who developed this disease during a naturally occurring hepatitis A outbreak in Massachusetts in 1969—the Holy Cross College football team outbreak (*65, 82*). We found, as shown in Table 3, that all 6 developed serological evidence of infection to the 27-nm hepatitis A particle by IEM, thereby suggesting strongly that the 27-nm particle was not an adventitious agent. Some of the heavily coated particles observed following reaction of the stool filtrate with a convalescent serum from the American Samoan and a convalescent serum from the Massachusetts outbreaks are shown in Figs. 20 and 21.

In further experiments designed to determine whether a serological relationship existed between the hepatitis A particle and hepatitis B antigens, we found, as shown in Table 3, that (*a*) none of the 12 seroresponders to the hepatitis A antigen developed a serological response

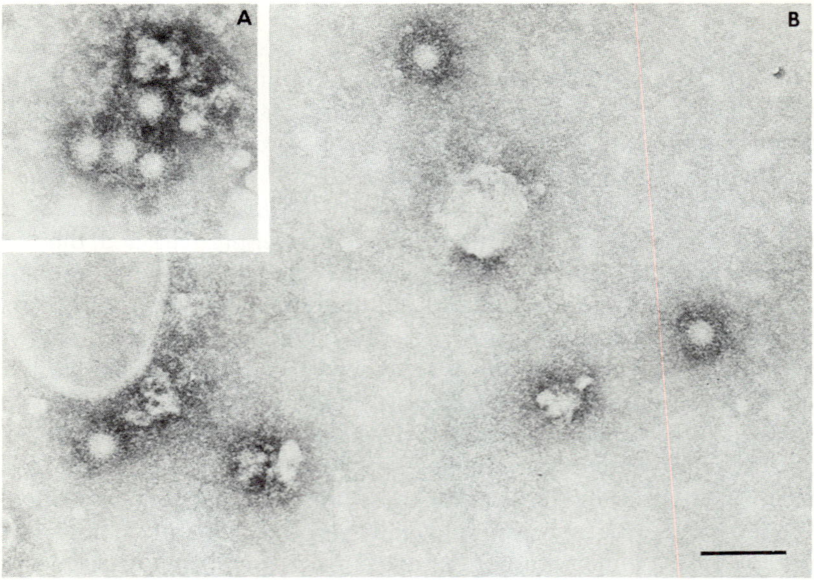

Fig. 20. A small aggregate of three particles plus a single particle (A) and three single particles (B) observed following (a) incubation of 0.9 ml of the stool filtrate with 0.1 ml of a 1:10 dilution of convalescent serum from a patient who developed hepatitis during the naturally occurring outbreak in American Samoa and (b) further preparation for electron microscopy. The particles are clearly covered with antibody. The ready recognition both of the antibody-coated particles in the aggregate and of the antibody-coated single particles illustrates clearly the use of the technique of immune electron microscopy in these studies. The bar = 100 nm and applies to A and B.

to hepatitis B antigen by radioimmunoassay (*57*); (*b*) 2 individuals with hepatitis B demonstrated a serological response to hepatitis B antigen by radioimmunoassay but did not develop a seroresponse to the hepatitis A antigen by IEM; (*c*) neither guinea pig hyperimmune serum to hepatitis B antigen nor guinea pig serum to the core of the Dane particle reacted with the hepatitis A antigen by IEM (*22, 42, 71*). Therefore, we could not detect a relationship between the hepatitis A antigen and hepatitis B antigens (*29*).

We also studied the possibility of an antigenic relationship between the hepatitis A and Norwalk particles by IEM since both were not only similar morphologically but also since infectious hepatitis A serum and infectious Norwalk stool filtrates have been shown by Provost *et al.* (*70*) and Dolin *et al.* (*26*), respectively, to be ether, acid, and relatively heat stable in infectivity studies. We could not demonstrate a serological response by IEM to the Norwalk particle in paired sera

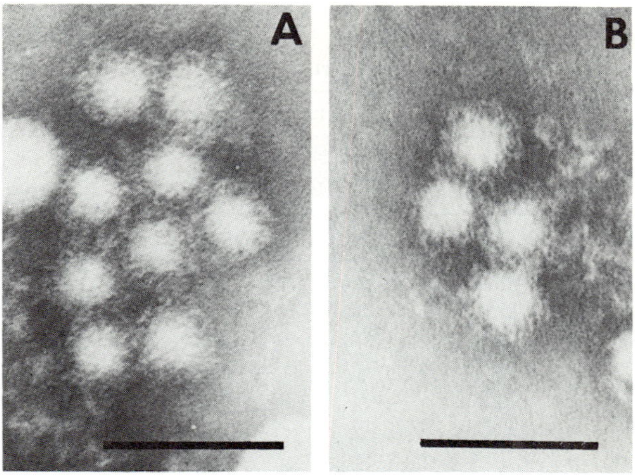

Fig. 21. Two aggregates observed following (a) incubation of the stool filtrate with 0.1 ml of a dilution of approximately 1:10 of a convalescent serum of a patient who developed hepatitis during the naturally occurring outbreak among members of the Holy Cross College football team and (b) further preparation for electron microscopy. The quantity of antibody on these particles was rated 3–4+. The bar in A and B = 100 nm. [Micrograph A after Feinstone *et al.* (*29*).

of 2 volunteers with experimentally induced hepatitis A (and with seroresponses to the hepatitis A antigen), nor could we demonstrate a serological response by IEM to the hepatitis A antigen in paired sera from 2 volunteers with experimentally induced Norwalk gastroenteritis (and with seroresponses to the Norwalk agent), indicating that the two agents were probably not serologically related.

In a previous study, a virus was recovered from cell cultures inoculated with specimens from volunteers in the Illinois study (*63*). This virus was later shown to be a parvovirus and to be related to the latent rat virus complex of agents (*64*). Although this isolate was most likely a contaminant unrelated to hepatitis A, the possibility of its association with the hepatitis A particle was examined with guinea pig hyperimmune serum to the Kilham strain of latent rat virus and rat virus antigen grown in cell culture. In a reciprocal cross-IEM study done under code, convalescent phase hepatitis A serum from a Joliet volunteer did not react with rat virus antigen but did with its homologous hepatitis A antigen, and the hyperimmune rat virus serum did not react with the hepatitis A antigen but did with its homologous rat virus antigen. Thus, it could be concluded that the hepatitis A antigen was not related to rat virus, and, in addition, the lack of an etiological relationship between rat virus and hepatitis A was confirmed.

TABLE 3. ANTIBODY TO ANTIGENS ASSOCIATED WITH HEPATITIS A, HEPATITIS B, AND NORWALK GASTROENTERITIS IN PAIRED SERA FROM PATIENTS WITH THESE DISEASES[a]

	Antibody to indicated antigen in first and second serum samples[b]					
	Hepatitis A antigen		Hepatitis B antigen		Norwalk gastroenteritis antigen	
Patient[c]	First	Second	First	Second	First	Second
	Experimental hepatitis A (MS-1), Illinois					
A [3]	0	3–4	–	–	1	1
B [4]	0	3–4	–	–	1	1
C [5]	0	1–2	–	–	NT	NT
D [6]	0	3–4	–	–	NT	NT
	Experimental hepatitis A (MS-1), New York					
E [1]	0	1–2	–	–	NT	NT
F [2]	0	1–2	+	+[d]	NT	NT
	Naturally acquired hepatitis A, American Samoa					
G [10]	0	2	–	–	NT	NT
H [11]	0	3	–	–	NT	NT
I [12]	1–2 (acute)	3–4	–	–	NT	NT
	Naturally acquired hepatitis A, Massachusetts					
J [7]	0–1 (acute)	3–4	–	–	NT	NT
K [8]	0 (acute)	3	–	–	NT	NT
L [9]	0 (acute)	3	–	–	NT	NT
	Naturally acquired hepatitis B					
M [13]	0	0	–	+	NT	NT
N [14]	0	0	–	+	NT	NT
	Experimental nonbacterial gastroenteritis					
O [15]	0	0	–	–	1	4
P [16]	2–3	2–3	–	–	1	4

[a] After Feinstone *et al.* (*29*).

[b] Antibody to the hepatitis A antigen was detected by immune electron microscopy (IEM) and was rated on a 0–4 scale, depending on the amount of antibody coating the particles. Antibody to the gastroenteritis antigen was determined in the same way. Antibody to hepatitis B antigen was measured by radioimmunoassay (RIA). The first serum of each serum pair listed was obtained prior to exposure, except in the case of four sera that were obtained during the acute phase of illness (acute). The second serum of each pair was obtained approximately 1–6 months after the onset of illness. Sera were tested by IEM at an initial dilution of 1:10 except for sera from patients J,K,L which were tested at approximately 1:10 and sera from patients E and F which were tested at 1:15. Sera were tested by RIA at a dilution of 1:4 or 1:10. Abbreviations: NT, not tested; –, negative; +, positive.

[c] Numbers in brackets indicate previous designation of patient in Feinstone *et al.* (*29*).

[d] No antibody increase.

We also tested another lot of immune human serum globulin for antibody to the hepatitis A antigen by IEM and found it to contain less antibody (2+) than the previously studied strongly positive (3–4+) lot (see Fig. 19). Such tests may prove useful in the potency standardization of immune serum globulin used for the prevention of hepatitis A.

The data from these IEM studies suggest that the 27-nm particle visualized in a stool filtrate derived from a patient with hepatitis A is the etiological agent of this disease. In addition, the detection of this agent in the stools of hepatitis A patients and the development of a serological technique with which to determine antibody to it provide a way to diagnose and study hepatitis A.

In collaboration with Dr. John Gerin, we attempted further to characterize the hepatitis A viruslike particle by determining its buoyant density in cesium chloride utilizing IEM to observe the distribution of particles in diluted fractions of the density gradient using minor modifications of the method employed previously for the Norwalk particle (*28, 47*). A convalescent serum known to contain antibody to the hepatitis A particle (*29*) from a patient with naturally acquired hepatitis A in American Samoa was used in the IEM studies (*28*). We studied original, approximately 20% stool extracts that had been shown, after ten-fold dilution and filtration, to be positive for the hepatitis A particle in studies by IEM as previously described (*29*). By this method we studied a stool extract derived from a volunteer during the acute phase of illness and found the peak fraction of the 27-nm particle to be at a buoyant density of approximately 1.41 gm/cm^3 in cesium chloride, as shown in Fig. 22. Figure 23 shows an aggregate of hepatitis A particles observed by IEM at a mean density in cesium chloride of about 1.4 gm/cm^3 (Fraction 4) in a previous isopycnic banding study of a stool extract from another volunteer in which only four consecutive fractions, 3–6, were reexamined in a repeat IEM study, and the peak number of particles was found in Fraction 4. (Details of the initial IEM study are described in the legend of Fig. 23.) The density of the hepatitis A particle of approximately 1.4 gm/cm^3 is consistent with that of a parvovirus or a rhinovirus but not an enterovirus (*37, 41, 44, 74, 79*). Although the antigen has a density similar to a rhinovirus, it can also be excluded from this subgroup of the picornaviruses, since rhinoviruses are acid-labile, whereas an infectious hepatitis A serum has been shown by Provost *et al.* to be acid-stable in studies in marmosets (*37, 44, 70, 74, 79*). Although the nucleic acid content of the hepatitis A particle has not yet been determined, it appears that this agent is parvovirus-like since it not only has a similar buoyant density in cesium chloride but also shares other properties with the parvoviruses such as morphology, and, as shown

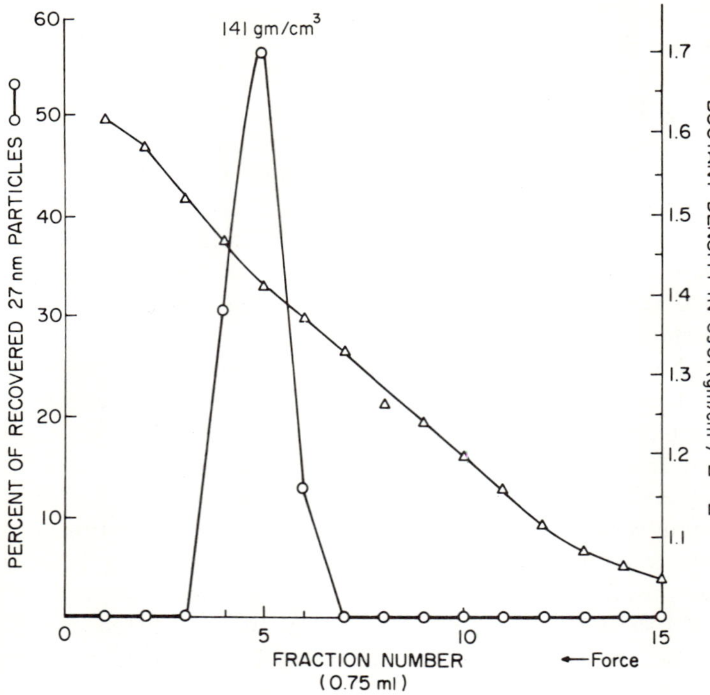

Fig. 22. Isopycnic banding of the hepatitis A viruslike particle in a cesium chloride density gradient. Quantitation of particle recovery was by IEM. [From Feinstone *et al.* (*28*).]

by Provost *et al.* in studies in marmosets with an infectious hepatitis A serum, acid, ether, and relative heat stability (*41, 70*).

VI. CONCLUSION

We have attempted to show the use of the technique of IEM for the detection of fastidious agents of human disease. For Norwalk gastroenteritis and hepatitis A, the technique enabled not only the definitive visualization and identification for the first time of viruslike particles from individuals with these diseases but also enabled us to demonstrate for the first time the serological association of an identifiable viruslike agent with each of these diseases. It is noteworthy that this was accomplished without benefit of any cell culture or animal system. I am certain that this technique will have many other applications, some practical, such as the typing of viruses directly after recovery or isolation, the latter having already been shown by Anderson and Doane (*9*) to be possible with certain enteroviruses. We have also recently been able to type a rhinovirus by IEM by reacting the cell culture isolate with

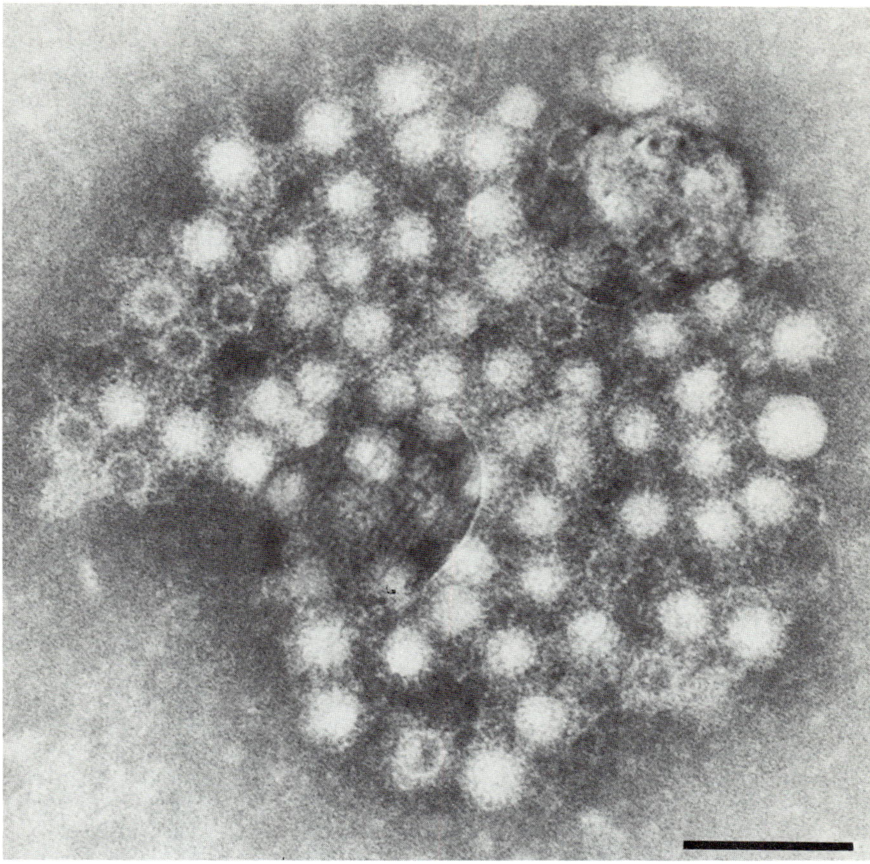

Fig. 23. An aggregate of hepatitis A particles observed by immune electron microscopy (IEM) at a mean density of about 1.4 gm/cm^3 in cesium chloride. The particles in the diluted fraction (Fraction 4) have been aggregated by a 1:10 dilution of a late convalescent serum from a patient with naturally acquired hepatitis A. Only four consecutive fractions, 3–6, of the gradient from this initial isopycnic banding experiment were reexamined in a repeat IEM study, and the peak fraction of particles was found in Fraction 4. [On initial examination of twelve fractions (2–13) by IEM, only 50 particles were seen: 6 in Fraction 3; 19 in Fraction 5; and 25 in Fraction 10. Fractions 4 and 11 could not be examined satisfactorily because of poor preparations.] The bar = 100 nm.

serum pools containing antibody to the eighty-nine numbered rhinoviruses plus the subtype (*1, 2*), followed by individual testing by IEM of the isolate with each serum in the pool that aggregated the virus. However, the most rewarding and exciting use of this technique will be in its application in the quest for the multitude of fastidious agents of human disease that cannot be cultivated or cannot be cultivated effectively in any *in vitro* or *in vivo* system.

ACKNOWLEDGMENTS

We thank Dr. R. F. Buscho for rhinovirus 13; Mr. H. D. James, Jr., for rhinovirus 1A and antisera to rhinoviruses 1A and 13; Drs. M. H. Hatch, J. L. Adler, and R. Zickl for serum specimens from the Norwalk, Ohio, gastroenteritis study; Dr. D. W. Gibson for the stool extracts for the hepatitis studies; Drs. J. Boggs, M. E. Conrad, S. Krugman, M. Oxman, and L. Rosen for serum specimens from hepatitis A patients; Dr. L Barker for antiserum to the core of the Dane particle; Dr. M. D. Hoggan for the rat virus reagents; and B. V. Armiger, D. B. Bertran, R. P. Chames, E. B. Compton, R. M. Faust, H. D. James, Jr., S. J. Kelly, L. P. Kendrick, A. L. Vaughn, and J. Valdesuso for assistance in these studies. A special debt of gratitude to Professor Anthony P. Waterson who agreed to my (A.Z.K.) working in his laboratory and to June D. Almeida who taught me immune electron microscopy.

REFERENCES

1. A collaborative report. Rhinoviruses. A numbering system *Nature* (*London*) **213**, 761–763 (1967).
2. A collaborative report. Rhinoviruses. Extension of the numbering system. *Virology* **43**, 524–526 (1971).
3. Adler, J. L., and Zickl, R., *J. Infec. Dis.* **119**, 668–673 (1969).
4. Almeida, J. D., Cinader, B., and Howatson, A., *J. Exp. Med.* **118**, 327–340 (1963).
5. Almeida, J. D., and Goffe, A. P., *Lancet* **2**, 1205–1207 (1965).
6. Almeida, J. D., and Morris, R., *J. Gen. Virol.* **20**, 369–375 (1973).
7. Almeida, J. D., Rubenstein, D., and Stott, E. J., *Lancet* **2**, 1225–1227 (1971).
8. Almeida, J. D., and Waterson, A. P., *Advan. Virus Res.* **15**, 307–338 (1969).
9. Anderson, N., and Doane, F. W., *Can. J. Microbiol.* **19**, 585-589 (1973).
10. Anderson, T. F., and Stanley, W. M., *J. Biol. Chem.* **139**, 339–344 (1941).
11. Ardenne, M. von, Friedrich-Freksa, H., and Schramm, G., *Arch. Gesamte Virusforsch.* **2**, 80–86 (1941).
12. Blacklow, N. R., Dolin, R., Fedson, D. S., DuPont, H., Northrup, R. S., Hornick, R. B., and Chanock, R. M., *Ann. Intern. Med.* **76**, 993–1008 (1972).
13. Blough, H. A., Tiffany, J. M., Gordon, G., and Fiala, M., *Virology* **38**, 694–712 (1964).
14. Boggs, J. D., Melnick, J. L., Conrad, M. E., and Felsher, B. F., *J. Amer. Med. Ass.* **214**, 1041–1046 (1970).
15. Bradburne, A. F., and Tyrrell, D. A. J., *Progr. Med. Virol.* **13**, 373–403 (1971).
16. Brenner, S., and Horne, R. W., *Biochim. Biophys. Acta* **34**, 103–110 (1959).
17. Bruckova, M., McIntosh, K., Kapikian, A. Z., and Chanock, R. M., *Proc. Soc. Exp. Biol. Med.* **135**, 431–435 (1970).
18. Candeias, J. A. N., de S. Carvalho, R. P., and Antonacio, F., *Rev. Inst. Med. Trop. Sao Paulo* **14**, 121–125 (1972).
19. Cavallaro, J. J., and Monto, A. S., *J. Infec. Dis.* **122**, 272–279 (1970).
20. Chapple, P. J., and Harris, W. J., *Nature* (*London*) **209**, 790–792 (1966).
21. Clarke, S. K. R., Cook, G. T., Egglestone, S. I., Hall, T. S., Miller, D. L., Reed, S. E., Rubenstein, D., Smith, A. J., and Tyrrell, D. A. J., *Brit. Med. J.* **3**, 86–89 (1972).
22. Dane, D. S., Cameron, C. H., and Briggs, M., *Lancet* **1**, 695–698 (1970).
23. Dingle, J. H., Badger, G. F., Feller, A. E., Hodges, R. G., Jordan, W. S., Jr., and Rammelkamp, C. H., Jr,, *Amer. J. Hyg.* **58**, 16–30 (1953).

24. Dingle, J. H., Badger, G. F., and Jordan, W. S., Jr., "Illness in the Home. A Study of 25,000 Illnesses in a Group of Cleveland Families," The Press of Western Reserve University, Cleveland, Ohio, 1964.
25. Dingle, J. H., McCorkle, L. P., Badger, G. F., Curtiss, C., Hodges, R. G., and Jordan, W. S., Jr., *Amer. J. Hyg.* **64,** 368–375 (1956).
26. Dolin, R., Blacklow, N. R., DuPont, H., Buscho, R. F., Wyatt, R. G., Kasel, J. A., Hornick, R., and Chanock, R. M., *Proc. Soc. Exp. Biol. Med.* **140,** 578–583 (1972).
27. Dolin, R., Blacklow, N. R., DuPont, H., Formal, S., Buscho, R. F., Kasel, J. A., Chames, R. P., Hornick, R., and Chanock, R. M., *J. Infec. Dis.* **123,** 307–312 (1971).
28. Feinstone, S. M., Kapikian, A. Z., Gerin, J. L., and Purcell, R. H., *J. Virol.* **13,** 1412–1414 (1974).
29. Feinstone, S. M., Kapikian, A. Z., and Purcell, R. H. Science **182,** 1026–1028 (1973).
30. Fukumi, H., Nakaya, R., Hatta, S., Noriki, H., Yumoki, K., Akagi, T., Saito, K., Vehiyama, K., Kobari, K., Nakanishi, R.. *Jap. J. Med. Sci. Biol.* **10,** 1–17 (1957).
31. Gerin, J. L,, Richter, W. R., Fenters, J. D., and Holper, J. C., *J. Virol.* **2,** 937–943 (1968).
32. Gibson, D. W., personal communication
33. Gordon, I., *Amer. J. Trop. Med. Hyg.* **4,** 739–755 (1955).
34. Gordon, I., Ingraham, H. S., and Korns, R. F., *J. Exp. Med.* **86,** 409–422 (1947).
35. Gordon, I., Meneely, J. K., Jr., Currie, G. D., and Chicoine, A., *J. Lab. Clin. Med.* **41,** 133–141 (1953).
36. Hamparian, V. V., Ketler, A., and Hilleman, M. R., *Proc. Soc. Exp. Biol. Med.* **108,** 444–453 (1961).
37. Hamre, D., *in* "Monographs in Virology" (J. L. Melnick, ed.), Vol. 1, Karger, Basel, Switzerland, New York. 1968.
38. Hamre, D., and Beem, M., *Amer. J. Epidemiol.* **96,** 94–106 (1972).
39. Hamre, D., and Procknow, J. J., *Proc. Soc. Exp. Biol. Med.* **121,** 190–293 (1966).
40. Hendley, J. O., Fishburne, H. B., and Gwaltney, J. M., Jr., *Amer. Rev. Resp.* Dis. **105,** 805–811 (1972).
41. Hoggan, M. D., *in* "Comparative Virology" (K. Maramorosch and E. Kurstak, eds.), Chapter 2, pp. 43–79 Academic Press, New York, 1971.
42. Hoofnagle, J. H., Gerety, R. J., and Barker, L. F., *Lancet* **2,** 869–873 (1973).
43. Jordan, W. S., Jr., Gordon, I., and Dorrance, W. R., *J. Exp. Med.* **98,** 461–475 (1953).
44. Kapikian, A. Z., *in* "Diagnostic Procedures for Virology and Rickettsial Diseases" (E. H. Lennette and N. J. Schmidt, eds.), 4th ed., pp. 603–640. Amer. Pub. Health Ass., New York, 1969.
45. Kapikian, A. Z., *in* "Diagnostic Procedures for Viral and Rickettsial Diseases" (E. H. Lennette and N. J. Schmidt, eds.), 4th ed., pp. 931–946. Amer. Pub. Health Ass., New York, 1969.
46. Kapikian, A. Z., Almeida, J. D., and Stott, E. J., *J. Virol.* **10,** 142–146 (1972).
47. Kapikian, A. Z., Gerin, J. L., Wyatt, R. G., Thornhill, T. S., and Chanock, R. M., *Proc. Soc. Exp. Biol. Med.* **142,** 874–877 (1973).
48. Kapikian, A. Z., James, H. D., Jr., Kelly, S. J., Dees, J. H., Turner, H. C., McIntosh, K., Kim, H. W., Parrott, R. H., Vincent, M. M., and Chanock, R. M., *J. Infec. Dis.* **119,** 282–290 (1972).
49. Kapikian, A. Z., James, H. D., Jr., Kelly, S. J., King, L. M., Vaughn, A. L., and Chanock, R. M., *Proc. Soc. Exp. Biol. Med.* **139,** 179–186 (1972).

50. Kapikian, A. Z., James, H. D., Jr., Kelly, S. J., and Vaughn, A. L., *Infec. Immunity* **7**, 111–116 (1973).
51. Kapikian, A. Z., Wyatt, R. G., Dolin, R., Thornhill, T. S., Kalica, A. R., and Chanock, R. M., *J. Virol.* **10**, 1075–1081 (1972).
52. Kawana, R., and Matsumoto, I., *Jap. J. Microbiol.* **15**, 207–217 (1971).
53. Kaye, H. S., Marsh, H. B., and Dowdle, W. R., *Amer. J. Epidemiol.* **94**, 43–49 (1971).
53a. Kelen, A. E., Hathaway, A. E., and McLeod, D. A., *Canad. J. Microbiol.* **17**, 993–1000 (1971).
54. Ketler, A., Hamparian, V. V., and Hilleman, M. R., *Proc. Soc. Exp. Biol. Med.* **110**, 821–831 (1962).
55. Kojima, S., Fukumi, H., Kusama, H., Yamamoto, S., Suzuki, S., Uchida, T., Ishimaru, T., Oka, T., Kuretani, S., Ohmura, K., Nishikawa, F., Fujimoto, S., Fujita, K., Nakano, A., and Sunakawa, S., *Jap. Med. J.* **1**, 467–476 (1948).
56. Krugman, S., Giles, J. P., and Hammond, J., *J. Amer. Med. Ass.* **200**, 365–373 (1967).
57. Lander, J. L., Giles, J. P., Purcell, R. H., and Krugman, S., *N. Engl. J. Med.* **285**, 303–307 (1971).
58. Mayor, H. D., *Virology* **22**, 158–160 (1964).
59. McGregor, S., and Mayor, H. D., *J. Virol.* **2**, 149–154 (1968).
60. McGregor, S, Phillips, C. A., and Mayor, H. D., *Proc. Soc. Exp. Biol. Med.* **122**, 118–121 (1966).
61. McIntosh, K., Becker, W. B., and Chanock, R. M., *Proc. Nat. Acad. Sci. U.S.* **58**, 2268–2273 (1967).
62. McIntosh, K., Kapikian, A. Z., Turner, H. C., Hartley, J. W., Parrott, R. H., and Chanock, R. M., *Amer. J. Epidemiol.* **91**, 585–592 (1970).
63. Melnick, J. L., Boucher, W., Craske, J., and Boggs, J., *J. Infec. Dis.* **124**, 76–85 (1971).
64. Mirkovic, R. P., Adamova, V., Boucher, W., and Melnick, J. L., *Proc. Soc. Exp. Biol. Med.* **138**, 626–631 (1971).
65. Morse, L. J., Bryan, J. A., Hurley, J. P., Murphy, J. F., O'Brien, T. F., and Wacker, W. F. C., *J. Amer. Med. Ass.* **219**, 706–708 (1972).
66. Morton, G. A., *RCA Rev.* **6**, 131–166 (1941).
67. Mudd, S., *J. Amer. Med. Ass.* **126**, 632–639 (1944).
68. Oshiro, L. S., Schieble, J. H., and Lennette, E. H., *J. Gen.. Virol.* **12**, 161–168 (1971).
69. Paver, W. K., Caul, E. O., Ashley, C. R., and Clarke, S. K. R., *Lancet* **1**, 237–240 (1973).
70. Provost, P. J., Ittensohn, O. L., Villarejos, V. M., Arguedas G, J. A., and Hilleman, M. R., *Proc. Soc. Exp. Biol. Med.* **142**, 1257–1267 (1973).
71. Purcell, R. H., Gerin, J. L., Holland, P. V., Cline, W. L., and Chanock, R. M., *J. Infec. Dis.* **121**, 222–226 (1970).
72. Reimann, H. A., Price, A. H., and Hodges, J. H., *Proc. Soc. Exp. Biol. Med.* **59**, 8–9 (1945).
73. Rosen, L., personal communication.
74. Rueckert, R. R., *in* "Comparative Virology" (K. Maramorosch and E. Kurstak, eds.), Chapter 8, pp. 255–306. Academic Press, New York, 1971.
75. Schaaf, D. M., Boswell, J., and Mufson, M. A., *Amer. Rev. Resp. Dis.* **104**, 41–51 (1971).
75a. Schieble, J. H., personal communication.
76. Tateno, I., Suzuki, S., Kagawa, S., Naito, H., and Goto, T., *Jap. J. Exp. Med.* **26**, 125–138 (1956).

77. Thomas, D. C., Conant, R. M., and Hamparian, V. V., *Proc. Soc. Exp. Biol. Med.* **133**, 62–65 (1970).
78. Thornhill, T. S., Kalica, A. R., Wyatt, R. G., Kapikian, A. Z., and Chanock, R. M., *J. Infec. Dis.* (submitted for publication).
79. Tyrrell, D. A. J., *Virol. Monogr.* **2**, 67–124 (1968).
80. Tyrrell, D. A. J., *J. Infec. Dis.* **121**, 561–571 (1970).
81. Tyrrell, D. A. J., and Bynoe, M. L., *Brit. Med. J.* **1**, 1467–1470 (1965).
82. Wacker, W. E. C., Riordan, J. F., Snodgrass, P. J., Chang, L. W., Morse, L. J., O'Brien, T. F., and Reddy, W. J., *Arch. Intern. Med.* **130**, 357–360 (1972).
83. Wyatt, R. G., Dolin, R., Blacklow, N. R., DuPont, H. L., Buscho, R. F., Thornhill, T. S., Kapikian, A. Z., and Chanock, R. M., *J. Infec. Dis.* **129**, 709–714 (1974).
84. Wyatt, R. G., Kapikian, A. Z., Thornhill, T. S., Dalgard, D. W., Allen, W. P., and Chanock, R. M., *J. Infec. Dis.* (submitted for publication).
85. Wyatt, R. G., Thornhill, T. S., DuPont, H., Kapikian, A. Z., and Chanock, R. M., unpublished studies.

DISCUSSION

DR. O'CONNOR: The elegant results reported by Dr. Kapikian prompt me to report that another technique besides immunoelectron microscopy is now available for this kind of study. Dr. Ralph deBlois and his associates at General Electric, Schenectady, have developed a new method for sizing and counting virus particles—the electrical resistance pulse analysis (ERP). In the ERP technique, virus particles in fluid flowing through a microcylindrical pore (made by General Electric Nucleopore process in a plastic sheet) are detected by the disturbance that they produce in an electrical voltage applied across the pore. Individual virus particles are sized and counted. The technique essentially involves an extension of the existing Coulter counter to particles of virus size. Viruses at concentrations $>10^8$/ml can be sized and counted in times of approximately 1 minute. To date the technique has been applied to polyomavirus, T phages, mouse leukemia virus, and mouse mammary tumor virus. Dr. Sam Mayassi and his associates at Charles Pfizer and Co., Maywood, have used this technique for sizing and counting virus particles routinely in production of oncornavirus stocks. They have further demonstrated that addition of specific antisera causes formation of aggregates that are detected in the resistance pulse analyzer. Although occasional plugging of the pore by virus–antibody complexes still presents mechanical difficulties, this problem appears solvable. This technique does appear amenable to applications in detection of viruses and virus–antibody complexes. The foregoing results by Dr. deBlois and associates and by Dr. Mayassi and associates were executed under the Molecular Control Program of the National Cancer Institute and will be reported at the spring meeting of the American Association for Cancer Research in Houston.

DR. KAPIKIAN: We reported in our original gastroenteritis study by immune electron microscopy that it took a median time of approximately 1 hour to examine a specimen in order to assign an antibody rating. It doesn't take us quite as long now. In the hepatitis A studies, such ratings took less time because the hepatitis antigen we used in the original serologic studies contained fewer particles than the Norwalk 8FIIa gastroenteritis antigen and therefore in the presence of specific antibody the hepatitis A particles were more randomly distributed as they occurred more often as antibody-coated single particles. In general, serological studies such as these by immune electron microscopy are quite time consuming; the system you mention would certainly be a great boon to us all if applicable for such serological studies.

DR. MELNICK:

1. Two distinct viruses have been passed in marmosets from hepatitis A patients, one by Deinhardt's group and a second by Hilleman's group. Have you examined paired sera from marmosets from these two laboratories?

2. From the acute sera of three of the Joliet volunteers, we have reported the isolation of a morphologically similar virus in the ill-fated Detroit-6 cells—at a time when they showed "hepatitis A susceptibility." On serial passage it appeared that the viruses were replaced with a contaminating rodent parvovirus. It would be of interest to examine the sera of the volunteers to determine if these sera were also positive for your virus by immune electron microscopy.

DR. KAPIKIAN: First, in response to the second question, Drs. Feinstone, Purcell, and I have not studied any of the sera from the Joliet study specifically for the purpose of trying to detect hepatitis A particles by immune electron microscopy as yet. We hope to do such studies soon. Second, we have not examined marmoset sera for antibody to the hepatitis A particle from the sources that you mentioned.

DR. KOCH: Can your method be used to get an estimate on the number of antibody molecules bound to one virus particle?

DR. KAPIKIAN: No. I don't think the technique enables such an estimate to be made. We can get a general idea of the amount of antibody coating a particle and assign a rating to it. Occasionally we see what we think is a bridging antibody molecule between two particles, but because of the limits of resolution of electron microscopy, I don't think we could estimate the number of such small molecules on a virus particle with the technique of immune electron microscopy as we use it.

DR. HALL: I would like to compliment Dr. Kapikian and his colleagues on their very beautiful work. Using fecal and serum samples obtained by Dr. David Madden (NINDS, NIH) during an outbreak of hepatitis A at the Lynchburg (Virginia) Training School and Hospital, Dr. H. E. Bond and I have been looking by immune electron microscopy at a naturally occurring epidemic of hepatitis A. We feel that we have found some particles similar to those described by Dr. Kapikian but require further studies before we can draw any firm conclusion about their distribution.

DR. KAPIKIAN: Drs. Feinstone, Purcell, and I recently collaborated with Dr. D. Madden, who provided paired serum specimens from patients who developed hepatitis during this Lynchburg outbreak, to determine if this hepatitis outbreak was associated serologically with the 27-nm hepatitis A particle. In a study under code of paired sera from each of 2 patients who developed hepatitis during this outbreak, it was found that both patients developed a significant serological response to the hepatitis A particle by immune electron microscopy.

DR. LENNETTE: My comments are impelled by the passing remark of the chairman concerning the potential applicability of the techniques described by Dr. Kapikian to practical problems of biomedicine.

An important example that almost immediately comes to mind is the rapid identification of variolavirus by the immunoelectron-microscopic approach described by Dr. Kapikian and the new techniques described by Dr. O'Connor.

Public health officials in this country are always on the alert for possible introduction of smallpox virus and its rapid identification, which is not always feasible under present circumstances. This is a matter of some importance since it is the word from the laboratory that determines whether or not the costly machinery of a vaccination campaign is to be put into motion. The need for an accurate, as well as a rapid, diagnosis is obvious.

As many of you are aware, vaccination against smallpox is no longer a requirement in this country, this action being based on the long-recognized fact that the number of adverse reactions to vaccination has outnumbered by far the incidence of the actual disease. Unless smallpox as a disease is completely eradicated, we will face the increasingly important problem of an increasing population of smallpox susceptibles and, hence, the need for a rapid method for recognizing and identifying the virus should a suspected case of variola be introduced into the country. Immunoelectron microscopy offers such a diagnostic aid (not only for smallpox but also for other viral diseases), and I take occasion to emphasize here that the electron microscope is no longer an instrument basic solely to the laboratory engaged in sophisticated or esoteric research, but is one that has now found, and will continue to find, pragmatic applications in the diagnostic public health and clinical virology laboratories.

DR. KAPIKIAN: Yes. I think the technique of immune electron microscopy could have practical applications in the rapid diagnosis of certain diseases, among them smallpox. Its use could assist in determining the etiology of certain vesicular lesions since with specific antisera an agent present in vesicular fluid could be identified. In addition, if the viral agent in a vesicle was too low titered to be visualized by conventional electron microscopy, the technique of immune electron microscopy might enable its visualization and with this, its identification.

CHAPTER 3

Current Understanding of Hepatitis B Virus Infection and Its Implications for Immunoprophylaxis

Robert H. Purcell

I. INTRODUCTION

Two papers summarizing the status of research on hepatitis B (serum hepatitis were presented at the Seventh Gustav Stern Symposium in 1970, an additional paper on hepatitis B was presented in 1972, and I am again summarizing the state of the art in 1974. That hepatitis B again appears on the program serves to emphasize the continuing rapid progress that has been made in this field since the discovery of Australia antigen by Dr. Blumberg in 1964 (*12*). This progress has been made despite the fact that the hepatitis B virus still has not been reproducibly isolated and passaged in cell or tissue culture. Some of the recent contributions to our understanding of hepatitis B are reviewed here, and how these may eventually lead to the successful immunoprophylaxis of hepatitis B infection is indicated.

II. NOMENCLATURE

The association of Australia antigen with hepatitis B led many to propose new names that more appropriately reflected its relationship to this disease. These included serum hepatitis antigen, hepatitis antigen, hepatitis-associated antigen, and hepatitis B antigen. The terms hepatitis B antigen (HB Ag) and antihepatitis B antigen (anti-HB Ag) or hepatitis B antibody (HB Ab) are the most commonly used terms at present and reflect the return to the use of the terms hepatitis B for serum hepatitis and hepatitis A for infectious hepatitis (*66*).

A fuller understanding of the complexity of the antigens associated with hepatitis B antigen has suggested to some the need for modification of this nomenclature (*20*). Thus, the term HB_s Ag has been proposed to denote the antigens found on the surface of hepatitis B antigen particles, and HB_c Ag has been proposed for certain internal "core" antigens found in one of the morphological forms of HB Ag (the Dane particle; see Section VI).

III. TESTS FOR HB_s Ag and anti-HB_s

Hepatitis B antigen and its antibody were first detected by the demonstration of a precipitin reaction in agar gel. However, subsequent studies demonstrated that this technique was relatively insensitive, particularly for the detection of antibody (anti-HB_s). The application of modern technology to the development of more sensitive tests has resulted in a bewildering array of sophisticated techniques that provide different degrees of specificity and sensitivity. At present the most sensitive tests for detecting HB_s Ag include radioimmunoassays, in which the antigen or its antibody is labeled with radioactive iodine, and passive hemagglutination techniques in which the antibody is coupled to red cells. Both types of tests yield false positive results and, therefore, require careful confirmation of specificity. These two types of tests are expected soon to replace the simpler but less sensitive counter-electrophoresis technique for the detection of antigen in the serum of prospective blood donors.

Hepatitis B antibody is not readily detected following primary exposure when techniques useful for the detection of antibody to other virus infections are employed. However, sensitive radioimmunoassays using radiolabeled HB_s Ag and passive hemagglutination techniques utilizing red cells coated with the antigen reveal anti-HB_s in the sera of most individuals convalescent from acute hepatitis B infection.

IV. EPIDEMIOLOGY

Application of the new techniques for detecting HB_s Ag and anti-HB_s to the epidemiology of hepatitis B infection has forced us to reevaluate our understanding of this disease. Paramount among these changing concepts is the realization that a significant proportion of hepatitis B infection must be acquired by nonparenteral or covert parenteral means. Although there was early epidemiologic evidence that such transmission may occur, this was overlooked in the face of the generally held belief that this type of hepatitis could be transmitted only through parenteral exposure to blood or blood products containing the virus, such as occurs during transfusion, the administration of blood components such as plasma and antihemophilic globulin, or the use of unsterile syringes contaminated with the blood of an infected individual.

Two factors served to shake our faith in this concept. One was the discovery by Krugman, Giles, and Hammond (*58*) that institutionalized retarded children were endemically exposed to two distinct types of hepatitis—one was epidemiologically similar to hepatitis A, and the other resembled hepatitis B. Both appeared to be transmitted by nonparenteral means, at least under the special conditions of living that pertained in such an institution.

The other finding resulted from surveys of various populations for the then recently discovered hepatitis B antigen. It was found that populations of certain countries, generally those with tropical climates, had frequencies of HB_s Ag as high as 10%, 10–100 times higher than the frequency of approximately 0.1% found throughout most of North America and Europe (*13*). The high frequency of antigen in such countries appears to be related to the developing nature of their cultures rather than their geographic locations, since similar high frequencies were discovered among populations living in sub-Arctic climates (*13, 94*). Furthermore, the prevalence of antigen was higher among the institutionalized, such as prisoners and the mentally retarded, and among populations of lower socioeconomic status than among "open" populations in the developed countries. Subsequent studies revealed that most of the individuals found to be positive for hepatitis B antigen were, in fact, chronic carriers of the antigen with or without concomitant evidence of chronic hepatitis.

Opportunities for parenteral exposure could not be identified in many antigen-positive individuals from the various populations studied. Furthermore, a high proportion of patients with sporadic "infectious" hepatitis not associated with documented parenteral exposure to blood or blood products was found to have HB_s Ag in their blood. These findings taken together, strongly suggested that a significant proportion of hepatitis B infection was transmitted by nonparenteral means.

Further evidence of such nonparenteral transmission was obtained from a population survey for antibody to HB_s Ag using a sensitive radioimmunoassay technique (*60*). In a collaborative study with Cherubin (*19*), antibody was found to increase in frequency with increasing age

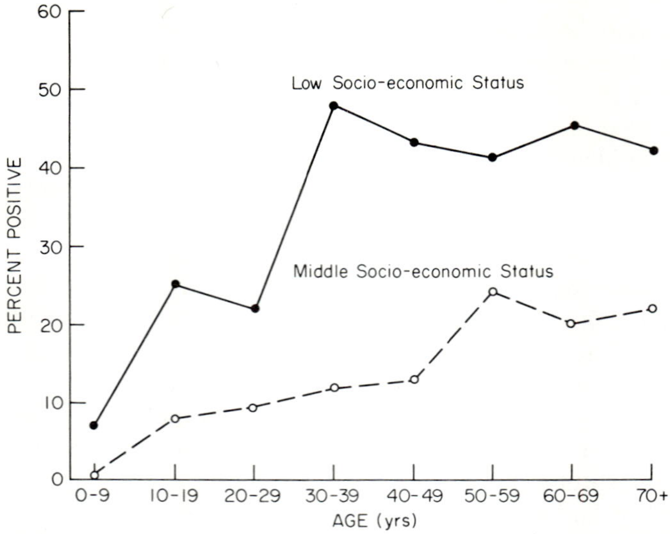

Fig. 1. Acquisition of antibody to hepatitis B surface antigen (anti-HB_s) in two socioeconomically diverse populations in New York City. The low socioeconomic status population consisted of individuals (mostly black) from Harlem. The middle socioeconomic status population consisted of working class whites from Staten Island. Sera from a high socioeconomic population were also tested; they had a lower frequency of antibody than either of the two populations. Sera were tested for antibody by a sensitive radioimmunoprecipitation technique (*19*).

of those sampled and to be found in inverse proportion to the socioeconomic status of the population studied (Fig. 1). In the population with the lowest socioeconomic standard studied, a maximum frequency of over 40% was found. In other studies with Reeves (*89*), we have found a similar high frequency of antibody among certain primitive populations, but not among other similar populations living under more hygienic conditions (Fig. 2). Similar patterns of anti-HB_s have been observed by others (*100–102*). Although 80–90% of experimentally infected individuals have been shown to develop anti-HB_s, such antibody is persistent in only approximately 50% (*9*). Therefore, the finding of a prevalence of anti-HB_s approaching or exceeding 50% in certain populations suggests an almost universal exposure to the agent in these populations, such infections usually are inapparent.

The means by which hepatitis B virus is maintained in the population is not presently understood. Several modes of transmission have been suggested. Krugman *et al.* (*58*) demonstrated in 1967 that the MS-2 strain of hepatitis B virus could be transmitted by ingestion of infectious serum, thus suggesting the possibility of fecal–oral spread. Hepatitis B

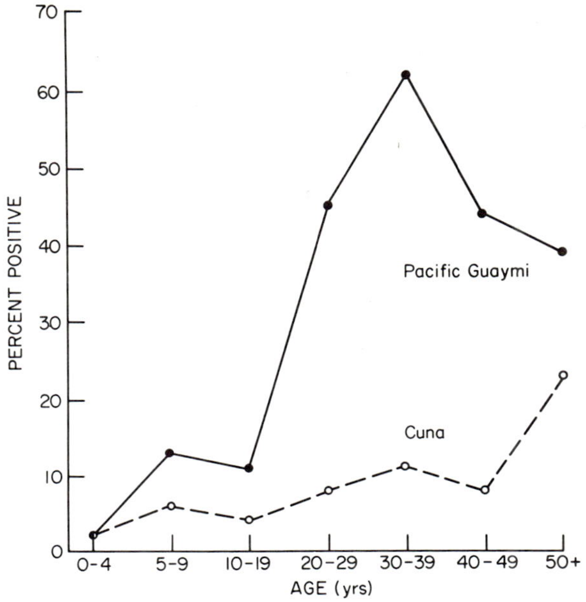

Fig. 2. Antibody to hepatitis B surface antigen (anti-HB_s) in two primitive Indian populations living in Panama. The Pacific Guaymi live in mountain forest regions along streams that they use for drinking, bathing, laundry, and defecation. The Cuna Indians live on islands off the Caribbean coast. They defecate in the ocean, but use freshwater from the mainland for drinking, bathing, and laundry. Anti-HB_s was measured with a sensitive radioimmunoprecipitation technique (*89*).

antigen has been detected in saliva, nasopharyngeal secretions, urine, and feces (*4, 38, 40, 79, 96, 105, 107*). Other laboratories have not always been able to confirm these findings, and there remains a degree of controversy over their significance. However, at least some of these observations appear to be bona fide. Whether such extracorporeal antigen is derived from contamination with blood or whether it represents locally produced antigen remains to be determined, but Ogra (*79*) has provided evidence of local production of antigen in the gastrointestinal and respiratory tracts followed by development of specific secretory IgA antibody.

Several reports have appeared in the literature documenting cases of hepatitis thought to be acquired by nonparenteral means, and aerosols have been suggested as one means of exposure (*1, 26, 29, 30, 41*). However, it is difficult to rule out unrecognized exposure to contaminated blood in these cases.

Another proposed means of transmission is via the venereal route. Higher than normal rates of HB_s Ag or anti-HB_s have been found among patients attending venereal disease clinics (*28, 47*), and transmission of hepatitis B from one spouse to the other has been observed. Mazzur (*70*) has cautioned against the possible infectivity of menstrual blood; such exposure might explain some of the apparently venerally transmitted hepatitis.

Vertical transmission from mother to offspring has been documented, particularly when the mothers have had acute hepatitis B in close temporal proximity to parturition (*5, 53, 71, 73, 91, 95, 108*). Such transmission occurs less frequently to offspring of the chronic carrier of hepatitis B antigen. It is not clear whether infection of the infant occurs more frequently *in utero*, at time of delivery, or in the immediate postnatal period.

Arthropods have been suggested as a possible means of covert parenteral spread of the virus, and HB_s Ag has been detected in wild-caught mosquitoes in the United States and Africa as well as in mosquitoes experimentally fed blood containing the antigen (*18, 23, 74, 81, 93, 97*). With the exception of one unconfirmed report, multiplication of HB_s Ag in mosquitoes has not been observed (*97*). However, mechanical transfer of the virus by arthropods sequentially feeding on two or more individuals has been postulated. Epidemiologic data suggest that arthropod-borne transmission, should it occur, is not a major cause of hepatitis B infection (*39*).

Clustering of HB_s Ag carriers within families is not a rare occurrence, and it seems to occur more frequently among genetically related family members than between spouses (*7, 11, 13, 80, 103*). No such association was seen when total infection rates (family members positive for HB_s Ag or anti-HB_s) were examined. Whether one of the suspected modes of transmission discussed above or some other environmental factor or, as suggested by Blumberg (*13*), a genetic factor is responsible is not understood at present.

In summary, hepatitis B virus appears to be endemic in many populations, to be spread by means not yet clearly understood, and to cause a very high ratio of subclinical infections.

V. ANTIGENIC DIVERSITY OF HBs Ag

It has now been well established that multiple antigenic specificities are associated with HB_s Ag (*6, 27, 55, 61–63, 67, 72, 76, 99, 106*). Our current understanding of these specificities is summarized in Table 1. Specificity *a* appears to be invariably associated with the various morphological forms of HB_s Ag and appears to be identical with the original Australia antigen described by Blumberg.

TABLE 1. SUBTYPES OF HEPATITIS B SURFACE ANTIGEN

Subtypes	Distribution	Relative frequency
Major		
adw	North, South America, western Europe, western, southern Africa	Common
ayw	Middle East, Mediterranean countries, eastern, northern Africa, Asian subcontinent (North America, Europe)	Common
adr	Far East	Common
ayr	Far East	Rare
Minor		
a_1, a_2, a_3	Not well characterized	Probably common

Two additional pairs of antigenic specificities, *d* and *y* (*61*) and *w* and *r* (*6*) have been described. Each pair behaves as though it were allelic and one member of each of these pairs of "subtype" antigens is found in association with the "group-reactive" specificity *a*. Thus, four groupings of specificities are possible (*adw, ayw, adr, ayr*), and all four have been identified.

The subtype specificities of HB_s Ag have epidemiologic significance. Antigen found in the serum of contact cases of hepatitis B invariably is of the same subtype as antigen found in the index cases. Such antigens have "bred true" through several serial passages in humans and chimpanzees.

In addition, each subtype specificity has a distinct geographic distribution (Table 1). Thus, *d* is found most commonly among populations in North and South America, Europe, and parts of Africa and the Far East, whereas subtype *y* predominates in the Middle East, northern Africa, southern Europe, and parts of Asia. Subtype *w* predominates throughout most of the world except in the Far East where subtype *r* is also found. In parts of the Far East where these zones of distribution overlap, all four subtype groupings have been found in the same populations.

Antigens that appear to have more than the three expected specificities have been described (*72, 77*); cross-absorption experiments suggest that, in some cases at least, all of the specificities reside on the same particle and, therefore, represent true "chimeras" and do not represent mixed infections with two or more subtype groupings, an occurrence that also has been reported (*106*). The significance of these infrequently found chimeras remains to be established, but they may represent dual infections with two or more hepatitis B viruses in which genotypic or phenotypic mixing has occurred (*77*).

Additional subspecificities of the group-reactive *a* antigen have been described by Soulier (*99*); these appear to be distinguishable from, but related to, *a* and have been designated a_1, a_2, and a_3. The significance of these subspecificities is unknown at present.

Two other antigenic specificities have been described and deserve mention. One, specificity *x*, does not appear to breed true and, therefore, may not be genetically determined by the virus. It has been suggested that this antigen may be derived from the host (*61*). The second antigenic specificity, designated *e*, has been found in patients with hepatitis B (*67*). It appears not to be physically associated with the morphological forms of HB_s Ag but is associated with a soluble protein found in the serum of some HB_s Ag carriers. Antibody to *e* (anti-*e*) has also been found in patients with hepatitis B. The significance of *e* and anti-*e* has not been determined.

Antibody to the subtype specificities *d, y, w,* and *r,* as well as to the group-reactive antigen *a* can be detected following primary or secondary exposure to HB_s Ag. This antibody response appears to be specific for the subtype of HB_s Ag associated with the infection, at least following primary exposure (*37*). Gold *et al.* (*37*) have shown that the relative frequency of subtype-specific antibodies in a population accurately reflects the relative frequency of subtypes of HB_s Ag in the same population.

VI. NATURE OF HEPATITIS B ANTIGENS

A. HBs Ag

The particulate nature of HB Ag was first recognized by Bayer *et al.* (*10*) who ascribed a viruslike morphology to the 22-nm spherical particles (Fig. 3). A filamentous form of the antigen, 22 nm in cross section and of varying length was also described.

The 22-nm form of HB_s Ag has a buoyant density in cesium chloride of approximately 1.20 and, in sucrose, a density of $1.16/cm^3$ (*24, 34, 56, 75, 83*). Sedimentation coefficients varying from 34 to 110 have been described (*14, 33, 34, 56, 83, 90*). The reason for the discrepancy is not clear, but may reflect differences in mean diameters of HB_s Ag derived from different sources as well as differences in technique.

This form of HB_s Ag consists of predominantly protein associated with small quantities of phospholipids and carbohydrates (*12, 17, 54, 75*). The proteins appear to account for most if not all of the antigenicity and immunogenicity of HB_s Ag, although the other components may also play a role.

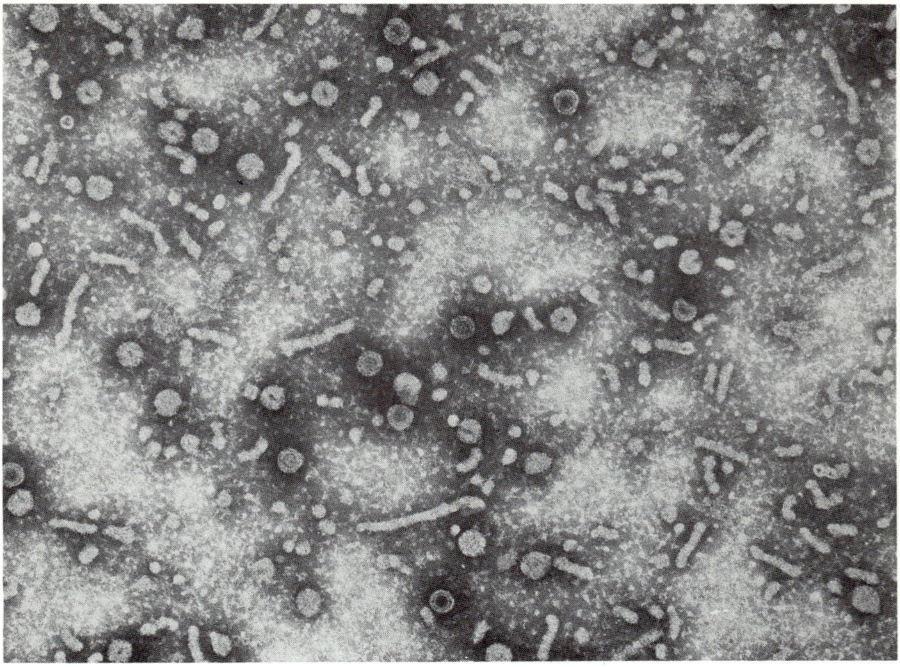

Fig. 3. Particulate hepatitis B antigens. This preparation consists of 22-nm spherical forms, filamentous forms, and 43-nm Dane particles. The 27-nm core is visible in several of the Dane particles. Electron micrograph taken by Mrs. Eugenie Ford and Dr. John Gerin. Magnification: × 100,000.

These proteins have been characterized by Gerin (*31*). Following solubilization with sodium dodecyl sulfate (SDS), urea, and mercaptoethanol, the proteins of HB_s Ag could be resolved into six or seven peptides with molecular weights of approximately 26,000 to 95,000 by acrylamide gel electrophoresis (*31, 33*). At least four of the peptides described by Gerin retained their antigenicity and immunogenicity following purification (*32, 92*). In addition, we have recently shown that individual peptides appear to contain group-reactive and type-specific determinants on the same peptide (*36*).

The detection of RNA in purified or partially purified preparations of HB_s Ag has been reported (*16, 49*). In addition, Hirschman *et al.* (*42*) reported the finding of DNA polymerase associated with partially purified HB Ag, and it was suggested that this enzyme was a reverse transcriptase. However, based on experiments performed by Gerin (*31*) and Kaplan *et al.* (*51*), it seems clear now that the 22-nm form of HB_s Ag contains neither nucleic acid nor DNA or RNA polymerase activity.

B. Dane Particle (HB_c Ag)

In 1970, Dane, Cameron, and Briggs (*21*) described a third morphological form associated with HB Ag. This form, now called the Dane particle, and present in by far the lowest concentration of any of the three morphological forms of HB Ag, has grown in significance as more and more evidence points to its identity as the hepatitis B virus. This particle is 43 nm in diameter and has a complex internal structure consisting of a viruslike core 27 or 28 nm in diameter and an outer shell or coat 7 or 8 nm in thickness (*21, 48*) (Fig. 3); HB_s Ag group and subtype specificities are present on its surface.

Almeida (*2*) demonstrated that when such particles were treated with detergent, the coat was disrupted, releasing the internal core. Serological evaluation, using immune electron microscopy, revealed that the core did not contain on its surface recognized group-reactive or type-specific antigenic specificities of HB_s Ag. Instead, a new antigenic specificity was revealed that reacted with antibodies in sera from patients convalescent from hepatitis B infection. Antibodies to both HB_s Ag and core antigen (HB_c Ag) were found in sera of hemophiliacs but only antibody to the core could be detected by immune electron microscopy in sera of patients convalescent from acute hepatitis B infection. The viruslike cores demonstrated by Almeida following detergent treatment of Dane particles were morphologically similar to particles she had observed previously in homogenates of livers from fatal cases of hepatitis B infection (*3*). Almeida employed the negative staining technique for the demonstration of the viruslike particles and, therefore, could not localize these entities in the liver cell. However, others, using the thin-sectioning technique, have demonstrated clusters of intranuclear viruslike particles in liver cells obtained from patients with hepatitis B (*45, 78*). If allowances are made for the differences in technique employed, the particles observed following negative staining of liver homogenates can be seen to resemble the intranuclear particles found in thin sections of liver. The viruslike particles have been found in the cytoplasm of liver cells but less frequently than in the nucleus. Some of the intracytoplasmic particles have been found to be enveloped with an outer coat, suggesting that these may be analogous to the Dane particles found in serum. Similarly, poorly resolved material found in the cytoplasm of liver cells from hepatitis B patients appears to consist of spherical and filamentous structures, and it has been suggested that this may be the spherical and filamentous forms of hepatitis B antigen.

Serological evidence that this is so has been provided by the immunofluorescence studies of Edgington and Ritt (*25*), Brzosko *et al.* (*15*),

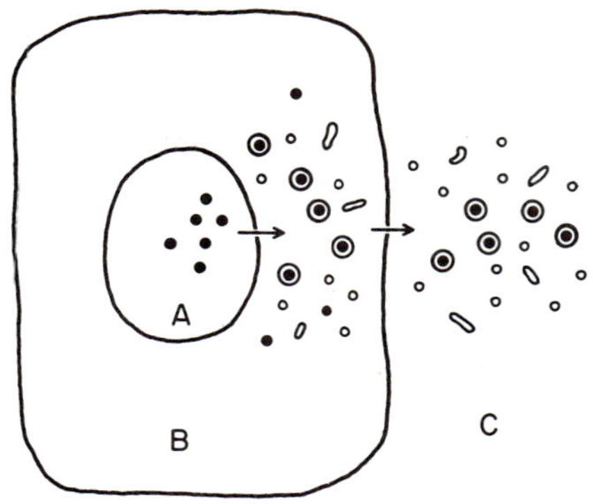

Fig. 4. Hypothesis for the synthesis of hepatitis B virus. Nucleocapsid is synthesized in the infected liver cell nucleus (A) and enveloped with the outer coat in the cytoplasm, where viral coat material is synthesized in excess (B). Enveloped nucleocapsids and excess viral coat material are released from the cell into the circulation (C) and are recognized as Dane particles and spherical and filamentous forms of hepatitis B antigen, respectively. Additional work must be done to determine if this hypothesis is correct. [Purcell (*85*); *in* Hepatitis and Blood Transfusion (G. N. Vyas, H. A. Perkins, R. Schmid, eds.) Grune & Stratton, New York, 1972, pp. 311–316, reprinted by permission.]

and Barker *et al.* (*8*). By use of immunofluorescence tests with antisera specific for, respectively, the 22-nm form of HB_s Ag and the core of the Dane particle, the former was localized to the cytoplasm and the latter was found in the nucleus.

These findings strongly suggest that the core of the Dane particle is, in fact, the capsid of the hepatitis B virus and, furthermore, that this capsid is synthesized in the nucleus of infected liver cells and then transferred to the cytoplasm where it is enveloped with cytoplasmically synthesized viral coat material. Such viral coat material appears to be synthesized in marked excess in the cytoplasm of infected cells and, when released with the complete virion, is recognized as the spherical and filamentous forms of HB Ag (*85*) (Fig. 4).

Strong support for this hypothesis was recently provided by the demonstration by Kaplan *et al.* (*50, 51*) of a DNA polymerase associated with the core of the Dane particle. The polymerase differed in many respects from a DNA polymerase detected previously by Hirshman *et al.* (*42*) in crude preparations of HBAg. It is possible that the polymerase activity detected by Hirshman was associated with Dane particles in preparations of HB Ag and that the discrepant results may have been related to the contamination of those preparations with other polymerases.

The core-associated polymerase synthesizes DNA from an endogenous template: activity cannot be stimulated by the addition of a number of natural or synthetic DNA's and RNA's. DNA polymerase activity could be detected only in a subpopulation of Dane particle cores that banded at a slightly greater density, suggesting that only those cores containing an endogenous nucleic acid template were capable of such synthesis.

The DNA polymerase activity was inhibited by actinomycin D, a compound known to inhibit DNA-dependent nucleic acid synthesis. Although not sufficient by itself, this finding, coupled with the localization of the Dane particle core in the nucleus of infected liver cells by immunofluorescence and electron microscopy, suggests that the endogenous template of this particle may be DNA. Additional studies must be carried out to determine if the hypothesis is correct.

Dane particle cores sufficiently pure for biophysical characterization have been prepared in several laboratories. Jackson *et al.* (*46*) recovered cores from the liver of an immunosuppressed chimpanzee infected with hepatitis B. These cores, which were found in great numbers within the nuclei of liver cells, were devoid of HB_s Ag and probably represented cores of Dane particles that had not yet been enveloped with coat material. The cores had a buoyant density in cesium chloride of 1.30 to 1.32. Gerin has also purified Dane particle cores from the plasma of HB Ag carriers. Cores prepared in such a manner had a density of approximately 1.31 and shared antigenic specificities with cores recovered from the nuclei of liver cells by Barker's group (*32, 86*). When negatively stained cores obtained from liver nuclei or Dane particles were examined by electron microscopy, most of the particles were penetrated by the stain, suggesting that they lacked a full complement of nucleic acid. However, a subpopulation of cores with a somewhat greater density was found to be associated with DNA polymerase activity, whereas cores with a density of 1.30 appeared to be devoid of such activity. Cores with similar morphology and biophysical characteristics have also been recovered from Dane particles by Lipman *et al.* (*64*).

Several serological tests, including immume electron microscopy (*2*), immunofluorescence (*15*), complement fixation (*44, 104*). and radioimmunoassay inhibition (*86*), have been developed for the detection of antibody to the core particle. Such tests have revealed an immune response following infection that is quantitatively and temporally distinct from the immume response to HB_s Ag. The temporal pattern of development of HB_s Ag, DNA polymerase activity, antibody to the core, and antibody to HB_s Ag in cases of acute and chronic viral hepatitis, Type B, is described by Krugman in the Discussion (p. 71) and elsewhere (*59*).

Individuals who undergo inapparent hepatitis B infection, characterized by the development of anti-HB_s in the absence of chemical or histological evidence of liver damage, generally do not develop anti-HB_c. Individuals apparently reexposed to hepatitis B virus, as evidenced by preexisting antibody to HB_s Ag and/or an anamnestic antibody response to this antigen, do not manifest an augmented immunological response to HB_c Ag. Hoofnagle *et al.* (*44*) have shown that, unlike anti-HB_s, anticore antibodies do not correlate positively with resistance to reinfection nor are they associated with recovery from infection, since all chronic carriers of the antigen have high titers of anti-HB_c.

In summary, these data, taken together, strongly suggest that the Dane particle is the virion of viral hepatitis, Type B, and that the other morphological forms of the antigen consist of noninfectious viral envelope material synthesized in excess by the infected cell. It has been widely suggested that such particulate material, which can be purified readily from human plasma obtained from chronic carriers of the antigen, may serve as a suitable vaccine for hepatitis B if appropriate means of assuring its safety can be found. Preliminary evidence for this has been obtained by Krugman (see Section VIII). The development of animal models for the study of hepatitis B offers promise of confirming and extending these observations.

VII. STUDIES OF HEPATITIS TYPE B IN NONHUMAN PRIMATES

Hepatitis B viruses were not successfully transmitted from man to laboratory animals until 1972, when London *et al.* (*65*) infected rhesus monkeys, and Maynard *et al.* (*68, 69*), Barker *et al.* (*8*), and Desmyter (*22*) infected chimpanzees. The success of these transmission experiments was due in large part to the development of very sensitive tests for the detection of HB_s Ag and anti-HB_s.

Hepatitis B infection in chimpanzees resembles infection in man and is characterized by a spectrum of host responses ranging from inapparent infection to frank hepatitis. Chimpanzees appear to be highly susceptible to HBV infection and serve as a sensitive indicator of the presence of the virus.

Because of the susceptibility of chimpanzees to hepatitis B infection, these animals are particularly useful for studies of the host response to the virus. However, several problems make the large-scale use of chimpanzees for the study of hepatitis B impractical: a large proportion of captive chimpanzees have serological evidence of prior exposure to hepatitis B; maintenance, particularly of large animals, is expensive and difficult; and populations of this species of Great Ape are rapidly dwindling. Thus, seronegative chimpanzees should be judiciously utilized for experiments that cannot be performed in other animals. To avoid duplication of experiments and wastage of seronegative chimpanzees, we are collaborating with Dr. Barker of the Bureau of Biologics, Food and Drug Administration, and Dr. Maynard of the Center for Disease Control, in an interagency study of hepatitis B infection in chimpanzees. Inocula for such studies are randomly distributed among the three participating laboratories and administered to laboratory-raised chimpanzees from the Holloman Air Force Base chimpanzee colony, currently under the supervision of Dr. Frederick Coulston of Albany Medical College.

A high priority of these studies is the establishment of pools of hepatitis B virus of known infectivity titer representing each of the subtypes of HB_s Ag. These preliminary infectivity studies, which are under the direction of Dr. Barker, are still in progress, but much useful information has already been obtained. Human serum or plasma containing HB_s Ag of subtype *adw, ayw, adr,* or *ayr* has been inoculated into chimpanzees and each has initiated hepatitis B infection as evidenced by the development of HB_s Ag, development of elevated transaminase levels, or development of antibody to HB_s Ag (Table 2). The inoculum chosen to represent subtype *ayw* was serum supplied by Dr. Saul Krugman from the MS-2 pool of hepatitis B virus (57, 58). When an infectivity titration of this pool was performed in seronegative chimpanzees, a titer of approximately $10^{7.5}$ chimpanzee infectious $dose_{50}$ (CID_{50}) per milliliter was obtained. Since Krugman's experiments with MS-2 virus involved inoculation of 0.1 ml of a 1:10 dilution of the MS-2 pool, we can now calculate that the challenge dose in those experiments was approximately $10^{5.5}$ CID_{50} of hepatitis B virus.

Infectivity titrations of hepatitis B virus associated with subtypes *adw, adr,* and *ayr* are currently in progress. These standardized pools will be used for characterization of the host response, cross-challenge experiments, and evaluation of various aspects of passive and active immunization against hepatitis B infection.

TABLE 2. EXPERIMENTAL HEPATITIS B INFECTION OF CHIMPANZEES

Incubation period to first detectable HB_s Ag	2–16 weeks
Duration of HB_s antigenemia	1–26 weeks
Chemical and histological evidence of hepatitis	Yes
Subtypes successfully transmitted	*adw, ayw, adr, ayr*
Infectivity titer of MS-2 pool[a]	$10^{7.5}$ CID_{50}/ml

[a] CID_{50} = chimpanzee infectious $dose_{50}$.

TABLE 3. EXPERIMENTAL HEPATITIS B INFECTION OF RHESUS MONKEYS

Incubation period to first detectable HB_s Ag	8–13 weeks
Duration of HB_s antigenemia	1–4 weeks
Chemical or histological evidence of hepatitis	No
Subtypes successfully transmitted	*adw* (one strain)
Infectivity titer of rhesus monkey-adapted pool[a]	$10^{4.2}$ $RMID_{50}$/ml

[a] $RMID_{50}$ = Rhesus monkey infectious $dose_{50}$.

In contrast to hepatitis B infection of chimpanzees, infection of rhesus monkeys with this virus is entirely inapparent: infection can be detected only by the most sensitive radioimmunoassays. This is followed, in some cases, by the development of anti-HB_s detectable by radioimmunoprecipitation and passive hemagglutination techniques. Neither histological evidence of liver damage nor elevated transaminase levels have been observed in infected rhesus monkeys, another indication of the mild nature of the infection (*87*).

Rhesus monkeys are quite resistant to hepatitis B virus infection, and only one strain of the virus has been successfully adapted to growth in this species if nonhuman primate. This strain, of subtype *adw*, has been serially transmitted to rhesus monkeys through seven passages. Despite numerous attempts, Dr. London, Dr. Alter, and I have been unable to adapt other strains of hepatitis B virus of the same subtype or other subtypes to the rhesus monkey. However, our rhesus monkey-adapted strain has provided us with a means of characterizing the agent by using rhesus monkeys as an indicator of the viability of the virus.

We have prepared a pool of hepatitis B virus passaged 5 times in rhesus monkeys; this pool has an infectivity titer of $10^{4.2}$ rhesus monkey infectious $dose_{50}$ ($RMID_{50}$) per milliliter as determined in repeated titrations (Table 3). Using this virus suspension we are currently in the process of determining the biological and biophysical characteristics of the hepatitis B virion. These studies are being carried out to determine the optimum conditions for inactivation of the hepatitis B virion with retention of immunogenicity of the protective antigen, believed to be HB_s Ag. Such information is a prerequisite for the successful development of an inactivated hepatitis B vaccine.

VIII. PROSPECTS FOR IMMUNIZATION AGAINST HEPATITIS B

A. PASSIVE

Numerous studies of the efficacy of immune serum globulin in preventing hepatitis B infection have yielded confusing and contradictory results (*43, 52*). However, most of these studies were designed to ascertain the protective effect of immune serum globulin in preventing post-transfusion hepatitis, but recent studies have indicated that a variable proportion (perhaps over 50%) of such hepatitis cannot be associated with hepatitis B infection. Furthermore, the immune serum globulin used in such studies was generally not characterized for anti-HB_s, and the challenge dose of hepatitis B virus (in the form of one or more blood transfusions) could not be controlled.

In recent studies, Krugman (*57*) carried out a controlled evaluation of the protective effect of standard immune serum globulin (ISG) with a titer of anti-HB_s of 1:16 as judged by passive hemagglutination and compared this with a hepatitis B immune serum globulin (HBIG) preparation prepared by Prince (*82, 84*) (passive hemagglutination titer of 1:256,000). The standard challenge dose of hepatitis B virus (0.1 ml of a 1:10 dilution of the MS-2 serum pool) was inoculated into seronegative children 4 hours before administration of 0.02 or 0.06 ml/kg body weight of globulin. All 25 children who received the MS-2 pool (but not globulin) developed HB_s Ag; 24 of these children developed hepatitis as evidenced by elevated transaminase levels. Thus, each of the 25 children was infected. Thirteen of the 15 children who received ISG following exposure to the MS-2 pool developed HB_s Ag and 12 of the 15 developed hepatitis. In contrast, only 4 of 10 children who received HBIG developed HB_s Ag and all 4 also developed hepatitis. Thus, a slight protective effect against illness was demonstrated with ISG, and a greater reduction in illness was achieved by prophylaxis with HBIG. However, at least 14 of the 15 children who received ISG and all 10 of the children who received HBIG had evidence of hepatitis B infection as judged by development of HB_s Ag and/or persistent anti-HB_s (passive–active immunization).

The challenge dose employed in these experiments was designed to simulate inadvertant exposure to a hypodermic needle contaminated with hepatitis B virus. Nevertheless, this inoculum contained $10^{5.5}$ CID_{50} of hepatitis virus. In addition, as pointed out by Prince (*82*), 0.01 ml of HB Ag-positive serum may contain as many as 10^{12} particles of HB_s Ag. Thus, for every hepatitis B virus there may be over a million noninfectious particles with hepatitis B antigen sites on their surface. If anti-HB_s is identical to neutralizing antibody to the virus, the noninfectious particles could greatly diminish the effectiveness of HBIG in neutralizing the hepatitis B virus. In the face of such potential problems, it is encouraging that a protective effect of ISG against hepatitis B has been demonstrated.

In addition to these studies utilizing controlled challenge with hepatitis B virus, epidemiologic evidence of protection against presumably non-parenterally acquired Type B hepatitis by standard ISG was presented by Ginsberg *et al.* (35). This lot of ISG had a somewhat higher titer of anti-HB_s (1:1280) than other lots tested (1:10 to 1:180) when measured by a radioimmunoassay procedure. This study provides additional evidence that anti-HB_s protects against hepatitis Type B when administered in sufficient quantity.

Double-blind studies of the efficacy of HBIG of known titer in preventing hepatitis B infection following presumed nonparenteral exposure, inadvertant "needle prick" exposure, or blood transfusion are currently in progress. When completed, these should help to answer questions regarding the magnitude of protection afforded by such preparations as well as provide estimates of optimum titers of anti-HB_s necessary for protection.

B. Active

In other studies, Krugman (57) vaccinated seronegative children with 0.1 ml of a 1:10 dilution of MS-2 serum that had been heat-inactivated (98°C for 1 minute). One to three subcutaneous inoculations were employed in this study. Following challenge with live hepatitis B virus (0.01 ml of the MS-2 pool) only 10 of 29 immunized children developed detectable HB_s Ag (compared to 25 of 25 nonimmunized children); 12 of the 29 children developed elevated transaminase levels indicative of hepatitis (compared to 24 of 25 in the nonimmunized group). However, at least 27 of the 29 children apparently experienced hepatitis B virus infection as evidenced by persistent anti-HB_s titers. Thus, the vaccine partially protected against illness but did not prevent hepatitis B infection.

Soulier (98) attempted similar studies of active immunization against hepatitis B with hepatitis B antigen-positive serum that had been heated at 60°C for 10 hours. However, HB_s Ag was detected in 6 of 7 vaccinees following an appropriate incubation period for hepatitis B, and 1 of these developed hepatitis. Thus, the heating conditions employed by Soulier were not sufficient to inactivate hepatitis B virus completely.

The studies carried out by Krugman involved the use of an inactivated vaccine containing HB_s Ag of subtype *ayw* and a challenge with the homologous virus. We do not yet know whether the various subtypes of HB_s Ag are reflections of different serotypes of hepatitis B virus or whether infection with virus associated with one subtype of antigen confers cross-protection against a virus associated with a different subtype of HB_s Ag. There is little epidemiologic evidence for sequential infections with hepatitis B virus containing different subtypes of HB_s Ag, but the high ratio of inapparent infections with hepatitis B virus makes it hard to interpret negative data in the absence of detailed serological evaluation. It is hoped that cross-challenge experiments in chimpanzees will provide important information about the significance of HB_s Ag subtypes. Thus the question of whether one or multiple vaccines will be needed to control hepatitis B cannot be answered at present. Furthermore, the inactivated vaccine used by Krugman in his studies contained, in addition to HB_s Ag, at least $10^{5.5}$ inactivated hepatitis B viruses, as well as, possibly, other immunizing antigens not yet characterized. It remains to be seen whether purified HB_s Ag stimulates antibody that is protective against hepatitis B infection. These studies are currently in progress in chimpanzees and rhesus monkeys.

It has been proposed that subunits of hepatitis B antigen or synthesized haptens containing the essential peptide sequences that define the antigen be used as the basis of a hepatitis B vaccine (*88, 109*). The rationale for such a vaccine is that it would be free of other potentially deleterious substances such as nucleic acids or human serum proteins. However, the former is not likely to be present if care is used in purifying the noninfectious 22-nm form of HB_s Ag. Although it is true that antibody to human IgG is often found following hyperimmunization of animals with purified HB_s Ag, this can be minimized by careful selection of HB_s Ag-positive plasma free of detectable immune complexes and careful purification to remove complexes from the final product.

The presence of immunogenic quantities of albumin in purified HB_s Ag is another problem, and this human protein may be intimately associated with the antigen. However, human albumin is generally considered to be a safe substance for administration to humans, and a synthetic HB_s Ag peptide vaccine would most probably be coupled to human albumin as a carrier protein. It seems to be premature at present to consider such subunit vaccines, at least until the nature of the protective antibody has been determined.

IX. CONCLUSION

Recent studies of hepatitis B infection, employing sensitive assays for HB_s Ag and anti-HB_s, have revealed a high subclinical attack rate for this virus. In some populations, exposure and subsequent infection probably approaches 100%. Exposure appears to be highest in populations with the lowest socioeconomic status and poorest hygienic conditions. The greatest rate of infection appears to occur during late childhood and early adulthood but the exact time varies from population to population, generally being earlier in those with the highest overall infection rate. This high subclinical attack rate and epidemiologic relationship to socioeconomic status is similar to that observed with other viruses, such as poliovirus, and suggests a high rate of nonparenteral transmission. However, failure to detect a high rate of infection in early life suggests that, unlike the situation with poliovirus, fecal–oral transmission is not readily accomplished. Additional studies must be carried out to determine how hepatitis B virus is maintained in the population.

Hepatitis B virus infection is associated with the production of large quantities of particulate antigen that is believed to consist of noninfectious viral coat materials. This particulate antigen can be purified in large quantities from the plasma of chronic carriers. Such antigen may provide a novel source of immunizing antigen for use as an inactivated vaccine.

Evidence for the feasibility of passive and active immunization against hepatitis B has been obtained in studies of limited scope. Large clinical trials of HBIG are currently in progress in an attempt to confirm the efficacy of this product in preventing hepatitis B. Attempts to confirm the efficacy of HB_s Ag in stimulating active immunity against hepatitis B virus are being carried out in studies with chimpanzees and rhesus monkeys.

Considerable progress has been made in recent years in our understanding of hepatitis B virus and the antigens associated with it. Part of this success has been due to the development of sophisticated biophysical and biochemical techniques that were not available until recently. It seems reasonable to conclude that immunoprophylaxis of hepatitis B infection may be achieved without having isolated the etiologic agent in cell or tissue culture.

REFERENCES

1. Almeida, J. D., Chisholm, G. D., Kulatilake, A. E., MacGregor, A. B., Mackay, D. H., O'Donoghue, E. P. N., Shackman, R., and Waterson, A. P., *Lancet* **2**, 849–850 (1971).
2. Almeida, J. D., Rubenstein, D., and Stott, E. J., *Lancet* **2**, 1225–1227 (1971).
3. Almeida, J. D., Waterson, A. P., Trowell, J. M., and Neale, G., *Microbios* **6**, 145–153 (1970).
4. Apostolov, K., Bauer, D. J., Selway, J. W. T., Fox, R. A., Dudley, F. J., and Sherlock, S., *Lancet* **1**, 1274–1275 (1971).
5. Aziz, M. A., Khan, G., Khanum, T., and Siddiqui A.-R., *J. Infec. Dis.* **127**, 110–112 (1973).
6. Bancroft, W. H., Mundon, F. K., and Russell, P. K., *J. Immunol.* **109**, 842–848 (1972).
7. Bancroft, W. H., Warkel, R. L., Talbert, A. A., and Russell, P. K., *J. Amer. Med. Ass.* **217**, 1817–1820 (1971).
8. Barker, L. F., Chisari, F. V., McGrath, P. P., Dalgard, D. W., Kirschstein, R. L., Almeida, J. D., Edgington, T. S., Sharp, D. G., and Peterson, M. R., *J. Infec. Dis.* **127**, 648–662 (1973).

9. Barker, L. F., Peterson, M. R., Shulman, R., and Murray, R., *J. Amer. Med. Ass.* **223**, 1005–1008 (1973).
10. Bayer, M. E., Blumberg, B. S., and Werner, B., *Nature (London)* **218**, 1057–1059 (1968).
11. Berris, B., Wrobel, D. M., Sinclair, J. C., and Feinman, S. V., *Ann. Intern. Med.* **79**, 690–693 (1973).
12. Blumberg, B. S., *Bull. N. Y. Acad. Med.* [2] **40**, 377–386 (1964).
13. Blumberg, B. S., Sutnick, A. I., London, W. T., and Millman, I., *in* "Perspectives in Virology" (M. Pollard, ed.), vol. 7, pp. 223-240. Academic Press, New York, 1971.
14. Bond, H. E., and Hall, W. T., *J. Infec. Dis.* **125**, 263–268 (1972).
15. Brzosko, W. J., Madalínski, K., Krawczýnski, K., and Nowaslawski, A., *J. Infec. Dis.* **127**, 424–428 (1973).
16. Brzosko, W. J., Mäntyjärvi, R. A., and Madalínski, K., *Acta Virol. (Prague)* **16**, 440–443 (1972).
17. Burrell, C. J., Proudfoot, E., Keen, G. A., and Marmion, B. P., *Nature (London), New Biol.* **243**, 260–262(1973).
18. Byrom, N. A., Davidson, G., Draper, C. C., and Zuckerman, A. J., *J. Infec. Dis.* **128**, 259–260 (1973).
19. Cherubin, C. E., Purcell, R. H., Lander, J. J., McGinn, T. G., and Cone, L. A, *Lancet* **2**, 149–151 (1972).
20. Committee on Viral Hepatitis of the National Research Council—National Academy of Sciences, *Morbidity Mortality* **23**, 29 (1974).
21. Dane, D. S., Cameron, C. H., and Briggs., M., *Lancet* **1**, 695–698 (1970).
22. Desmyter, J., Liu, W. T., De Somer, P., and Mortelmans, J., *Vox Sang.* **24**, Suppl., 17–26 (1973).
23. Dick, S. J., Tamburro, C. H., and Leevy, C. H., *J. Amer. Med. Ass.* **229**, 1627–1629 (1974).
24. Dreesman, G. R., Hollinger, F. B., Suriano, J. R, Fujioka, R. S., Brunschwig, J. P., and Melnick, J. L., *J. Virol.* **10**, 469–476 (1972).
25. Edgington, T. S., and Ritt, D. J., *J. Exp. Med.* **134**, 871–885 (1971).
26. Favero, M. S., Maynard, J. E., Peterson, N. J., Boyer, K. M., Bond, W. W., Berquist, K. R., and Szmuness, W., *Lancet* **2**, 1455 (1973).
27. Feinman, S. V., Berris, B., Sinclair, J. C., Wrobel, D. M., Alter, H. J., and Holland, P. V., *Lancet* **2**, 867–869 (1973).
28. Fulford, K. W. M., Dane, D. S., Catterall, R. D., Woof, R., and Denning, J. V., *Lancet* **1**, 1470–1473 (1973).
29. Garibaldi, R. A., Hatch, F. E., Bisno, A. L., Hatch, M. H., and Gregg, M. B., *J. Amer. Med. Ass.* **220**, 963–966 (1972).
30. Garibaldi, R. A., Rasmussen, C. M., Holmes, A. W., and Gregg, M. B., *J. Amer. Med. Ass.* **219**, 1577–1580 (1972).
31. Gerin, J. L., *in* "Hepatitis and Blood Transfusion" (G. N. Vyas, H. A. Perkins, and R. Schmid, eds.), pp. 205–219. Grune & Stratton, New York, 1972.
32. Gerin, J. L., personal communication.
33. Gerin, J. L., Holland, P. V., and Purcell, R. H., *J. Virol.* **7**, 569–576 (1971).
34. Gerin, J. L., Purcell, R. H., Hoggan, M. D., Holland, P. V., and Chanock, R. M., *J. Virol.* **4**, 763–768 (1969).
35. Ginsberg, A. L., Conrad, M. E., Bancroft, W. H., Ling, C. M., and Overby, L. R.. *N. Engl. J. Med.* **286**, 562–566 (1972).
36. Gold, J. W. M., personal communication.
37. Gold, J. W. M., Alter, H. J., Holland, P. V., Gerin, J. L., and Purcell, R. H., *J. Immunol.* **112**, 1100-1106 (1974).

38. Grob, P. J., and Jemelka, H., *Lancet* **1**, 206–208 (1971).
39. Hawkes, R. A., Vale, T. G., Marshall, I. D., and Maclennan, R., *Amer. J. Epidemiol.* **95**, 228–237 (1972).
40. Heathcote, J., Tsianides, A., and Sherlock, S., *Lancet* **2**, 593–595 (1973).
41. Hersh, T., Melnick, J. L., Goyal, R. K., and Hollinger, F. B., *N. Engl. J. Med.* **285**, 1363–1364 (1971).
42. Hirschman, S. Z., Vernace, S. J., and Schaffner, F., *Lancet* **1**, 1099–1103 (1971).
43. Holland, P. V., *in* "Hepatitis and Blood Transfusion" (G. N. Vyas, H. A. Perkins, and R. Schmid, eds.), pp. 331–333. Grune & Stratton, New York, 1972.
44. Hoofnagle, J. H., Gerety, R. J., and Barker, L. F., *Lancet* **2**, 869–873 (1973).
45. Huang, S., *Amer. J. Pathol.* **64**, 483–500 (1971).
46. Jackson, D. R., McGrath, P. P., and Barker, L. F., *Proc. Electron Microsc. Soc. Amer.* **31**, 588 (1973).
47. Jeffries, D. J., James, W. H., Jefferiss, F. J. G., MacLeod, K. G., and Willcox, R. R., *Brit. Med. J.* **2**, 455–456 (1973).
48. Jokelainen, P. T., Krohn, K., Prince, A. M., and Finlayson, N. D. C., *J. Virol.* **6**, 685–689 (1970).
49. Józwiak, W., Koŝcielak, J., Madalínski, K., Brzosko, W. J., Nowoslawski, A., and Kloczewiak, M., *Nature (London), New Biol.* **229**, 92–94 (1971).
50. Kaplan, P. M., Gerin, J. L., and Alter, H. J., *Nature (London), New Biol.* **249**, 762–764 (1974).
51. Kaplan, P. M., Greenman, R. L., Gerin, J. L., Purcell, R. H., and Robinson, W. S., *J. Virol.* **12**, 995–1005 (1973).
52. Kerr, D. N. S., *Lancet* **2**, 627 (1973).
53. Keys, T. F., Sever, J. L., Hewitt, W. L., and Gitnick, G. L., *J. Pediat.* **80**, 650–653 (1972).
54. Kim, C. Y., and Bissell, D. M., *J. Infec. Dis.* **123**, 470–476 (1971).
55. Kim, C. Y., and Tilles, J. G., *J. Infec. Dis.* **123**, 618–628 (1971)
56. Kim, C. Y., and Tilles, J. G., *J. Clin. Invest.* **52**, 1176–1186 (1973).
57. Krugman, S., and Giles, J. P., *N. Engl. J. Med.* **288**, 755–760 (1973).
58. Krugman, S., Giles, J. P., and Hammond, J., *J. Amer. Med. Ass.* **200**, 365–373 (1967).
59. Krugman, S., Hoofnagle, J. H., Gerety, R. J., Kaplan, P. M., and Gerin, J. L., *N. Engl. J. Med.*, **290**, 1331–1335 (1974).
60. Lander, J. J., Holland, P. V., Alter, H. J., Chanock, R. M., and Purcell, R. H., *J. Amer. Med. Ass.* **220**, 1079–1082 (1972).
61. Le Bouvier, G. L., *J. Infec. Dis.* **123**, 671–675 (1971).
62. Le Bouvier, G. L., McCollum, R. W., Hierholzer, W. J., Irwin, G. R., Krugman, S., and Giles, J. P., *J. Amer. Med. Ass.* **222**, 928–930 (1972).
63. Levene, C., and Blumberg, B. S., *Nature (London)* **221**, 195–196 (1969).
64. Lipman, M. B., Hierholzer, W. J., Jr., and Schluederberg, A., *J. Infec. Dis.* **128**, 664–667 (1973).
65. London, W. T., Alter, H. J., Lander, J., and Purcell, R. H., *J. Infec. Dis.* **125**, 382–389 (1972).
66. MacCallum, F. O., *Lancet* **2**, 435 (1947).
67. Magnius, L. O., and Espmark, J. A., *J. Immunol.* **109**, 1017–1021 (1972).
68. Maynard, J. E., Berquist, K. R., Krushak, D. H., and Purcell, R. H., *Nature (London)* **237**, 514–515 (1972).
69. Maynard, J. E., Hartwell, W. V., and Berquist, K. R., *J. Infec. Dis.* **123**, 660–664 (1971).
70. Mazzur, S., *Lancet* **1**, 749 (1973).
71. Mazzur, S., Blumberg, B. S., and Friedlaender, J. S., *Nature (London)* **247**, 41–43 (1974).

72. Mazzur, S., Burgert, S., and Blumberg, B. S., *Nature (London)* **247**, 38–40 (1974).
73. Merrill, D. A., Dubois, R. S., and Kohler, P. F., *N. Engl. J. Med.* **287**, 1280–1282 (1972).
74. Metselaar, D., Blumberg, B. S., Millman, I., Parker, A. M., and Bagshawe, A. F., *Lancet* **2**, 758–760 (1973).
75. Millman, I., Loeb, L. A., Bayer, M. E., and Blumberg, B. S., *J. Exp. Med.* **131**, 1190–1199 (1970).
76. Mosley, J. W., Edwards, V. M., Meihaus, J. E., and Redeker, A. G., *J. Epidemiol.* **95**, 529–535 (1972).
77. Nordenfelt, E., and Le Bouvier, G., *Intervirology* **2**, 65–74 (1973/74).
78. Nowoslawski, A., Brzosko, W. J., Madalínski, K., and Krawczýnski, *Lancet* **1**, 494–498 (1970).
79. Ogra, P. L., *J. Immunol.* **110**, 1197–1205 (1973).
80. Ohbayashi, A., Okochi, K., and Mayumi, M., *Gastroenterology* **62**, 618–625 (1972).
81. Prince, A. M., Metselaar, D., Kafuko, G. W., Mukwaya, L. G., Ling, C. M., and Overby, L. R., *Lancet* **2**, 247–250 (1972).
82. Prince, A. M., Szmuness, W., Brotman, B., Ikram, H., Lippin, A., Stryker, M., Ehrich, C., and Finlayson, N. C. D. C., *in* "Hepatitis And Blood Transfusion" (G. N. Vyas, H. A. Perkins, and R. Schmid, eds.), pp. 335–347. Grune & Stratton, New York, 1972.
83. Prince, A. M., Szmuness, W., Hargrove, R. L., Jeffries, G. H., Cherubin, C. E., and Kellner, A., *in* "Perspectives in Virology" (M. Pollard, ed.), vol. 7, pp. 241–292. Academic Press, New York, 1971.
84. Prince, A. M., Szmuness, W., Woods, K. R., and Grady, G. F., *N. Engl. J. Med.* **285**, 933–938 (1971).
85. Purcell, R. H., *in* "Hepatitis And Blood Transfusion" (G. N. Vyas, H. A. Perkins, and R. Schmid, eds.), pp. 311–316. Grune & Stratton, New York, 1972.
86. Purcell, R. H., Gerin, J. L., Almeida, J. D., and Holland, P. V., *Intervirology* **2**, 231–243 (1973/74).
87. Purcell, R. H., London, W. T., Alter, H. J., and Edgington, T. S., *J. Infec. Dis.* (in press).
88. Rao, K. R., and Vyas, G. N., *Nature (London), New Biol.* **241**, 240–241 (1973).
89. Reeves, W. C., Peters, C. J., Lander, J. J., Purcell, R. H., and Johnson, K. M., *J. Infec. Dis.* **128**, 265–270 (1973).
90. Schober, A., Sundmacher, R., Thomssen, R., and Drees, O., *Arch Gesamte Virusforsch.* **35**, 314–316 (1971).
91. Schweitzer, I. L., Wing, A., McPeak, C., and Spears, R. L., *J. Amer. Med. Ass.* **220**, 1092–1095 (1972).
92. Shih, J., personal communication.
93. Skinhoj, P., *Lancet* **2**, 715 (1972).
94. Skinhoj, P., McNair, A., and Andersen, S. T., *Amer. J. Epidemiol.* **99**, 50–57 (1974).
95. Skinhoj, P., Olesen, H., Cohn, J., and Mikkelsen, M., *Acta Pathol. Microbiol. Scand.* **80**, 362–366 (1972).
96. Skinhoj, P., and Steiness, I., *Acta Pathol. Microbiol. Scand.* **80**, 801–805 (1972).
97. Smith, J. A., Ogunba, E. O., and Francis, T. I., *Nature (London)* **237**, 231–232 (1972).
98. Soulier, J. P., Blatix, C., Courouce, A. M., Benamon, D., Amouch, P., and Drouet, J., *Amer. J. Dis. Child.* **123**, 429–434 (1972).

99. Soulier, J. P., and Couroucie-Pauty, A. M., *Vox Sang.* **25,** 212–234 (1973).
100. Szmuness, W., Pick, R., and Prince, A. M., *Amer. J. Epidemiol.* **92,** 51–61 (1970).
101. Szmuness, W., Prince, A. M., Brotman, B., and Hirsch, R. L., *J. Infec. Dis.* **127,** 17–25 (1973).
102. Szmuness, W., Prince, A. M., Etling, G. F., and Pick, R., *J. Infec. Dis.* **126,** 498–506 (1972).
103. Szmuness, W., Prince, A. M., Hirsch, R. L., and Brotman, B., *N. Engl. J. Med.* **289,** 1162–1166 (1973).
104. Thiry, L., Clinet, G., Toussaint, C., and Vereerstraeten, P., *Vox. Sang.* **24,** Suppl., 36–43 (1973).
105. Tripatzis, I., and Horst, H. G., *Nature (London)* **231,** 266–267 (1971).
106. Van Kooten Kok-Doorschodt, H. J., Van den Akker, R., and Gispen, R., *J. Infec. Dis.* **126,** 117–122 (1972).
107. Ward, R., Borchert, P., Wright, A., and Kline, E., *Lancet* **2,** 726–727 (1972).
108. Wright, R., Perkins, J. R., Bower, B. D., and Jerrome, D. W., *Brit. Med. J.* **4,** 719–721 (1970).
109. A. J. Z. (A. J. Zuckerman), and C. R. H. (C. R. Howard) *Nature (London)* **246,** 445–447 (1973).

DISCUSSION

DR. KRUGMAN: During the course of his excellent presentation, Dr. Purcell referred to two new tests which are now available for the study of hepatitis B virus infection. Detection of hepatitis B-specific DNA polymerase activity was described by Kaplan, Gerin, and associates and a complement fixation test for detection of antibody to hepatitis B core antigen was described by Hoofnagle, Gerety, and Barker.

Recently, we [Krugman, S., Hoofnagle, J. H., Gerety, R. J., Kaplan, P. M., and Gerin, J. L., *N. Engl. J. Med.* **290,** 1331–1334 (1974)] tested serial serum specimens from immunized and unimmunized individuals who were exposed to the MS-2 strain of hepatitis B virus for DNA polymerase activity and for antibody to the hepatitis B core antigen (anti-HB_c). The following additional tests were performed: serum glutamic oxalacetic transaminase (SGOT), hepatitis B surface antigen (HB_s Ag), and antibody to hepatitis B surface antigen (anti-HB_s).

Deoxyribonucleic acid polymerase activity was detected in unimmunized individuals; (1) it appeared 1–2 weeks *after* HB_s Ag was detectable and several days *before* or at the time of abnormal SGOT activity, and (2) it was present for several days or weeks in those with transient antigenemia and for many months or years in chronic carriers (Figs. 1 and 2). The DNA polymerase activity was not detected in the serum of immunized individuals; neither in those who were passively immunized with hepatitis B immune serum globulin (Fig. 3), nor in those who were immunized with heat-inactivated MS-2 serum (Figs. 4 and 5).

Antibody to hepatitis B core antigen was detected in the serum of unimmunized individuals: (1) it appeared about 4–6 weeks *after* detection of HB_s Ag, about 2 weeks *after* DNA polymerase activity, and about 4 months *before* detection of anti-HB_s (Fig. 1), and (2) it persisted for years, especially in chronic carriers (Fig. 2). Antibody to hepatitis B core antigen was detected transiently (about 6 weeks) in the serum of one individual who was passively immunized with hepatitis B immune serum globulin (Fig. 3); it was not detected in the serum of two individuals who were immunized with heat-inactivated MS-2 serum.

The results of these studies provide support for Dr. Purcell's conclusion that immunoprophylaxis of hepatitis B infection may be achieved without having isolated the etiological agent in cell or tissue culture.

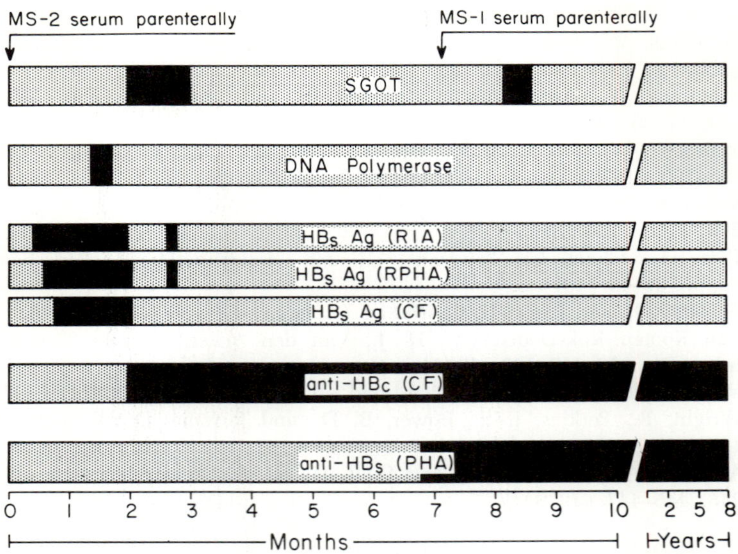

Fig. 1. Viral hepatitis, type B, following parenteral exposure to MS-2 serum and viral hepatitis, type A, following parenteral exposure to MS-1 serum. Solid black area denotes abnormal or detectable. Stippled area denotes normal or not detectable. White area denotes not tested. SGOT = serum glutamic oxalacetic transminase; HB_s Ag = hepatitis B surface antigen; anti-HB_s = antibody to HB_s Ag; anti-HB_c = antibody to hepatitis B core antigen; RIA = radioimmunoassay; CF = complement fixation; PHA = passive hemagglutination; RPHA = reversed passive hemagglutination.

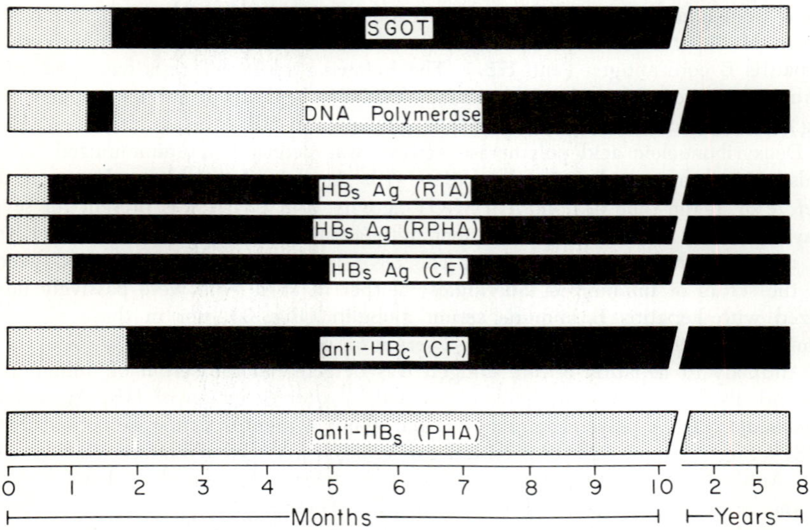

Fig. 2. Occurrence of viral hepatitis, type B, and subsequent chronic carrier state after a parenteral exposure to MS-2 serum. Black, abnormal or detectable; stippled, normal or not detectable; white, not tested. Abbreviations same as in Fig. 1.

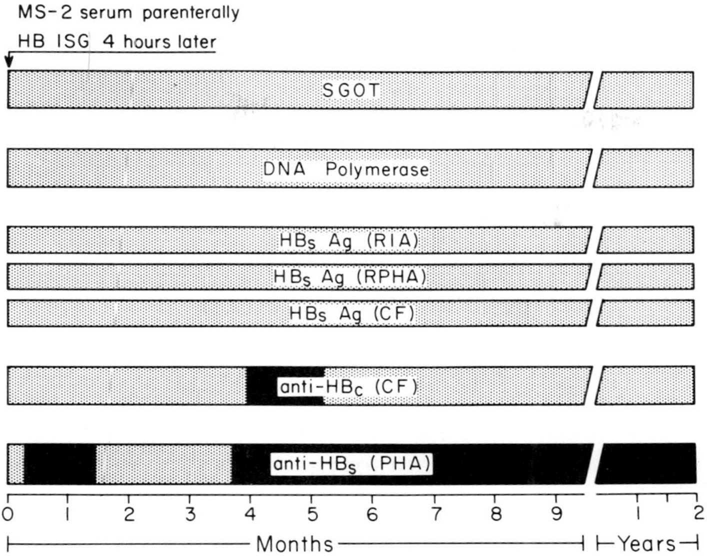

Fig. 3. Prevention of viral hepatitis, type B, with hepatitis B immune serum globulin (HBISG). Results of tests for SGOT, DNA polymerase activity, HB_s Ag, anti-HB_s, and anti-HB_c. Black, abnormal or detectable; stippled, normal or not detectable; white, not tested. Abbreviations same as in Fig. 1.

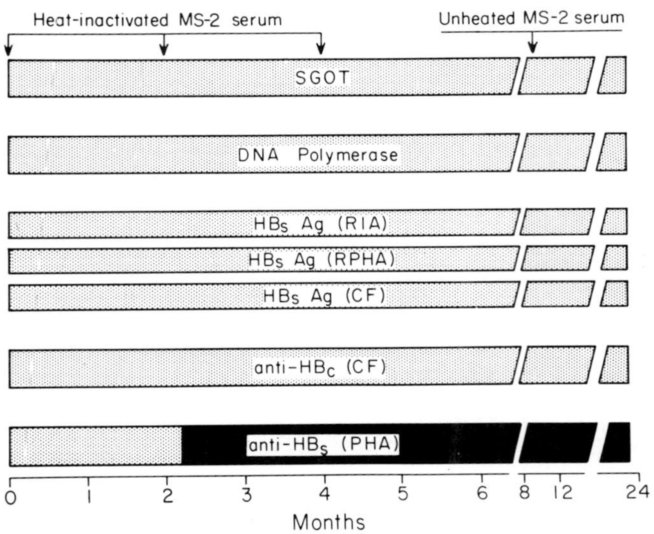

Fig. 4. Prevention of viral hepatitis, type B, by active immunization with heat-inactivated MS-2 serum. Results of tests for SGOT, DNA polymerase activity, HB_s Ag, anti-HB_s, and anti-HB_c. Black, abnormal or detectable; stippled, normal or not detectable; white, not tested. Abbreviations same as in Fig. 1.

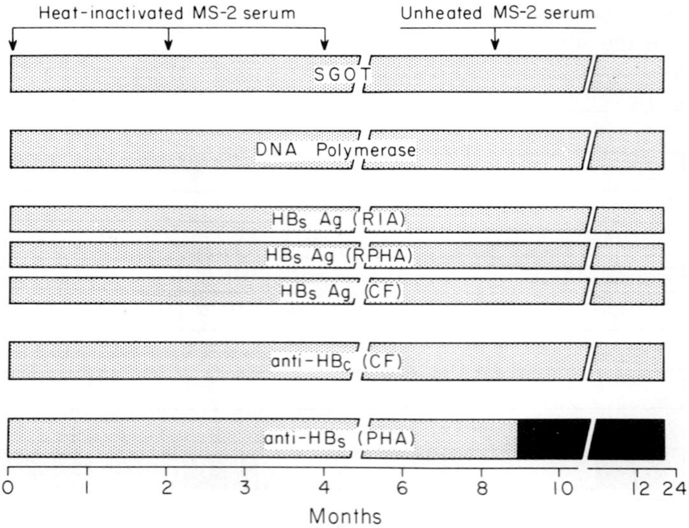

Fig. 5. Prevention of viral hepatitis, type B, by active immunization with heat-inactivated MS-2 serum. Results of tests for SGOT, DNA polymerase activity, HB_s Ag, anti-HB_s, and anti-HB_c. Black, abnormal or detectable; stippled, normal or not detectable; while not tested. Abbreviations same as in Fig. 1. Figures 1–5 from Krugman *et al. N. Engl. J. Med.* **290,** 1331–1334 (1974).

DR. STEWART: Will the rhesus monkey-adapted virus grow in monkey cells in culture?

DR. PURCELL: Well, human strains do not grow in human cells in culture, and the rhesus strain does not seem to grow in rhesus cells in culture either. We still have the problem of not isolating reproducibly this virus in any known culture.

DR. GINSBERG: This is a very beautiful demonstration, but I am still somewhat confused by my old memory that ether does not inactivate hepatitis virus, yet these apparently have envelopes and I was wondering if the core was infectious after removal of the envelope.

DR. PURCELL: It is difficult to say definitely. There is some evidence from chimpanzee studies and some evidence from our rhesus monkey studies that the core without its envelope is not infectious, and this would make it analogous to herpes virus and a number of other viruses. We do have additional evidence from the rhesus monkey model that the virus does, indeed, seem to be sensitive to lipid solvents and detergents. I do not know how to rationalize this in light of the previous studies, except that those previous studies were carried out with pools of virus of unknown titer, which may have contained a tremendous amount of virus, and generally in the presence of high concentrations of serum or plasma which have been shown previously, in studies by Liebhaber, H., Krugman, S., Giles, J. P., and McGregor, D. M. [*Virology* **24,** 109 (1964)] to protect at least partially some lipid-sensitive viruses.

DR. CASALS: On listing the various possibilities of transmission of hepatitis B virus by nonparenteral or by cryptoparenteral routes, was that discussion put there as an exercise to cover all possibilities, or do you have some evidence to indicate that there is arthropod-borne transmission? The mechanical transmission would offer no problem, but biological transmission is a different matter. Is there any evidence that you know of to indicate that there is actual biological transmission?

DR. PURCELL: One paper from Africa reported what appeared to be multiplication of the antigen, as judged by immunofluorescent detection in mosquitoes following their ingestion of the antigen. Also, there was evidence in that paper of transmission to chickens, but that work has not yet been confirmed. Most individuals who have examined mosquitoes that have either fed on artificial preparations or on antigen-positive volunteers have failed to demonstrate any evidence of multiplication. I think most people now believe that mechanical transmission is more likely to occur, if in fact, arthropod-mediated transmission occurs at all.

DR. WELLER: Are there further comments on this question of biological transmission as contrasted to mechanical, because this is an important point; or do you accept Dr. Purcell's summary which is essentially negative?

DR. PURCELL: Antigen certainly can be detected in mosquitoes that have been trapped both in the United States, in Africa, and elsewhere, but detecting antigen in mosquitoes may simply be a way of sampling the population. It's clear that these populations often have up to 10% antigenemia. So it is not surprising that if you detect blood in mosquitoes, you will detect antigen in some of them.

DR. MERIGAN: I believe that Almeida *et al.* [Almeida, J. D., Rubenstein, D., and Stott, E. J., *Lancet* **2**, 1225 (1971)], has identified a patient who has primarily core antibody. I wonder whether you or Dr. Krugman have any information as to whether core antibody is protective?

DR. PURCELL: Most of the evidence would indicate that core antibody is probably not protective. Core antibody as Dr. Krugman showed, appears first, sometimes months before antibody to the surface antigen appears. It has been demonstrated by Hoofnagle *et al.* [Hoofnagle, J. H., Gerety, R. J., and Barker, L. F., *Lancet* **2**, 869 (1973)] with their complement fixation tests, and we found the same with our radioimmunoassay blocking test, that every chronic carrier of the antigen tested has high levels of anticore antibody but little, if any, antisurface antibody. In the face of these high levels of anticore antibody, infection progresses. It certainly has no effect in stopping the infection. Moreover, Hoofnagle has some evidence, from the reanalysis of Dr. Murray's earlier studies in volunteers, that antisurface antibody does appear to be protective, whereas anticore antibody does not seem to correlate with protection.

DR. BLUMBERG: I wanted to comment on the question about the mosquitoes. We have reported that the antigen has been detected in the eggs of mosquitoes which could be due to some kind of mechanical transmission but is also consistent with some sort of biological action.

DR. PURCELL: I believe, Dr. Blumberg, that you have also found persistence of the antigen in mosquitoes and that Dr. Maynard has recently found the same also. It is difficult to know whether that's just a slow leeching out of trapped antigen or whether there is actual multiplication.

DR. BLUMBERG: I think that your general conclusion is sound. So far, it is not clear that there is any replication in the mosquitoes but I think it's still a possibility and warrants continued study. We found a persistence of antigen in mosquitoes up to 2–3 days after they were experimentally fed on blood containing the antigen.

DR. KOCH: Two comments concerning the infectivity of core particles. Recently, it has been reported that isolated DNA from herpes virus is infectious. In most virus host cell systems, one finds that the infectivities of subvirus particles and of isolated viral nucleic acids are several orders of magnitude lower than the infectivity of intact virus particles. Therefore, we might expect that it will be more difficult to demonstrate the infectivity of core particles than of complete virus.

DR. PURCELL: Yes. I'm aware that the nucleic acid has been shown to be infective, but I don't believe there's any good evidence that the capsid itself, containing nucleic acid but free of the envelope, is infectious.

DR. CROWELL: I should like to encourage those attempting to culture the hepatitis viruses *in vitro* by calling attention to our recent findings [Landau, B. J., Crowell, R. L., and Boclair, C. W., *Proc. Soc. Exp. Biol. Med.* **141**, 753–758 (1972)] that tissue cultures of differentiating fetal mouse muscle cells serve as susceptible hosts for Coxsackie viruses of group A. Several of these A viruses had not been cultured *in vitro* prior to this finding, even though many attempts to do so had failed. Thus, since the hepatitis viruses share with Coxsackie viruses of group A a highly specific tissue tropism, the liver for the former and mucles for the latter, continued efforts should be made to obtain differentiating hepatocytes *in vitro*. It is reasonable that such cells may serve as susceptible hosts for replication of the hepatitis viruses.

DR. WELLER: It's an interesting comment, but I sort of feel that the Coxsackie viruses and hepatitis B agents are different orders of magnitude in terms of their challenge.

CHAPTER 4

Defective Interfering Particles As Antiviral Agents

Alice S. Huang
Eduardo L. Palma

I. INTRODUCTION

The properties of defective interfering (DI) particles, the extent of their presence among virus groups, and their molecular biology have been recently reviewed (*4*). In this paper the antiviral properties of DI particles are examined through a series of experiments that demonstrate their ability to affect the outcome of virus–cell interactions. Also, the molecular basis of interference is presented.

Before discussing data on the role of DI viruses as antiviral agents, let us first examine the natural history of DI particles and the theoretical implications of their presence for viral persistence.

The modern molecular virologist's understanding of viruses naively stems largely from studies on laboratory strains. Laboratory isolates of viruses are selected for their rapid growth and cytopathogenicity. It is often assumed that laboratory strains behave in their hosts identically as the original isolates and that the ameliorating effects on viral infections in animals are almost completely due to host defenses. This can only be part of the truth and, perhaps, not even the truth. For it is presumptuous of us to assume that such absolute parasites as viruses would depend so completely on their hosts to prevent their own extermination. Without any checks on viral growth, viruses might be expected to rampage, killing every last cell and every available host; thus, causing not only our death, but also the extinction of the animal viruses themselves. Even without host controls on viruses, such a final outcome is unlikely to happen because viruses have a built-in self-limiting mechanism programmed within their growth cycle. This control resides in their ability to generate DI particles.

DI particles are, therefore, natural antiviral agents. Because they are found in most preparations of animal viruses, their presence may explain how agents of acute viral disease may possibly, also, be etiological agents of chronic infections. The DI particles provide a mechanism for explaining how viruses persist without killing their hosts. It is important to note that such a mechanism does not necessarily explain the causes for the symptoms of chronic viral disease.

II. CONTINUOUS INFECTION OF CELLS BY VESICULAR STOMATITIS VIRUS

The system chosen for these studies is vesicular stomatitis virus (VSV), the prototype of the bullet-shaped, RNA-containing rhabdoviruses. The strain was selected for its rapid cytopathogenicity (*19, 20*). In order to examine virus–host interactions in the absence of the immune response and other host defenses, a reconstruction experiment was initiated. Chinese hamster ovary (CHO) cells in suspension (*18*) were infected with standard, plaque-forming B particles and defective interfering T particles (*6*) in a ratio of 7:1. Such an infection was allowed to proceed in the presence of a constantly increasing number of uninfected CHO cells by diluting the infected cells 1:9 with fresh CHO cells every six successive 24-hour periods. Aliquots were taken during each 24-hour period for the determination of three parameters: (1) the percent of surviving cells by dye exclusion, (2) the production of B particles, and (3) the production of T particles.

Figure 1 shows that the percentage of cells surviving a 24-hour infection with a mixture of B and T particles reached a value of 50% by the fourth successive 24-hour period and then declined to 30% at the sixth period. When cells from the fourth period were examined, they were found to be resistant to superinfection and were synthesizing a low level of virus-specific messenger RNA's (*10*). These cells eventually died because they were infected by a very cytopathic virus. Nevertheless, their increased survival time indicated a variability in host response due to the presence of T particles.

When the production of B and T particles were examined by the incorporation of radioactive precursor into the nucleic acids of the two particles, there was a nonoverlapping cyclic production of B and T

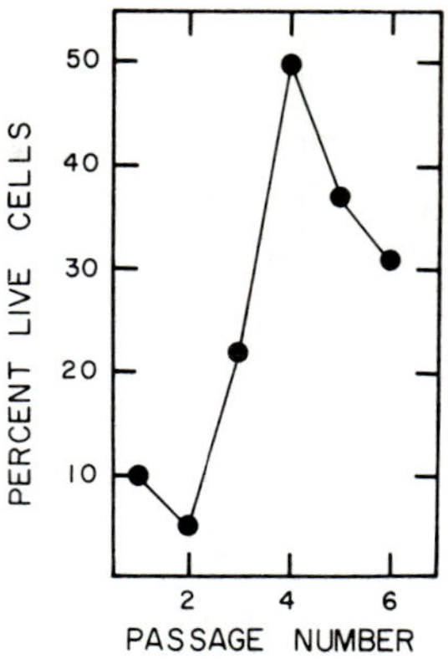

Fig. 1. Survival of host cells during successive passages of B and T particles (*10*). A first passage of a mixture of B and T particles was made by infecting 4×10^7 Chinese hamster ovary cells suspended in 10 ml of medium with B particles at a multiplicity of 20 and T particles at an estimated multiplicity of 3 (*7*). After 30 minutes of attachment at room temperature the cells were diluted to 100 ml and incubated at 37°C for 20 hours. Successive passages were made by resuspending 4×10^7 fresh CHO cells with 10 ml of infected cells from the previous passage. After attachment at 4°C for 30 minutes the cells were diluted to 100 ml and further incubated at 37°C for 20 hours. At the end of each passage, an aliquot of cells was mixed in equal volume with undiluted trypan blue and the cells were counted in a hemocytometer chamber. Percent survival is expressed as

$$\frac{\text{No. of unstained cells}}{\text{No. of unstained cells} + \text{No. of stained cells}} \times 100$$

The total number of cells expressed in the denominator did not vary significantly from passage to passage.

particles (Fig. 2). This cyclic pattern was predicted from the known interdependence of B and T particles. The interdependence results from the fact that B particles are necessary helpers for the synthesis of T particles and, at the same time, T particles inhibit the synthesis of B particles. The significance of this interactive cyclic pattern for viral pathogenesis will be discussed.

When B particles were assayed by plaque formation, the amount produced during each of the 24-hour periods varied over a range of 10^4 to 10^8 plaque-forming units per milliliter (Fig. 3). A one-step growth cycle for VSV occurs within 8 hours. Therefore, such a variable production of B particles could not be due to harvesting the virus at the wrong time during viral replication. Such variation in the production of B particles was solely attributable to the presence or absence of T particles. The total number of viral particles and their relative distribution into

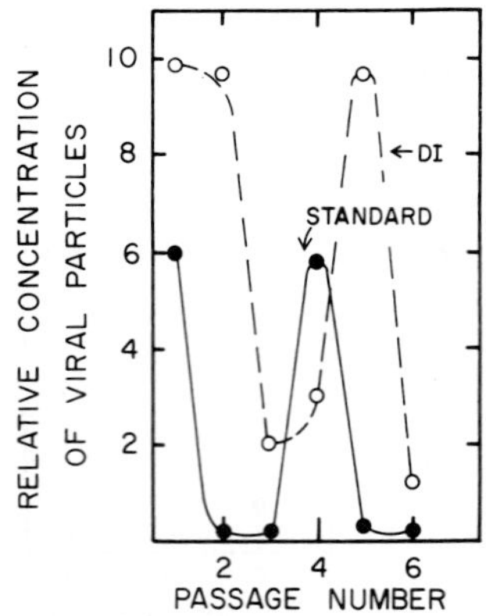

Fig. 2. Relative concentrations of B and T particles during successive passages (*10*). Aliquots of Chinese hamster ovary cells were infected identically to the cells shown in Fig. 1, except that actinomycin at a final concentration of 5 μg/ml was added to the concentrated cells (10 ml) during the attachment period and later uridine-^{3}H at a final concentration of 10 μCi/ml was added to cells at the beginning of incubation at 37°C. The extracellular viral particles from each passage were pelleted from the medium by centrifugation at 106,000 *g* for 30 minutes at 4°C and resuspended in 1 ml of Earle's saline. Extracellular virus-specific material represents 100% of the total pelleted viral particles. Extracellular ^{3}H-labeled-uridine viral particles were deproteinized with sodium dodecyl sulfate (SDS) and separated on a 15–30% sucrose gradient containing SDS (*18*). The relative radioactivity sedimenting at 40 S and at 19 S representing RNA's of B particles and T particles, respectively, are graphed here. The relative counts per minute in T RNA was not multiplied by 3 and, therefore, reflects the minimum number of viral particles that contain one-third the RNA contained in B particles (*8*). Raw data appeared similar to those shown in Fig. 4b. DI, defective interfering (particles).

B and T particles did account for the very different outcomes resulting from the VSV–CHO cell interactions.

III. NUCLEOCAPSID ACCUMULATION

Because many viral infections, including persistent ones, are diagnosed by the presence of inclusion bodies and because some inclusion bodies have been characterized to contain viral nucleic acid and protein, the infected cells from each of the 24-hour periods were examined for intracellular accumulation of viral structures. Only cells producing a maximum amount of T particles retained any viral structures. These cells were from the second and fifth periods shown in Fig. 2. The structures TNC were characterized to be nucleocapsids sedimenting at 70 S (Fig. 4a). They contained RNA the size of T-particle RNA and protein sub-

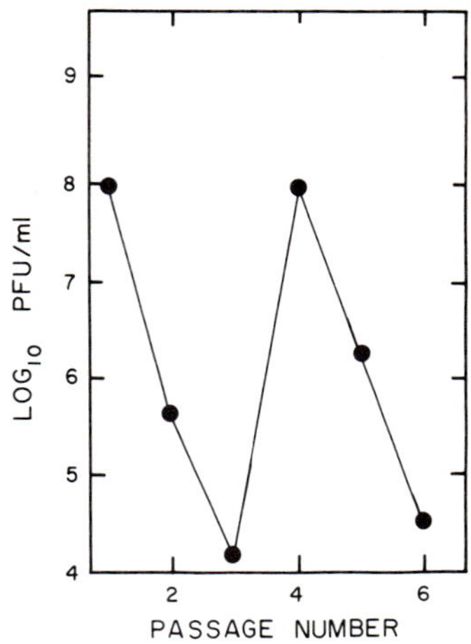

Fig. 3. Plaque assay of B particles produced during successive passages of B and T particles in Chinese hamster ovary (CHO) cells (*10*). At the end of each passage shown in Fig. 1, an aliquot was withdrawn, the cells pelleted by centrifugation at 600 *g* for 7 minutes, and the medium assayed on monolayers of CHO cells for plaque-forming B particles as previously described (*18*).

units that protected the RNA from nuclease digestion. A search for intracellular nucleocapsids similar in size to nucleocapsids from B particles (BNC) revealed that the ratio of intracellular T to B nucleocapsids was 24:1 (Fig. 4a), whereas extracellular RNA from viral particles produced by these same cells was in ratio of 6:1 for T to B RNA (Fig. 4b).

In an attempt to determine why T nucleocapsids accumulated intracellularly, the RNA inside the nucleocapsids was examined by nucleic acid hybridization. It was annealed to B virion RNA as well as to complementary messenger RNA from VSV-infected cells. Purified intracellular T nucleocapsids contained 19 S RNA, 60% of which had nucleotide sequences identical to extracellular virion RNA, and 40% of which had complementary sequences (Table 1). If the 19 S RNA were taken from an extract of total cytoplasmic RNA, the distribution of virionlike and complementary RNA remained 60:40. These results suggest that all of the intracellular single-stranded 19 S RNA was found in nucleocapsid structures. Similar results have been found for the intracellular B virion 40 S RNA and its complete 40 S complement (*21*).

For controls, RNA from purified extracellular B and T particles was shown to be of only one type and almost completely complementary to messenger RNA (Table 1).

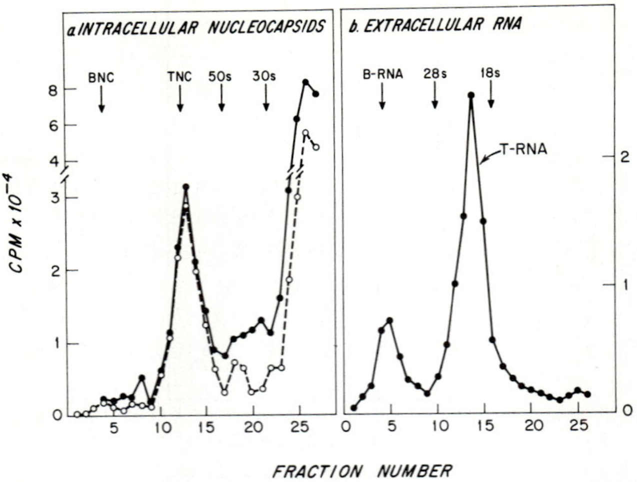

Fig. 4. Sucrose gradient profiles of intracellular RNA–protein complexes and extracellular viral RNA produced by cells coinfected with B and T particles. One 4 ml sample of 1.6×10^6 Chinese hamster ovary cells containing 10 μg/ml of actinomycin D was coinfected with B particles at a multiplicity of 25 and T particles at an estimated multiplicity of 60. After 30 minutes at 35°C, 4 ml of medium containing 50 μCi/ml of uridine-^{3}H was added. The cells were incubated at 35°C for 4 hours. Then the cells were separated from the supernatant by centrifugation at 600 *g* for 5 minutes and fractionated. The pellet containing cells was processed for the cytoplasmic extract as previously described (7). After sodium deoxycholate at a final concentration of 1% was added to the extract it was layered on a 15–30% sucrose gradient containing buffer without divalent cations—NEB (7) and centrifuged in the SW 27 rotor at 4°C for 15 hours at 50,000 *g*. The collected fractions from the gradients were divided in half and one half was treated with ribonuclease (7). (○---○), acid-precipitable ribonuclease-resistant RNA; (●—●), total acid-precipitable RNA. The supernatant containing extracellular virus particles was processed identically to the extracellular particles shown in Fig. 2. The RNA sedimenting at 40 S is from B particles and the RNA sedimenting at 19 S is from T particles (8). BNC, B-particle nucleocapsids; TNC, T-particle nucleocapsids.

Nucleocapsids containing RNA with nucleotide sequences complementary to that found in T particles accounted for the inability of some nucleocapsids to mature into extracellular T particles. However, other mechanisms must be operative. One such example was the preferential maturation of B nucleocapsids over T nucleocapsids into extracellular particles (cf. Fig. 4a and b).

IV. MOLECULAR BASIS OF INTERFERENCE

How T particles inhibit the synthesis of B particles is now understood on a molecular basis. The RNA of T particles maintains a selective

TABLE 1. ANNEALING OF 19 S RNA FROM T PARTICLES AND FROM INTRACELLULAR NUCLEOCAPSIDS TO VIRION RNA OR TO MESSENGER RNA[a]

Source of labeled RNA	Ribonuclease resistance (%)
19 S RNA from intracellular nucleocapsids[b]	
Annealed to virion RNA	41.9
Annealed to 13–28 S mRNA	67.0
Not annealed	5.7
19 S RNA from a cytoplasmic extract[b]	
Annealed to virion RNA	40.8
Annealed to 13–28 S mRNA	71.9
Not annealed	3.2
19 S RNA from T particles[c]	
Annealed to virion RNA	8.5
Annealed to 13–28 S mRNA	106.7
Not annealed	5.4
40 S RNA from B virions[c]	
Annealed to virion RNA	6.4
Not annealed	2.8
13–15 S mRNA from a cytoplasmic extract[c]	
Annealed to virion RNA	101.0
Annealed to 13–28 S mRNA	3.1
Not annealed	2.4

[a] Different samples of ^{3}H-labeled RNA, containing 3000 cpm or greater, were resuspended in 0.3 ml of 2 × SSC, containing 20% sucrose and 0.1% sodium dodecyl sulfate (SDS). The samples were annealed either with 8 μg/ml of nonradioactive 40 S virion RNA or with 300 μg/ml of unlabeled 13–28 S intracellular RNA or not annealed. The annealing was performed by incubation at 70°C for 3.5 hours. Then, all the samples were diluted with 2 × SSC to 2 ml and divided into two aliquots, one of which was treated with ribonuclease as previously described (*7*).

[b] Intracellular 19 S RNA was prepared either from 70 S nucleocapsid material which was labeled and isolated (as shown for Fig. 4a) or from whole cytoplasmic extracts (as shown for Fig. 4a). The RNA was extracted by treatment with 1% SDS and centrifuged in a 15–30% sucrose gradient in SDS buffer containing 0.1% SDS (*18*). Analysis of the 70 S nucleocapsid in parallel experiments showed that all of the RNA was ribonuclease-resistant prior to SDS treatment. After deproteinization, all of the RNA migrated as a single 19 S species on polyacrylamide gels and was completely sensitive to ribonuclease.

[c] Preparation of 19 S RNA from T particles, 40 S RNA from B particles, and messenger RNA species were done as described (*7*).

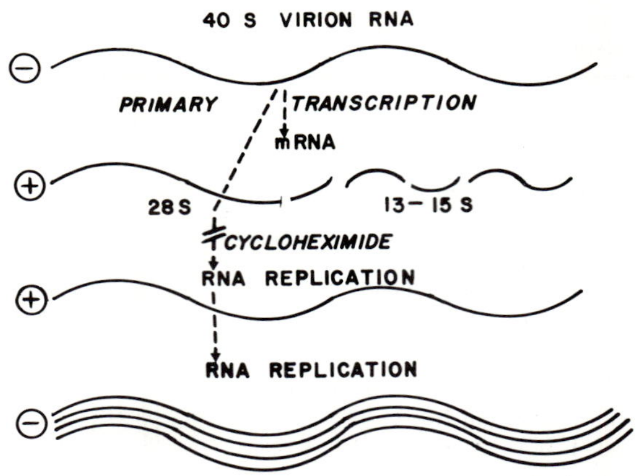

Fig. 5. Virus-specific single-stranded RNA species synthesized in cells infected only by B particles of vesicular stomatitis virus. The identification of RNA species is from Stampfer *et al.* (*18*), Baltimore *et al.* (*1*), Huang *et al.* (*5*), Huang and Manders (*7*), and Stampfer (*17*).

advantage during RNA synthesis (*18*). To understand this competition in detail let us first summarize VSV RNA synthesis (Fig. 5). There are only a limited number of virus-specific RNA species. These are virion 40 S RNA, complementary 40 S RNA, 28 S messenger RNA, and 13–15 S messenger (*5, 9, 17*). The intermediates for the synthesis of these RNA's sediment heterogeneously from 25 to 35 S and represent only a small proportion of the total virus-specific RNA (*16, 18*).

During viral growth, two separate pathways exist for RNA synthesis: transcription is the synthesis of multiple complementary smaller messenger RNA species from the virion RNA template (*1*); replication is the synthesis of progeny virion RNA from a complementary template the full size of virion RNA. This template is synthesized directly from the virion RNA and is not a product resulting from the processing of smaller messenger RNA pieces (*12, 13*).

During coinfection of cells with B and T particles, transcription by B particles proceeds normally (*7, 14*), and interference occurs at the level of replication (*11*). Therefore, the synthesis of T RNA molecules competes only with the enzyme(s) catalyzing replication and not transcription. The reduction of overall virus-specific RNA synthesis by T particles is due to this direct inhibition of B RNA replication and subsequent indirect inhibition of secondary transcription resulting from newly replicated templates.

This rather specific step of interference by T RNA on B RNA replication is best demonstrated by coinfecting cells with T particles derived

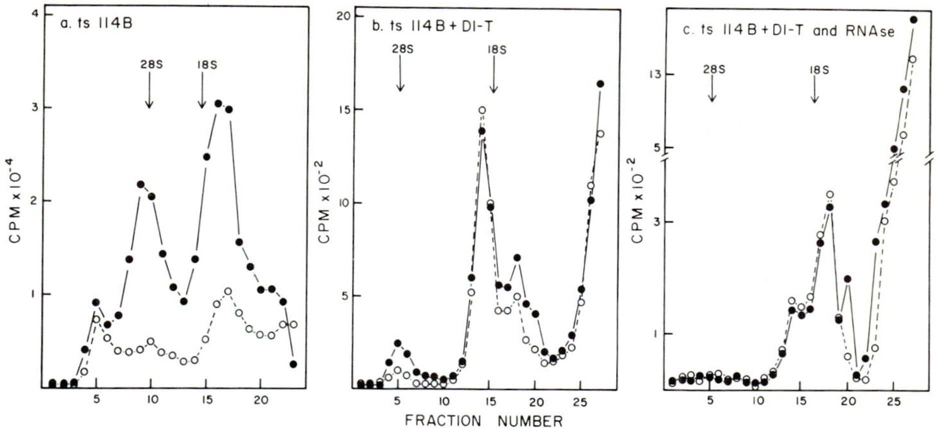

Fig. 6. Sucrose gradient patterns of RNA made by T particles and ts 114 at the permissive and nonpermissive temperatures (*11*). Two 1.5-ml samples of 6×10^6 Chinese hamster ovary cells were infected with either ts 114 B particles alone at a multiplicity of 50 or with ts 114 B particles at a multiplicity of 30 plus T particles at a multiplicity of 100. Each sample received actinomycin D to a final concentration of 10 μg/ml. After adsorption for 30 minutes at 22°C the samples were diluted two-fold with medium and the samples further incubated at 31°C for 2.5 hours. Then half of each sample was transferred to 37°C. After another 5 minutes uridine-^{3}H was added to each sample to a final concentration of 33 μCi/ml. After 1 hour of labeling, cytoplasmic extracts were made (*7*) and layered onto separate 15–30% sucrose gradients containing sodium dodecyl sulfate. After centrifugation these gradients were collected and fractionated. Half of the fractions collected from cells infected with ts 114 B particles alone were acid-precipitated; fractions from coinfected cells were divided in half, and one-half was digested with ribonuclease prior to acid-precipitation of both halves (*7*). (●) Ribonucleic acid synthesized at 31°C; (○) at 37°C. DI, defective interfering (particles).

from wild-type virus and helper B particles derived from a temperature-sensitive mutant ts 114 (*15*). At nonpermissive temperatures in ts 114-infected cells, transcription is reduced considerably and replication is completely unaffected (*13*). The same mutant-infected cell can be treated in such a way so that an opposite dissociation is achieved. By inhibiting protein synthesis at the nonpermissive temperature, replication is completely inhibited and transcription is restored as the only functional RNA synthetic pathway (*12*).

By using the two conditions to uncouple transcription from replication, it was shown that T RNA synthesis was dependent only on the replicative and not the transcriptive pathway. Figure 6 shows the sucrose gradient patterns of RNA species synthesized at the permissive and nonpermissive temperatures by cells infected with ts 114 B particles alone or coinfected with ts 114 B particles and T particles. The inhibition of transcription

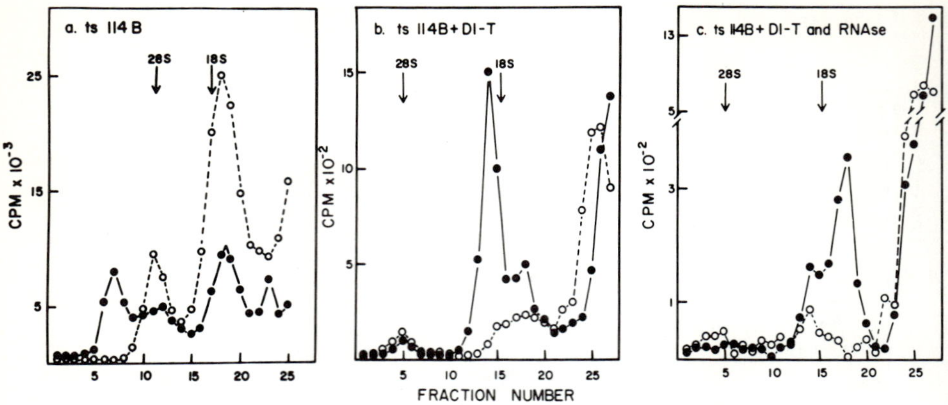

Fig. 7. Sucrose gradient patterns of RNA synthesized by ts 114 B and T particles at the nonpermissive temperature in the presence and absence of cycloheximide (*11*). Two 1.5-ml samples containing 6×10^6 Chinese hamster ovary cells were infected with either ts 114 B particles alone at a multiplicity of 50 or with ts 114 B particles at the same multiplicity plus T particles at a multiplicity of 100. The conditions of incubation were the same as for Fig. 6 except that whole samples were transferred to 37°C and half of each sample was treated with 100 μg/ml of cycloheximide at this temperature. After 5 minutes of incubation, 50 μCi of uridine-^{3}H was added and the samples were labeled for 1 hour. Cytoplasmic extracts were processed exactly as indicated for Fig. 6. (●) RNA synthesized at 37°C in the absence of cycloheximide; (○) in the presence of cycloheximide. DI, defective interfering (particles).

by a shift in temperature does not affect the synthesis of T RNA and the replicative intermediates for its synthesis. By contrast, the inhibition of replication abolishes T RNA synthesis (Fig. 7). Therefore, these studies permit the conclusion that interference by T particles on the growth of B particles occurs during the step of RNA replication [for details, see Palma *et al.* (*11*)].

V. DISCUSSION

In summary, these results show that decreased synthesis of B particles is due to the ability of T RNA to compete successfully during RNA replication. Also, increased cell survival and intracellular accumulation of viral nucleocapsids during infection can be correlated to the presence of T particles. Extrapolation of these observations on VSV to other viruses would suggest that interference by DI particles occurs during replication of viral genomes and that the presence of DI particles can be suspected (1) when viable cells are resistant to superinfection by

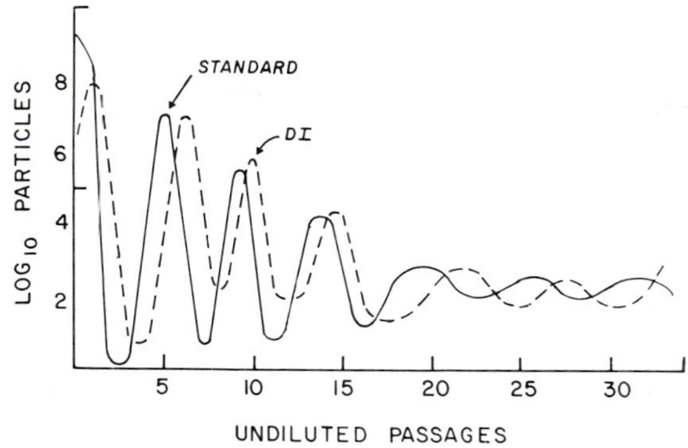

Fig. 8. A model for the oscillating interactions between standard virus and DI particles during continuous passages. Reproduced with permission from A. S. Huang, *Annu. Rev. Microbiol.* **27.** Copyright © 1973 by Annual Reviews, Inc. All rights reserved.

homologous virus, (2) when difficulties initially arise in isolating an infectious agent but not when cells are diluted or passaged, and (3) when intracellular viral antigens accumulate. Moreover, as shown by the cyclic patterns these findings can vary considerably within the infected animal, as well as from animal to animal.

Considering the cyclic pattern again, the wide ranges in the production of infectious standard particles within a period of six 24-hour periods may be interpreted to represent viral infections. The first peak of standard virus production (Fig. 8) may represent acute viral disease. In 3 days when DI particles appear and the level of standard virus is reduced, host defenses will have been marshalled and the infection controlled. If not, another burst of synthesis of standard virus occurs and a biphasic pattern for the acute disease emerges. Persistent infection may be represented by a continuation of this cyclic pattern with an overall decreased production of both kinds of viral particles or just decreased production of standard virus. This may be accomplished by host immune responses or by a selective mutation in the virus itself. Disease symptoms then would result from an accumulation of viral antigens or by anything that would upset the balance between standard and DI particles. A localized burst in the production of standard virus may cause the exacerbations that accompany chronic infections. These are testable hypotheses which are presented here as perspectives in viral pathogenesis.

Lastly, the question of whether DI particles are antiviral agents should be considered. Available evidence indicates that DI particles are extremely specific inhibitory agents of the viruses from which they are derived. Their ability to inhibit, however, is modified by viral strain differences and by the genetic makeup of the host [reviewed by Huang

(*4*)]. Other factors are suspected to affect the inhibitory properties of DI particles, but until they are understood, use of DI particles as antiviral agents is contraindicated. Preliminary studies by Holland and his colleagues (*2, 3*) demonstrating the prophylactic properties of DI particles in animals have shown that with the wrong combination of DI and standard particles, the disease may become prolonged, go underground, and even increase in severity. Therefore, only after further studies will we know if DI particles are potentially useful antiviral agents. The potential of DI particles lies not only in their antiviral properties but they will undoubtedly, also, provide much information on our understanding of viral pathogenesis.

ACKNOWLEDGMENTS

This work is supported by research grants from the American Cancer Society VC-63A and the U. S. Public Health Service research grant AI-10100. Author A. S. H. is a Research Career Development Awardee of the U. S. Public Health Servcie, and E. L. P. is a World Health Organization–Pan American Health Organization Fellow on leave from the University of Buenos Aires. We thank Craig Pringle for sending us ts 114 and Norma Hewlett for excellent technical support. Also, provocative discussions were provided by Drs. Stanley M. Perlman and Marco Soria.

REFERENCES

1. Baltimore, D., Huang, A. S., and Stampfer, M., *Proc. Nat. Acad. Sci. U. S.* **66,** 572–576 (1970).
2. Doyle, M., and Holland, J. J., *Proc. Nat. Acad. Sci. U. S.* **70,** 2105–2108 (1973).
3. Holland, J. J., and Doyle, M., *Infec. Immunity* **7,** 526–531 (1973).
4. Huang, A. S., *Annu. Rev. Microbiol.* **27,** 101–117 (1973).
5. Huang, A. S., Baltimore, D., and Stampfer, M., *Virology* **42,** 946–957 (1970).
6. Huang, A. S., Greenwalt, J. W., and Wagner, R. R., *Virology* **30,** 161–172 (1966).
7. Huang, A. S., and Manders, E. K., *J. Virol.* **9,** 909–916 (1972).
8. Huang, A. S., and Wagner, R. R., *J. Mol. Biol.* **22,** 381–384 (1966).
9. Mudd, J. A., and Summers, D. F., *Virology* **42,** 958–968 (1970).
10. Palma, E. L., and Huang, A. S., *J. Infec. Dis.* **129,** 402–410 (1974).
11. Palma, E. L., Perlman, S. M., and Huang, A. S. *J. Mol. Biol.* **85,** 127–136 (1974).
12. Perlman, S. M., and Huang, A. S., *J. Virol.* **12,** 1395–1400 (1973).
13. Perlman, S. M., and Huang, A. S., *Intervirology* **2,** 312–325 (1974).
14. Perrault, J., and Holland, J. J., *Virology* **50,** 159–170 (1972).
15. Pringle, C. R., *J. Virol.* **5,** 559–567 (1970).
16. Schincariol, A. L., and Howatson, A. F., *Virology* **42,** 732–743 (1970).
17. Stampfer, M., Ph.D. Thesis, Massachusetts Institute of Technology, Cambridge, Massachusetts (1972).
18. Stampfer, M., Baltimore, D., and Huang, A. S., *J .Virol.* **4,** 154–161 (1969).
19. Stampfer, M., Baltimore, D., and Huang, A. S., *J. Virol.* **7,** 409–411 (1971).
20. Wagner, R. R., Levy, A. H., Snyder, R. M., Ratcliff, G. A., and Hyatt, D. F., *J. Immunol.* **91,** 112–122 (1963).
21. Soria, M., Little, S. P., and Huang, A. S. *Virology* **61,** 270–280 (1974).

DISCUSSION

DR. PFAU: I would like to make some comments about the arenaviruses. If you take lymphocytic choriomeningitis (LCM), the prototype of the arenavirus group, and carry infected cells for long periods of time in tissue culture, you get the almost exclusive production of what is undoubtedly defective interfering (DI) virus. These types of cultures were studied extensively by Dr. Lehmann-Grube in 1969 [Lehmann-Grube, F., Slenczka, W., and Tees, R. *J. Gen. Virol.* **5**, 63–81 (1969).] and he showed that these DI viruses could be immunogenic in mice and protect against lethal challenge with standard virus. There are three strong human pathogens in the arenavirus group: Lassa fever, Junin, and Machupo. Two other viruses in this group, Parana and Pichinde, will also produce DI virus, and it is probable that Junin virus will also. I, therefore, think that there is a real possibility for using these as antiviral agents.

One short comment on regulation. As you know, LCM causes a classic persistent infection in mice and there is a very good regulation mechanism there. If you follow an individual mouse that has this tolerant infection for its whole life-span, there will be an almost constant titer of virus in that mouse, and immunological mechanisms do not seem to control this titer.

DR. HUANG: I think when we discuss the role of the DI particles in animals, it is important to differentiate between the maintenance of viral antigen in the cell and in the organism, i.e., the maintenance of such antigen of virus particles without killing the animal, together with the actual direct causes of disease symptoms such as immune complexes.

DR. LEVY: Dr. Huang's model system has been actually tried with the C-type viruses. If you infect cell cultures with leukemia viruses, you are unable to superinfect them with leukemia viruses of the same type or sarcoma viruses carrying the same type-specific antigen. Actually this was first described in the chicken system by Rubin and in the mouse system by Sarma. Rowe showed in 1963 that if you infected mice with Moloney virus and then challenged them with Friend virus, which shared very similar type of coats, Friend disease has a much shorter latent period. You could prevent the onset of Friend disease. Recently, Barski and his co-workers have shown this as well in other mouse systems using other leukemia viruses. In fact, a few years ago, Latajet commented that if you could get a nontransforming nonleukemogenic virus and infect an animal with this virus and have every cell infected, this would be an ideal method of preventing cancer by propagating viruses. And, one last point, in the Finkel, Biskis, and Jenkins virus (the FBJ sarcoma virus), we have found autoinhibition, in that there are very high titers of nontransforming nonleukemogenic virus associated with the sarcoma virus, which, if you were to test for the sarcoma virus directly, is difficult to show pathogenicity because of its interfering virus. However, as Dr. Huang commented, if you titrate out the virus you'll certainly find the sarcoma virus particle.

DR. HUANG: I would like to comment on this interference that has been so well characterized among the RNA tumor viruses especially in the avian system. In this regard, when we look at our cells that were surviving fairly well during passage 4, there was primary transcription which was unaffected by the presence of T particles. Now, primary transcription permits primary translation which allows all proteins specific to virus infection. And we hypothesize that G protein, which is the surface antigen of the virus, is produced and reaches the outside of the cells, interacts with the receptor site on the cells, and, therefore, causes the interference for homologous superinfecting virus. We are presently testing this hypothesis, and our preliminary studies indicate that, indeed, this resistance to superinfection is a surface one and not an intracellular one.

Synthetic Vaccines of the Future*

Michael Sela†

I. INTRODUCTION

I think that it is nice to relax and dream, and it is in this spirit that, although I may have some limited competence in chemistry and immunology, I am propelling into futurology. My sole excuse is that I shall try to extrapolate from known data.

Louis Pasteur, who was probably the first great experimental immunologist, introduced the attenuation of virulent organisms, and used them for immunization. He called this treatment vaccination to honor Jenner and his method of preventive treatment against smallpox. Not much has changed in our methods of immunization since Louis Pasteur. I would like to draw attention to dreams concerning the synthetic vaccines of the future, and to certain approaches that might possibly convert these dreams, and I insist on the word *dreams*, into reality.

Now, with the increase in the number of vaccines used, it is obvious that fundamentally new approaches are needed to cope with the problem in the future. One such possibility is to use for immunization synthetic substances containing several immunopotent determinants, each one characteristic of a different disease, on the same macromolecule. I want also to make clear that, if the attenuated rather than the killed viruses are going to be the universal answer to all cases of vaccinations, then the present approach is of no interest. Conversely, if killed vaccines are going to be an important factor for immunization also in the future, the approach described here might be ultimately the correct answer.

* After-luncheon address.

† This paper was prepared while the author was a Fogarty Scholar-in-Residence, National Institutes of Health, Bethesda, Maryland.

II. SPECIFICITY DETERMINANTS

In discussing the synthetic vaccines of the future, I would really like to mention several components, and I shall try to illustrate them with a few examples. First, you will recall that when you have antibodies against a bacterium or against a virus, or against a tissue, the antibodies are really always against a very limited juxtaposition of atoms in a certain macromolecule on the surface. One iron rule of immunology is that *whatever sticks out more is more immunopotent;* another one is that one has to think in terms of molecules rather than complete "beasts."

Another point, to which I shall come back, is that some antibodies are formed against certain sequences of amino acids in a polypeptide chain, or of sugars in an oligosaccharide chain, and these I call *sequential determinants,* but the great majority of antigenic determinants that are of interest exist in a unique conformation in three dimensions, and these I shall call *conformational determinants.* The triggering of the synthesis of antibodies against such antigenic determinants must occur while the immunogenic macromolecule is still intact. I am not referring here to a different level of organization, namely to the shattering of a bacterial cell wall or a virus into its component proteins or other macromolecules, but many arguments can be brought for the intactness of the antigenic macromolecule in the initial steps of antibody biosynthesis, and the most obvious one is the role of conformation (*18, 19*). Thus, in most cases the destruction of the native conformation of a protein will also abolish the cross-reaction with antibodies to the native molecule. It has always been a puzzle to me that making vaccines by heat-killing or chemical change is such a good procedure, whereas it seems that the antigenic determinants should under these circumstances lose their native form. I cannot help thinking that during this process one perhaps loses 80–90% of wonderful native determinants and uses essentially just the small percentage of those that still survive.

III. ANTIGENIC COMPETITION

The specificity of the antibodies formed, however, is just one ingredient. If one thinks in terms of this dream of a synthetic vaccine in the future, one should also think about antigenic competition, i.e., competition between antigenic determinants. It would be very efficient to put forty different vaccines onto one molecule, but it will never work: it is probably not possible to put within one macromolecule more than four or five determinants and get a satisfactory immune response to all of them. The few review articles on the subject of antigenic competition usually start by mentioning Leonor Michaelis who described in 1904 the phenomenon and gave it its name (*12*). What is less known, and usually not mentioned, is that Friedberger (*5*) described the same phenomenon at approximately the same time. He did not give it a name, and he did not claim that it is a new phenomenon, but rather quoted several earlier papers to illustrate that it has been recognized and described before, and is a known immunological observation!

Antigenic competition is a phenomenon that has to be taken in consideration in the kind of plan we are discussing here. One cannot envisage one macromolecule containing all the desirable determinants, each one capable of leading to a different specific response. But by studying this phenomenon, being aware of it, one might find easily an interplay of as many as four or five different antigenic determinants, each of which could express itself.

IV. SYNTHETIC ANTIGENS AND ROLE OF CONFORMATION

Certainly, what induced me to think in terms of synthetic vaccines of the future was our initial work on synthetic antigens (*16–18*) and the fact that we can make in a test tube—starting essentially from atoms, at least conceptually—efficient immunogens leading to an immune response that is cross-reactive with various natural substances.

I shall not discuss here our earlier work concerned with synthesizing an antigen that leads to antibodies depending entirely on the steric conformation of immunogen, and another antigen of similar peptide composition, but provoking antibodies recognizing exclusively the peptide sequence (*18*). What I would like to describe here is our work with a native protein. We did not start outright with a viral coat protein, about whose chemistry and three-dimensional structure very little is known until this day. We chose as our model a very well-studied molecule, namely hen egg-white lysozyme. We know the chemical sequence of its amino acids, as well as its three-dimensional structure which was solved crystallographically several years ago by Phillips and his colleagues (*4*).

By means of chemical scissors (pepsin), it is possible to cut out one portion of the lysozyme molecule. It is a chain of around twenty amino acids containing a disulfide bridge. We denote it the *loop* (*2*)—in the three-dimensional model of lysozyme it sticks out more than any other area. The loop can be used to isolate immunospecifically antibodies against that region from an antiserum containing the totality of antilysozyme antibodies. The loop can be also attached to synthetic polymers such as multichain polyalanine (*2, 11*) and polyproline (*10*), and the resulting immunogens provoke antibodies that cross-react very well with native hen egg-white lysozyme. The antibodies are against conformation-dependent determinants as they do not react either with the loop or with lysozyme after disulfide bridges (one in the case of the loop, four in the case of lysozyme) are opened. Thus, we have obtained what I call a semisynthetic immunogen. More recently, in collaboration with Dr. Anfinsen, we have synthesized the loop (*1*) with one disulfide bridge, attached it to branched poly-DL-alanine, and used the resulting material for immunization of rabbits and goats. The resulting antibodies, provoked with a completely synthetic immunogen, react specifically with a native protein (lysozyme) via a unique region (the loop) which is conformation-dependent.

Now, this was the moment when I felt that the conceptual road to a synthetic vaccine has been opened, because if it can be done for one protein, it should be possible to do it for another one. In view of the quick progress in the methodology of the organic chemistry of proteins as well as their X-ray crystallography, it is possible to envisage that one day in the future, and maybe not in the very far future, a similar approach could be used for some viral coat proteins.

V. SYNTHETIC ANTIGENS CROSS-REACTIVE WITH NATURAL PRODUCTS

Several other synthetic antigens have been described, leading to immune responses cross-reactive with natural products. One other example, studied in detail in our laboratory, is that of collagen and its synthetic model, the periodic polymer of the tripeptide L-Pro-Gly-L-Pro (*9*). The synthetic model possesses a three-dimensional structure similar to collagen. Both exist in a triple-stranded helical form. Antibodies prepared against the synthetic polymer cross-react with collagens of different sources. Anticollagen antibodies cross-react with the synthetic polymeric model. Guinea pigs showing delayed hypersensitivity to the polymer exhibit it also to collagen (*9*). Antibodies to both collagen and its synthetic model are cytotoxic, in the presence of complement, to rat embryo fibroblasts, which are cells producing collagen (*8*).

We have shown (*23*) already years ago that one can make synthetic antigens containing nucleosides that provoke antibodies reacting with denatured DNA and RNA. We have produced synthetic antigens leading to antibodies that react with cytolipin H (*3*) as well as with brain lipids devoid of sugar moieties (*22*). The sphingomyelin-specific antibodies reacted with the natural sphingomyelin present in sheep red blood cell membrane. On the other hand, guinea pig red blood cells, which have a low level of sphingomyelin in their membranes, did not undergo hemolysis with our antisera. Synthetic antigens have been also described capable of provoking antibodies cross-reacting with a bacterial cell wall (*25*) and with the basic protein of the myelin sheath (*24*). It is thus possible to conclude that one can prepare materials that will lead to antibodies capable of reacting with all kinds of natural substances.

VI. GENETIC CONTROL OF IMMUNE RESPONSE

After discussing the possibility of preparing immunogens leading to specific immune responses and after mentioning antigenic competition which serves as a warning as to the upper limit of different specificities hooked to one immunogen, I would like to come now to another point of importance for our dreams of the synthetic vaccines of the future. I refer to the genetic control of immune response (*6, 14*). By the way, this is one area which has been developed, largely by making use of synthetic antigens.

Using simple synthetic antigens and inbred strains of guinea pigs, mice, and rats, it was possible to show that some strains are good antibody producers and other are poor producers and that this is under genetic control. I think the most exciting finding in this whole area was the one made by McDevitt and Chinitz (*7*) when they found that the capacity of various strains of mice to respond poorly or well to a series of synthetic multichain polypeptides was closely correlated with the genotype for the major histocompatibility locus. This is not true for all genetic controls of immune response but certainly is valid for many of them.

Now, the point I want to stress here is that the first observation of the genetic control of immune response, made by McDevitt and Benacerraf (*6*), was that strain 13 of guinea pigs is a poor responder to any determinant attached to polylysine, even though it could make antibodies to the same determinant attached to another carrier. Thus, the polylysine gene is carrier-specific. The genetic control is at the level of the carrier. Although cases are known in which the genetic control is at the level of antigenic determinant (*6, 14*), we have described recently two cases in which mice, strains SJL and SWR, are defective, respectively, to multichain poly-DL-alanine (*14*) and to multichain poly-L-proline (*13*). In other words, SJL mice are incapable of responding efficiently to any determinant attached to branched polyalanine, whereas SWR mice cannot make antibodies to anything attached to branched polyproline.

Moving to futurology, I shall not be surprised if some of these carrier-specific genetic controls of immune response will be linked in the human to the major histocompatibility locus. In a few years we may all know our main transplantation antigen the way we know today our blood groups. It is possible that we shall reach a day when the right carrier for the specificity determinants against which we wish to be immune, will be found depending on our histocompatibility background. I know that this sounds like a pure dream, but, if we start to think in terms of such biomolecular engineering, we might as well think about all the aspects.

VII. BUILT-IN ADJUVANTICITY

The last point I would like to mention is that these synthetic vaccines of the future should be good immunogens. Rather than mixing them with extraneous adjuvants, it would be preferable to introduce into the immunogenic macromolecule ingredients that will considerably enhance its immunogenicity without significantly contributing to its specificities. This is the way I stumbled into immunology 18 years ago, showing that enrichment of gelatin with tyrosine strongly enhanced its immunogenicity (*16*). Attachment of lauric acid has been also used quite successfully (*15*).

VIII. VACCINE SPECIFICITIES?

Making antibodies against various determinants on the virus particle might not always be the desirable answer. Under some circumstances viruses may, as a matter of fact, lead to the expression of new antigenic determinants on the cell, and such cellular neoantigens may be characteristic for various viruses. Elucidation of the chemical nature of such neoantigens could lead, in turn, to the synthesis of macromolecules capable of triggering an immune response specific for the cell possessing the new antigenic determinants.

There are at least two recent examples in which synthetic macromolecules were used for protective immunization, and they add to my confidence that the synthetic approach to vaccination may become promising in the future. In one case the synthetic immunogen was a macromolecule containing a sugar component of *Salmonella typhi*, *O*-acetylabequose, and it was found capable of provoking antibodies that agglutinate certain strains of *Salmonella* (*21*). Studies of the effect of such antigen on infectivity, and of its possible vaccination effectivity, could serve as a step forward in reaching the goal of synthetic vaccines.

In the second reported case, a synthetic antigen containing the presumed receptor site of wheat germ agglutinin (a lectin capable of specifically agglutinating tumor cells), conjugated with methylated serum albumin, was found to elicit an immune response in mice, capable of cross-reacting with the receptor sites on tumor cell surface (*20*). It was claimed that mice immunized against this antigen rejected 5 times as many transplanted myeloma tumor cells as are rejected by otherwise identically treated control mice. Without going into detailed criticism of the last approach, I would only say here that the synthetic approach to vaccination is not limited to viruses and may be envisaged in other areas, including cancer.

IX. CONCLUDING REMARKS

The problem that I discussed may not seem urgent today but it will inevitably become of great urgency in some 10–20 years, and we usually tend to underestimate rather than overestimate.

It is obvious that the approach suggested here may have several drawbacks. First, no information is as yet available about the persistence of an antibody response to synthetic immunogens designed on the lines discussed here. Second, it is not certain whether it will be successful for all viruses. Last, the achievement of large-scale applicability of vaccines produced by methods based on this approach may take a long time.

There are, on the other hand, several advantages to the synthetic approach over the presently used methods of vaccination. First, each adequately designed macromolecule will be "multivalent" and, thus, be able to replace several narrowly specific vaccines of today. Second, it will eliminate immunization against many irrelevant antigenic determinants of the virus or irrelevant proteins that contaminate the viral preparation in the vaccine. Thus, the frequently occurring undesired side reaction may be avoided.

By adequate molecular engineering, many variations of vaccines might possibly be designed, enabling us to have on the shelf a series of different multivalent synthetic vaccines to be used according to a preprogrammed key. This certainly sounds today like a dream. The road will be a long and difficult one. However, with the increasing pace of accumulation of knowledge concerning the structure of proteins and their synthesis, it may be hoped that, if the proper effort will be invested in research on these lines, then this dream may become a reality.

REFERENCES

1. Arnon, R., Maron, E., Sela, M., and Anfinsen, C. B., *Proc. Nat. Acad. Sci. U. S.* **68,** 1450 (1971).
2. Arnon, R., and Sela, M., *Proc. Nat. Acad. Sci. U. S.* **62,** 163 (1969).
3. Arnon, R., Sela, M., Rachaman, E. S., and Shapiro, D., *Eur. J. Biochem.* **2,** 79 (1967).
4. Blake, C. C. F., Koenig, D. F., Mair, G. A., North, A. C .T., Phillips, D. C., and Sarma, V. R., *Nature (London)* **206,** 757 (1965).
5. Friedberger, E., *Berlin. Klin. Wochenschr.* **41,** 242 (1904).
6. McDevitt, H. O., and Benacerraf, B., *Advan. Immunol.* **11,** 31 (1969).
7. McDevitt, H. O., and Chinitz, A., *Science* **163,** 1207 (1969).
8. Maoz, A., Dym, H., Fuchs, S., and Sela, M., *Eur. J. Immunol.* **3,** 839 (1973).
9. Maoz, A., Fuchs, S., and Sela, M., *Biochemistry* **12,** 4246 (1973).
10. Maron, E., Scher, H. I., Mozes, E., Arnon, R., and Sela, M., *J. Immunol.* **111,** 101 (1973).
11. Maron, E., Shiozawa, C., Arnon, R., and Sela, M., *Biochemistry* **10,** 763 (1971).
12. Michaelis, M., *Deut. Med. Wochenschr.* **30,** 1240 (1904).
13. Mozes, E., and Sela, M., *Proc. Nat. Acad. Sci. U. S.* **71,** 1574 (1974).
14. Mozes, E., and Shearer, G. M., *Curr. Top. Microbiol. Immunol.* **59,** 167 (1972).
15. Rüde, E., Meyer-Delius, M., and Gundelach, M.-L., *Eur. J. Immunol.* **1,** 113 (1971).
16. Sela, M., *Advan. Immunol.* **5,** 29 (1966).
17. Sela, M., *Science* **166,** 1365 (1969).
18. Sela, M., *Harvey Lect.* **67,** 213 (1973).
19. Sela, M., Schechter, B., Schechter, I., and Borek, F., *Cold Spring Harbor Symp. Quant. Biol.* **32,** 537 (1967).
20. Shier, W. T., *Proc. Nat. Acad. Sci. U. S.* **68,** 2078 (1971).
21. Stellner, K., Lüderitz, O., Westphal, O., Staub, A.-M., Leluc, B., Coynault, C., and Le Minor, L., *Ann. Inst. Pasteur, Paris* **123,** 43 (1972).
22. Teitelbaum, D., Arnon, R., Sela, M., Rabinsohn, Y., and Shapiro, D., *Immunochemistry* **10,** 735 (1973).
23. Ungar-Waron, H., Hurwitz, E., Jaton, J.-C., and Sela, M., *Biochim. Biophys. Acta* **138,** 513 (1967).
24. Webb, C., Teitelbaum, D., Arnon, R., and Sela, M., *Eur. J. Immunol.* **3,** 279 (1973).
25. Zeiger, A. R., and Maurer, P. H. *Biochemistry* **12,** 3387 (1973).

CHAPTER 6

Inhibition of Viral Neuraminidase As a New Approach to the Prevention of Influenza*

Edwin D. Kilbourne
Peter Palese
Jerome L. Schulman

I. INTRODUCTION

The usual objective of viral chemotherapy or immunoprophylaxis is the complete inhibition or restriction of infection. Another stratagem that has received insufficient attention is to achieve a lesser victory through a quantitative and partial limitation of infection sufficient to prevent disease but insufficient to inhibit the desirable immunizing effects of that infection.

Table 1 summarizes the basic approaches available for antiviral prophylaxis. Those categorized here as restrictive are obvious, and for the most part are already in use. The second category (B) of approaches may be less obvious. In diseases with long incubation periods such as rabies, or even measles, administration of neutralizing antibody *after*

* This work was monitored by the Commission on influenza of the Armed Forces Epidemiological Board, and was supported (in part) by the U. S. Army Medical Research and Development Command under research contract No. DADA1769-C-9137, by Public Health Service research grant No. AI09304 from the National Institute of Allergy and Infectious Diseases, and by contract No. U-2076 from the Health Research Council of the City of New York.

TABLE 1. STRATEGIES OF ANTIVIRAL PROPHYLAXIS

A. Nonpermissive (restrictive). Prevention of *infection* by:
   1. Prevention of virus–host contact
   2. Inhibition of viral replication (early events)
   3. Preexisting neutralizing antibody

B. Prevention of *disease* by quantitative limitation of infection by:
   1. Postinfection administration of neutralizing antibody
   2. Postinfection chemotherapy
   3. Preinfection administration or induction of suppressive chemotherapy or antibody

infection can prevent or modify the evolution of disease. It is conceivable that sharp inhibition of viral replication by chemotherapy *after* the initiation of disease could have a similar effect. Still another, virtually untried approach (B3), is to immunize with minor antigenic components of the virus or to inhibit viral biological activities that are not essential for initiation of infection. For reasons that will be presented, influenza viruses lend themselves ideally to this latter stratagem.

In Fig. 1 the structure of influenza virus is somewhat simplistically schematized. Emphasized here is the external anatomy of the virion which consists of spikelike projections of hemagglutinin and neuraminidase polypeptides as the biologically active surface antigens through which the virus encounters its environment. It is these proteins and not those beneath the host-derived lipid bilayer that are peculiarly subject to antigenic mutation. The hemagglutinin, the most abundant surface protein, functions in attachment to cells, whereas most evidence indicates that the neuraminidase is involved in viral release.

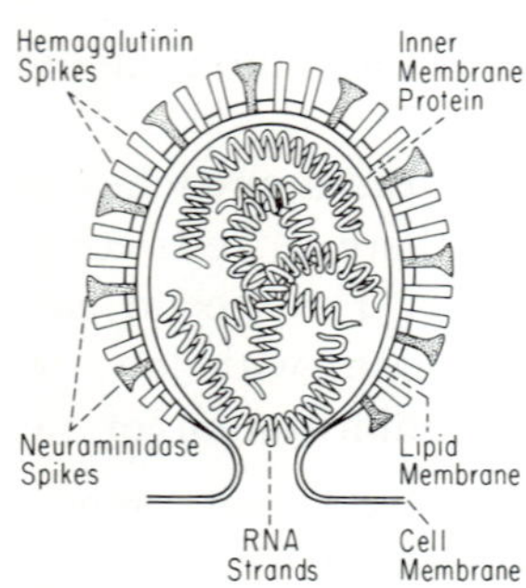

Fig. 1. Schematic diagram of influenza virion showing hemagglutinin and neuraminidase as external glycoprotein "spikes" susceptible to interaction of antibody and to antigenic mutation.

II. COMPARISON OF ANTIHEMAGGLUTININ AND ANTINEURAMINIDASE IMMUNIZATION

Hemagglutinin and neuraminidase are coded by different genes (7) and hence are segregatable by genetic recombination or reassortment into antigenic hybrids (Fig. 2). The creation of such antigenic hybrids, as reported in *Perspectives in Virology* in 1967,[a] has facilitated analysis of the function of these proteins and the production of antisera monospecific for them. Figure 3 characterizes the qualitatively different effects of antihemagglutinin and antineuraminidase antibody. Classic and complete plaque inhibition is seen when infected cell monolayers are overlaid with antihemagglutinin. In contrast, no decrease in plaque number is observed with antineuraminidase; rather, a decrease in plaque size (PSR) is characteristically seen.

Extensive experiments in mice have paralleled studies in cell culture in that antihemagglutinin clearly prevents infection and antineuraminidase is "infection permissive," but depresses viral replication and, therefore, disease. In Fig. 4 are illustrated the results of immunization of mice with virus (H0N1) completely heterologous to the challenge virus (H2N2), or with virus (H0N2) antigenically related to the challenge virus only through the neuraminidase antigen, N2. The latter (effectively monospecific) immunization does not prevent infection but depresses pulmonary virus replication and, consequently, the development of gross pulmonary lesions.

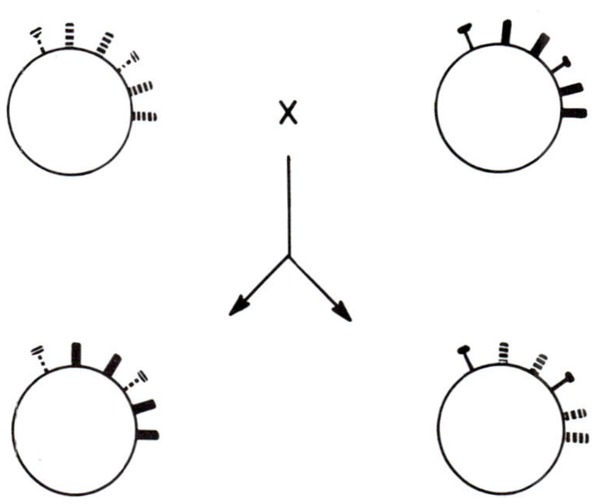

Fig. 2. Segregation by genetic recombination of hemagglutinin (H) and neuraminidase (N) antigens into antigenically hybrid viruses. Rod-shaped spikes are H, the major external antigen; "toadstools" are N, the minor antigen. [From Kilbourne (5).]

[a] Kilbourne *et al.*, 5, 87–106.

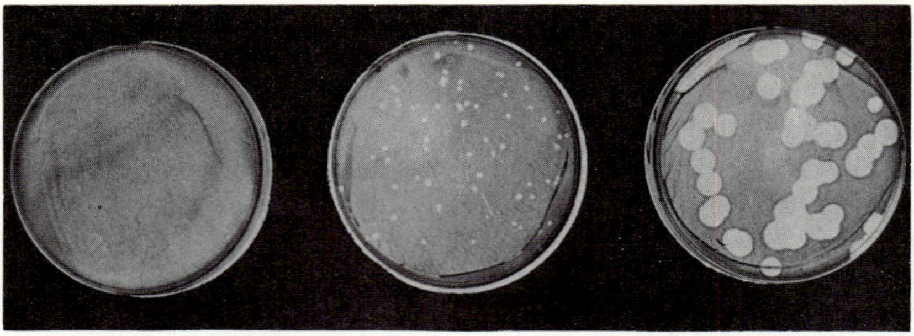

Fig. 3. Differing effects of antihemagglutinin (anti-H) and antineuraminidase (anti-N) antibodies on influenza virus plaque formation in clone 1-5C-4 monolayer cultures. Antiserum is incorporated into the agar overlay. Complete inhibition of plaque formation occurs with anti-H (left); reduction in plaque size (PSR) with no reduction in plaque number occurs in the presence of anti-N. Plaque development in the absence of specific antibody is shown at the right. [From Kilbourne and Schulman (*6*).]

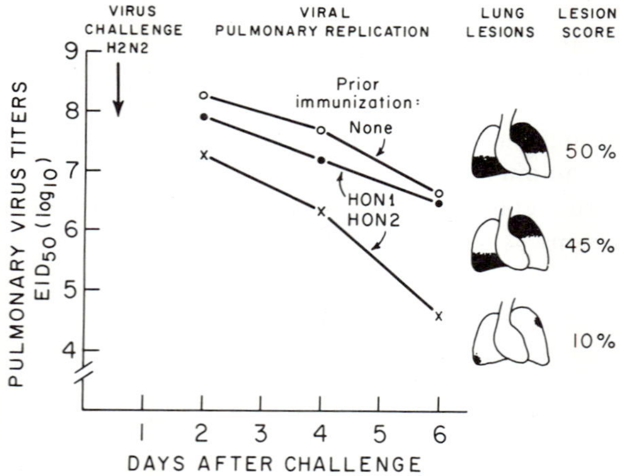

Fig. 4. Protective effect of neuraminidase-specific immunization in mice. Mice were injected intraperitoneally with UV-irradiated A/NWS/33 (H0N1) virus or with X-7 (H0N2) virus. Control mice were injected with saline. Challenge, 4 weeks later, was by aerosol with A/Japan/305/57 (H2N2) virus. Pulmonary virus titers were measured 2 and 4 days after infection and lung lesions were assessed 7 days after infection.

The efficiency of antineuraminidase immunization has also been demonstrated in man in preliminary collaborative studies with Dr. Robert Couch and his associates (*2*). The results of studies in several systems of dissociated hemagglutinin-specific and neuraminidase-specific immunization are summarized in Table 2.

TABLE 2. COMPARATIVE EFFECTS OF ANTIHEMAGGLUTININ AND ANTINEURAMINIDASE ANTIBODIES

Effect	Antihemagglutinin	Antineuraminidase
In vitro		
Attach to RBC	+	±[a]
Release from RBC	±[b]	+
Attach to cells supporting replication	+	0
Release from cells supporting replication	±[c]	+
Plaque inhibition (neutralization)	+	0
Plaque size reduction	0	+
In vivo		
Neutralization	+	0
Prevent infection	+	0
Prevent disease	+	+[d]

[a] Will cause hemagglutination inhibition of certain viruses.
[b] May *increase* elution rate.
[c] May form lattices but primary effect is neutralizing.
[d] If level of viral replication is sufficiently diminished.

It will be seen that the primary effect of antihemagglutinin is the prevention of primary attachment of virus to cells (neutralization), whereas the action of antineuraminidase is at the later stage of multicycle infection in inhibiting the release or detachment of virus. Therefore, antihemagglutinin but not antineuraminidase will prevent infection, but antineuraminidase can modify it by inhibition of the spread of virus.

The qualitatively different effects of encounter of the host with the two antigens suggested to us the new approach of infection-permissive immunization against influenza briefly mentioned in the Introduction of this paper. This approach is outlined in Fig. 5. It is clear that effective protection against influenza can be achieved with inactivated vaccines but this protection is only briefly maintained. Thus, infection is prevented through the medium of antihemagglutinin (anti-H) antibody soon after immunization as shown in the upper portion of the figure, but the subject is vulnerable to infection and disease a year or two later. In contrast, as we have seen, neuraminidase-specific (anti-N) immunization does not prevent infection (i.e., is infection-permissive) but does diminish viral replication to a point below that threshold necessary for the development of disease (Fig. 5, lower panel). A legacy of the partially suppressed infection will be the development of the more enduring immunity

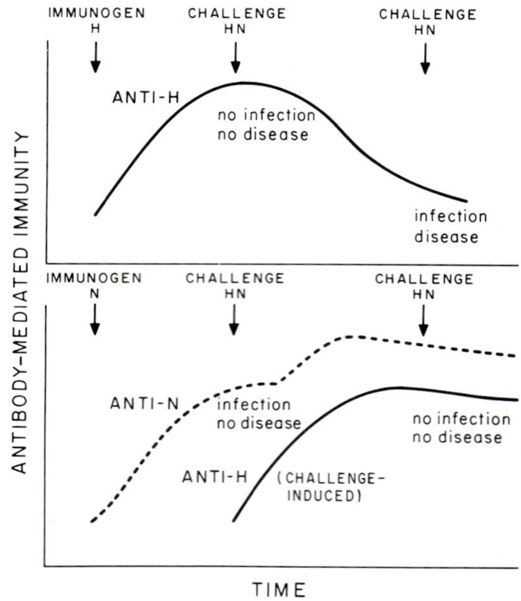

Fig. 5. Schematic representation of contrasting strategies of immunization with conventional inactivated vaccine (H) or neuraminidase-specific vaccine (N). (Strictly speaking, conventional vaccine is HN, but effect of H is overriding.) Vaccine H induces totally effective (infection-prohibitive) but briefly sustained immunity (upper panel) that leaves subject vulnerable to infection (challenge HN) after waning of immunity. As an immunogen, N induces the formation of antibodies ineffective in preventing infection (infection-permissive) but effective in restricting viral replication (after challenge HN) below the threshold required for induction of disease. This partial replication of virus after initial (HN) challenge will in itself induce primary antibody formation to H and secondary response to N. This immunity of natural infection should be of sufficient duration and quality to withstand future HN challenges.

characteristic of live virus administration—immunity now to *both* antigens.

III. MECHANISMS OF INFLUENZA VIRUS SUPPRESSION

A. Antineuramidase Antibody

That antineuraminidase antibody acts in the suppression of influenza virus infection is indisputable; however, the mechanism of this action has not been resolved. *In vitro* the access of the enzyme of substrate is sterically blocked, as illustrated in Fig. 6, and the extent of antineuraminidase inhibition is dependent on substrate size. Thus, the antigenic

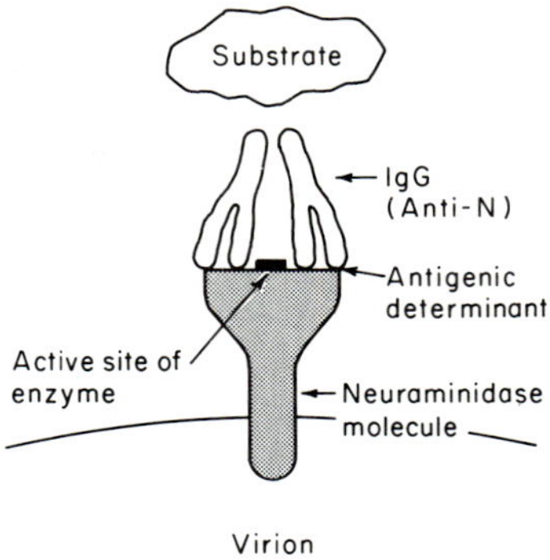

Fig. 6. Steric inhibition of neuraminidase function by antibody. A single hypothetical neuraminidase molecule is shown with two bivalent antibody molecules attached. Experimental evidence indicates that attachment of IgG will sterically block access of high molecular weight substrates to the enzymically active site which is discrete from the antigenic determinants. Anti-N, anti-neuraminidase.

determinant occupies a different site on the neuraminidase molecule than does the enzymically functional site as originally emphasized by Fazekas de St. Groth (*4*).

It is evident that the bivalent antibody of conventional antiserum preparations might act in any of several ways in influencing the course of multicycle infection, as is schematically depicted in Fig. 7. Such antibody might cause (1) steric inhibition of enzyme function, thus delaying viral release; (2) aggregation (cross-linking) of particles, thus reducing the number of effective infective units; (3) cementing of free particles to cell-associated neuraminidase antigen; or (4) inhibition of final detachment of the budding particle.

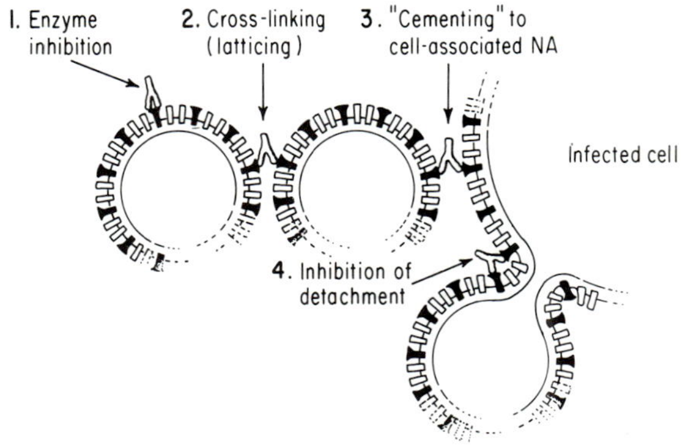

Fig. 7. Putative mechanisms of effect of antineuraminidase antibody. NA—neuraminidase.

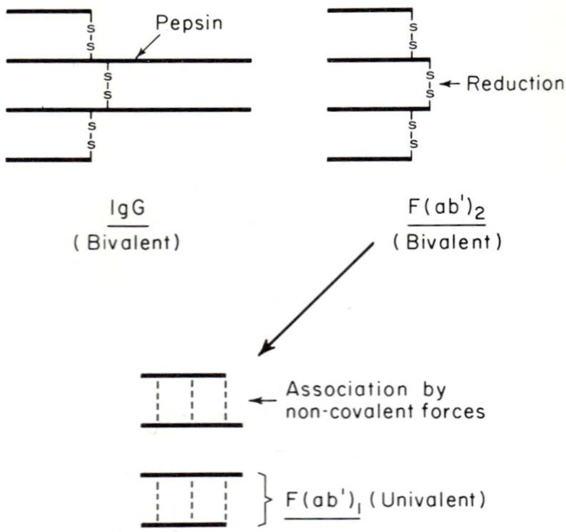

Fig. 8. Progressive fractionation of IgG. Pepsin cleavage splits off Fc portion of heavy chains to yield $F(ab')_2$ which still possesses bivalent activity. Reduction of disulfide bonds produces F(ab′) fragments that are functionally univalent.

We have approached the problem by the use of purified IgG fractionated to yield bivalent and univalent F(ab′) fragments, as independently suggested by Becht and colleagues (*1*), and by Dowdle *et al.* (*3*).

Pepsin hydrolysis of IgG splits off the Fc portion of the molecule leaving the bivalent antigenically reactive sites, $F(ab')_2$ (Fig. 8). Reduction breaks disulfide bonds to yield univalent F(ab′) fragments in which light and heavy chain pieces are associated by noncovalent forces. Such univalent molecules are obviously incapable of bridging or latticing effects and must act solely through direct enzyme inhibition.

The comparative activity of the IgG and its fractionated components was studied with respect to inhibition of elution of virus from red blood cells (Fig. 9). In the test illustrated, the end point is disagglutination of red blood cells previously agglutinated by influenza virus (M. Stern and E. D. Kilbourne, unpublished data). The substantial although lesser activity of univalent F(ab′) is to be noted. Indeed, as summarized in Table 3, in neuraminidase inhibition, elution inhibition, disagglutination, and plaque size reduction assays the univalent antibody was found to be active, although activity was diminished 4–16-fold. We conclude that infection can be modified, at least in part, by direct enzyme inhibition (see mechanism 1 of Fig. 7) but admit the possibility of any or all of the other mechanisms (*1, 2, 4*) being operative as well.

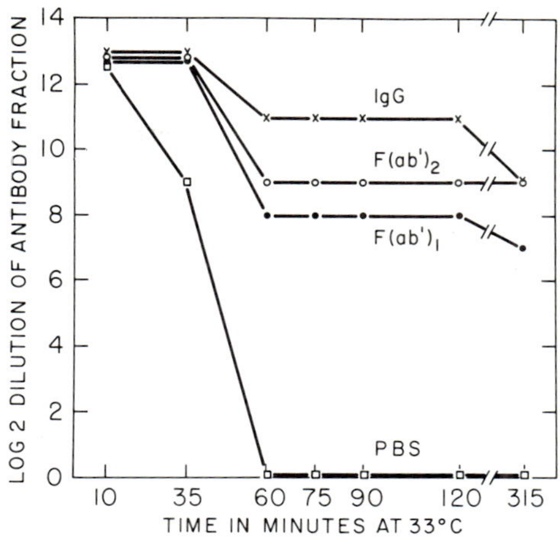

Fig. 9. Elution (disagglutination) inhibition of virus–red blood cell (RBC) patterns by antibody fractions. The RBC are agglutinated at 4°C with multiple aliquots of fast eluting X-32 (H equi/N2) virus. Dilutions of antibody preparations are then added to the tubes and incubation in a 33°C water bath is begun. Serial readings at the time intervals indicated determine the stability of hemagglutination patterns in the presence of antibody. Thus, loss of classic pattern with time is read as disagglutination with reference to RBC–virus controls in the absence of antibody. PBS, phosphate-buffered saline.

B. Chemical Inhibition of Viral Function

We have examined another tactic, but with the same strategic objective, namely inhibition of viral neuraminidase function.

The influenza viral neuraminidase requires a substrate with an α-glycosidic linkage from the second carbon atom of *N*-acetylneuraminic acid to a second compound such as lactose (Fig. 10). A synthetic fluorinated analog of neuraminic acid, 2-deoxy-2,3-dehydro-*N*-trifluoroacetylneura-

TABLE 3. ACTIVITY OF NEURAMINIDASE-SPECIFIC ANTIBODY FRACTIONS[a]

Antibody vs:	Antibody fraction	NI	Elut. inhib.	Disaggl.	PSR
N2	IgG	6400	160	8192	6400
(Isolated antigen)	$F(ab')_2$	6400	160	8192	6400
	F(ab′)	400	40	1024	800
Activity reduction	$F(ab')_2/F(ab')$	16×	4×	8×	8×

[a] Derived by immunization of rabbits with N2 neuraminidase isolated from X-7 (H0N2) virus. NI = neuraminidase inhibition; Disaggl. = disagglutination titer as defined in Fig. 9; PSR = plaque size reduction (as in Fig. 3). Values are expressed as reciprocal of arithmetic dilution at end point (except for activity reduction).

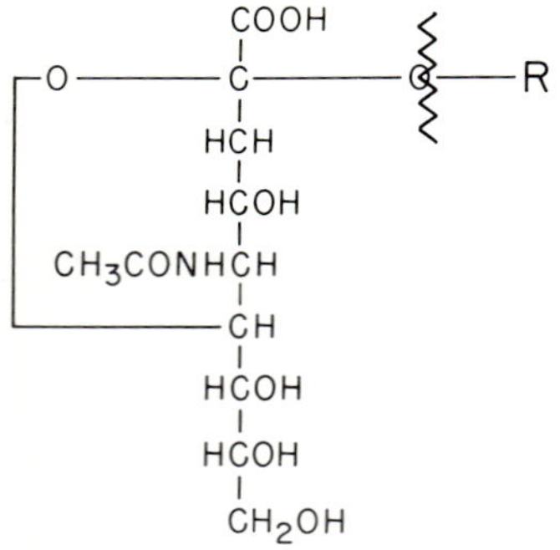

Fig. 10. Hydrolysis of neuraminidase substrates. Neuraminidase splits off sugars and alcohols that are bound ketosidically to the second carbon atom of neuraminic acid.

minic acid (FANA) (Fig. 11) has been found to have marked *in vitro* inhibitory activity against bacterial and viral neuraminidases—this activity mediated by competition with substrate (*8*). A summary of results obtained with this compound in our laboratory is presented in Table 4.

It is evident that activity against a variety of enzymes is demonstrable at low molar concentrations of FANA. The activity of this compound in inhibiting elution of virus from red blood cells, a phenomenon principally dependent on viral neuraminidase function, is illustrated in Fig. 12.

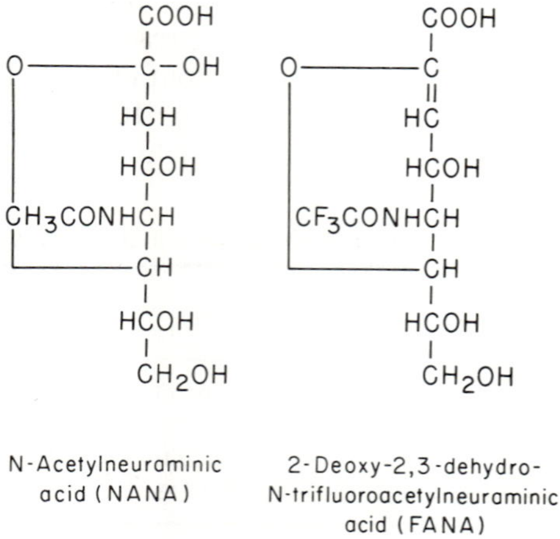

Fig. 11. Comparison of structures of neuraminic acid and its synthetic analog. Compound FANA differs from *N*-acetylneuraminic acid by one double bond between carbon atom 2 and carbon atom 3. The amino group is blocked with a trifluoroacetyl group instead of the acetyl group.

TABLE 4. *In Vitro* INHIBITION OF NEURAMINIDASE ACTIVITY BY FANA[a]

Neuraminidase	Concentration (in moles/liter) for 50% neuraminidase inhibition[b]
RDE[c]	2.5×10^{-6}
NDV/California	1.3×10^{-6}
X-7 (HON2)	4.8×10^{-6}
A/Melbourne/35 (HON1)	4×10^{-6}

[a] FANA = 2-deoxy-2,3-dehydro-*N*-trifluoroacetylneuraminic acid.
[b] Neuraminidase activity was measured using a standard thiobarbituric acid assay, the neuraminic acid content, of the fetuin used in the test was 1.1×10^{-3} *M*. All samples contained 0.1 *M* phosphate buffer pH 5.9; the test for the bacterial enzyme contained 2 m*M* $CaCl_2$ in addition.
[c] RDE = neuraminidase from *Vibrio cholerae* (Behring Diagnostics).

Experiments in cell culture showed little effect of FANA on viral replication if single-cycle growth was assured by high (plaque-forming units per cell) multiplicity of infection (Fig. 13); however, significant inhibition was shown under conditions in which successive cycles of replication occurred (low multiplicity of infection).

Of great interest to us was the demonstration that plaque size of influenza viruses but not of VSV (which lacks neuraminidase) was re-

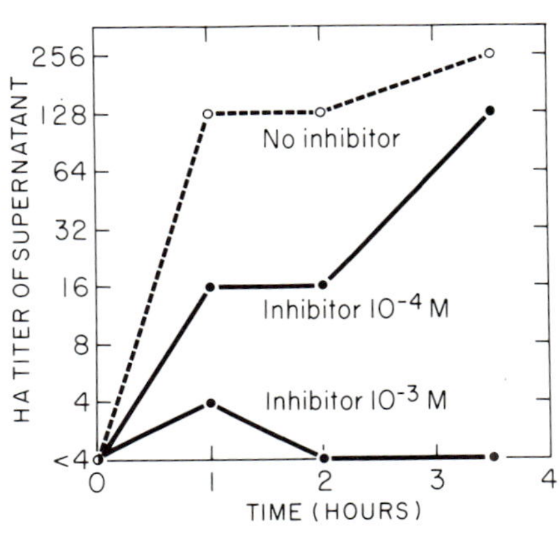

Fig. 12. Effect of FANA on elution of X-7 (H0N2) virus from red blood cells. Human red blood cells type *O* at a final concentration of 0.25% and X-7 (H0N2) virus were mixed at 4°C in the presence and absence of FANA in a total volume of 2 ml. After 1 hour the mixtures were transferred to 37°C, and 0.025-ml samples of the supernatant were measured for hemagglutinin (HA) activity at appropriate intervals.

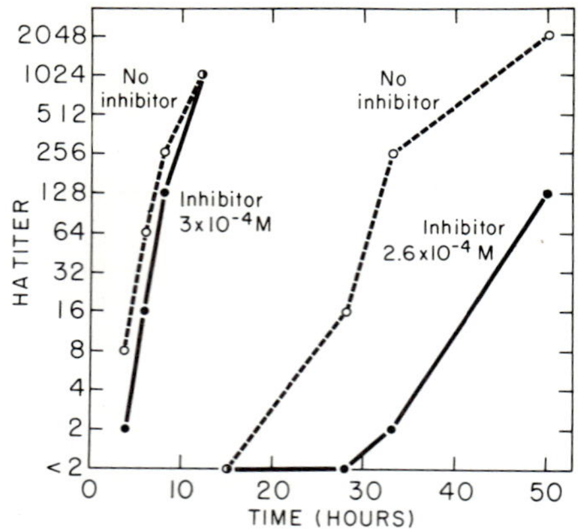

Fig. 13. Effect of FANA on replication of influenza A/WSN(H0N1) virus during single and multiple step growth cycles. The MDBK cells were inoculated with WSN virus at a multiplicity of 8 PFU/cell (curves on left). After 2 hours excess input virus was removed by WSN antiserum [(HI) titer 2048] diluted 1:100. Cells were kept under liquid medium. At intervals, virus yields were measured from cultures incubated with and without the neuraminidase inhibitor FANA. Similarly, a multicycle growth curve with WSN was obtained (curves on right side of graph), 10^{-4} PFU/cell were inoculated and dishes incubated with and without FANA were harvested at appropriate intervals and measured for virus yields.

duced when FANA was incorporated into overlays of semisolid media—a parallel to our earlier observations with neuraminidase antibody (Fig. 14).

Neuraminidase inhibitory compounds used previously to inhibit influenza viral replication have had concomitant nonspecific effects that have

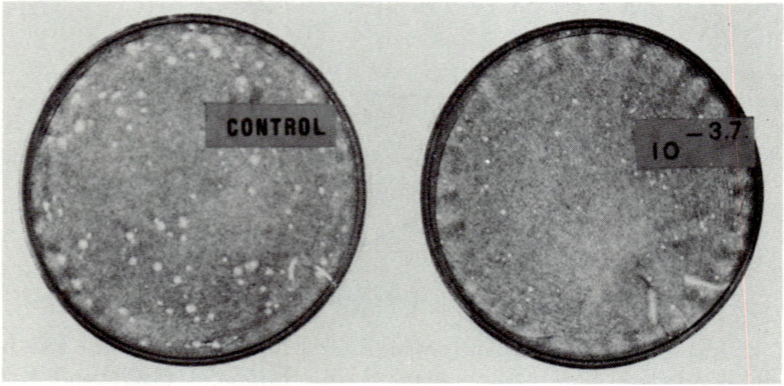

Fig. 14. Effect of FANA on plaques of X-7 (H0N2) virus in clone 1-5C-4 cells. Monolayers were infected with X-7 virus. The agar overlay medium contained FANA in a concentration of 2×10^{-4} *M* (right side); control (left side). Four days after infection dishes were stained with crystal violet.

TABLE 5. EVIDENCE FOR SPECIFICITY OF FANA IN RELATION TO NEURAMINIDASE INHIBITION[a]

1. No effect on attachment of influenza viruses to red blood cells
2. No effect on viral attachment in cell culture
3. Relative effectiveness of related analogs in neuraminidase inhibition parallels inhibitory effects on viral replication
4. No effect on replication of another RNA-enveloped virus (VSV) that does not contain neuraminidase
5. In recombinants, differing susceptibility to FANA is related to neuraminidase, not hemagglutinin phenotype (e.g., H0N1 and H0N2 differ 100×).

[a] FANA = 2-deoxy-2,3-dehydro-*N*-trifluoroacetylneuraminic acid.

obscured interpretation of results. Evidence that the antiviral effects of FANA are mediated through enzyme inhibition is summarized in Table 5. Of special significance is the demonstration that the differing susceptibility of recombinant viruses to FANA varies with neuraminidase rather than hemagglutinin phenotype.

In conclusion, comparison of the effects of FANA and antineuraminidase antibody shows them to be remarkably concordant (Table 6). Neither neutralizes nor inactivates virus nor prevents attachment of virus to cells. Both either directly or indirectly inhibit neuraminidase activity *in vitro* as well as viral release from cells. As a consequence, multicycle viral replication dependent on cell-to-cell spread of virus is depressed.

From these and other data, we conclude that neuraminidase has an important role in the total viral replicative process. Furthermore, we now see two independent approaches to our goal of partial inhibition of influenza viral replication; immunoprophylactic and chemoprophylactic. Both the tactics and the basic stratagem require much more study.

TABLE 6. COMPARATIVE EFFECTS OF FANA AND ANTINEURAMINIDASE (ANTI-N) ANTIBODY[a]

Effect	FANA	Anti-N
Viral neutralization or inactivation	0	0
Viral attachment	0	0
Single-cycle replication	±	±[b]
Neuraminidase inhibition	+	+
Viral release (inhibition)	+	+
Multicycle replication (inhibition)	+	+
Disease prevention	?	+

[a] FANA = 2-deoxy-2,3-dehydro-*N*-trifluoroacetylneuraminic acid.
[b] Plus (+) for bivalent antibody.

REFERENCES

1. Becht, H., Hammerling, U., and Rott, R., *Virology* **46**, 337 (1971).
2. Couch, R. B., Kasel, J. A., Knight, V., Gerin, J. L., Schulman, J. L., and Kilbourne, E. D., *J. Infec. Dis.* **129**, 411 (1974).
3. Dowdle, W. R., Downie, J. C., and Laver, W. G., *J. Virol.* **13**, 269 (1974).
4. Fazekas de St. Groth, S., *Ann. N. Y. Acad. Sci.* **103**, 674 (1963).
5. Kilbourne, E. D., *J. Infec. Dis.* **127**, 478–487 (1973).
6. Kilbourne, E. D., and Schulman, J. L., *Trans. Ass. Amer. Physicians* **78**, 323–333 (1965).
7. Laver, W. G., and Kilbourne, E. D., *Virology* **30**, 493 (1966).
8. Meindl, P., Bodo, G., Palese, P., Schulman, J., and Tuppy, H., *Virology* **58**, 457 (1974).

DISCUSSION

DR. ANDREWES: Dr. Kilbourne and I look at things in much the same way. Two concepts seem important to me in understanding the epidemiology of flu. First, many of us escape disease if we inhale a small dose and get an inapparent infection that protects against a big dose. Anything Kilbourne does to achieve this is all to the good. Incidentally, it is alleged that amantadine is particularly effective in stopping subclinical infections. Insofar as it does this, it may do harm rather than good.

I feel that the individuals who shed very big doses of virus are particularly dangerous. George Hirst showed long ago that there might be a millon-fold difference between the amounts of virus shed by individuals. If so, antineuraminidase might work out of control virus shedding. Its activity could be greatest if the appropriate vaccine were given to whole groups. But Fred Davenport knows well how difficult it is to do controlled trials on a basis that will satisfy critical statisticians.

DR. DAVENPORT: Interpretation of Dr. Kilbourne's slide depicting the events following infection and vaccination should be undertaken with caution. The duration of immunity following natural infection or vaccination with either live or killed vaccines is unknown. The few observations made indicate that in either case it can extend for a few years when the challenge comes from a virus of the same subtype. Since in all outbreaks of influenza the rate of subclinical cases exceeds the rate of clinical cases, Dr. Kilbourne's strategy is to extend nature's experiment.

DR. KILBOURNE: I'd just like to answer Dr. Davenport by pointing out that I agree with him with minor qualifications. That is why I didn't put any numbers on the chart.

DR. LOCKART: There is considerable evidence from mouse experiments that amantadine acts by reducing the concentration of virus in infected lungs about ten-fold and permitting immune conversions to occur. In so doing, it effectively prevents or ameliorates clinical symptoms and prevents death.

DR. SELA: You might consider one other explanation for the mechanism of action of your antineuraminidase antibodies. The antiserum most probably contains a mixture of antibodies against several different antigenic determinants of neuraminidase. Some of these antibodies may be directed against determinants to the enzymically active site, and they might inhibit the enzymic activity. Other antibodies might be against a region close to the one producing the inhibiting antibodies, but they might not be inhibiting the enzymic activity. On the other hand, they might compete with the inhibiting antibodies. Thus, as long as the antiserum contains a mixture of these two kinds of antibodies: (1) it will not inactivate completely the virus, but will cause only, as found, a decrease in infectivity; and (2) the monomeric antibody fragments obtained after peptic splitting, followed by reduction, will inhibit neuraminidase activity only if they are derived from the truly "inhibiting" antibodies, whereas the divalent antibody species will be more efficient, because it will act both through direct enzymic inhibition and through steric hindrance.

DR. KILBOURNE: I'm delighted you brought that point up, because again I've been simplistic, and certainly we don't believe that there is a single antigenic determinant on the molecule. We have just a small suggestion of the possibility that we may get different results with antisera produced against the isolated enzyme and antiserum produced against the whole virus recombinants, in terms of the efficiency of inhibition. This might expand what you are saying by introducing the idea of multiple antigenic determinants.

DR. BANG: In your diagram of neuraminidase–antineuraminidase activity, you have indicated various possible places in the development of the virus at which the antibody may act. I wonder, however, whether the activity which is most prominent, that is the plaque reduction size, might not be explained by inhibition of a cell-to-cell spread of virus through the activity of neuraminidase. This would be different from release, as such, but would involve both release and penetration of neighboring cells.

DR. KILBOURNE: We thought of this of course as being a microinfection with the radial spread of such infection as occurs with the evolution of a plaque, and I don't see that it is necessary to invoke anything happening within the cell or by direct cell-to-cell spread. We don't know of any such mechanisms with influenza virus as occur with herpes for example. Just parenthetically, I might say that plaque size reduction seems to be unique for neuraminidase-containing viruses in relation to their antiserum, except for some experiments that were done by Frank Ennis in which he studied herpes virus which, of course, can escape antibody by intracellular bridges. In that kind of situation he has been able to show that what ostensibly would be a neutralizing antibody is able to cause this plaque size effect.

CHAPTER 7

Cellular Immune Response in Viral Infections

Abner Louis Notkins
Donald Louis Lodmell

I. INTRODUCTION

It has been suggested for a number of years that cellular immunity plays a role in the defense against certain viral infections. Support for this thesis comes from several lines of evidence. First, it is known that certain viruses, such as herpes simplex, can persist *in vivo* and *in vitro* in the presence of high concentrations of neutralizing antibody (*14*). Second, from experiments of nature, it has been shown that patients with deficiencies in cellular immunity have difficulty in coping with certain viruses such as herpes simplex, cytomegalo, varicella zoster, measles, and vaccinia, but do not have difficulty with a number of other common viral infections, such as polio (*1, 7, 23*). Third, experiments in animals have shown that suppression of the cellular immune response by neonatal thymectomy or antilymphocyte serum makes animals more susceptible to only certain viruses (*23*). Fourth, if both cellular and humoral immune responses are experimentally suppressed in animals by cyclophosphamide, reconstitution with antibody can prevent death by viruses such as Coxsackie or encephalomyocarditis (*13, 17*), but immune cells are needed to protect against other viruses such as mousepox (*2*). The precise mechanism by which cell-mediated immunity operates to protect the host, however, is far from clear.

TABLE 1. RELATIONSHIP BETWEEN ROUTE OF SPREAD OF VIRUSES AND IMMUNE MECHANISMS FOR THEIR ERADICATION

Type	Route of spread	Example	Proposed sites of action and mechanisms by which the immune response stops viral spread
I	Extracellular	Poliovirus	Neutralization of extracellular virus
II	Cell to contiguous cell	Herpes simplex	Destruction of infected and/or surrounding uninfected cells; breaking of cell-to-cell contact; inhibition of viral replication by mediators
III	Parent cell to progeny cells	Murine leukemia	Destruction of cells carrying viral genome; inhibition of viral replication by mediators

II. DIFFERENT ROUTES OF SPREAD OF VIRUS INFECTION

I would like to review some of the experiments performed in our laboratory over the last couple of years which I believe shed some light on this problem. I will discuss the various immunological responses of the host to viral infections, with emphasis on lymphokines, and will try to show that the effectiveness of each type of immune response in eradicating a particular infection is intimately dependent on the mode by which that virus spreads from one cell to another. Viruses can spread from one cell to another by three different routes (Table 1) which we refer to as Type I, Type II, or Type III spread (*15*). In Type I spread, the virus infects nearby or distant cells by the extracellular route (e.g., polio). In Type II, the virus spreads from cell to contiguous cell presumably as a result of cell fusion. In this case the virus is protected from antiviral antibody (e.g., herpes simplex virus). In Type III spread the virus or viral genome is transmitted from the parent to the progeny cells during cell division (e.g., murine leukemia viruses). When viral infections are analyzed in terms of their mode of spread, it becomes quite clear that different immunological processes are required to stop the different modes of spread. Type I spread is stopped primarily by the neutralization of extracellular virus by antiviral antibody. This has been extensively studied and provides the basis for most of our modern concepts in viral immunology. In comparison, very little is known about how the host stops Type II and Type III spread.

III. IMMUNE MECHANISMS PREVENTING TYPE II SPREAD

I will concentrate primarily on Type II spread, as exemplified by herpes simplex virus (HSV). Theoretically, the immune response of

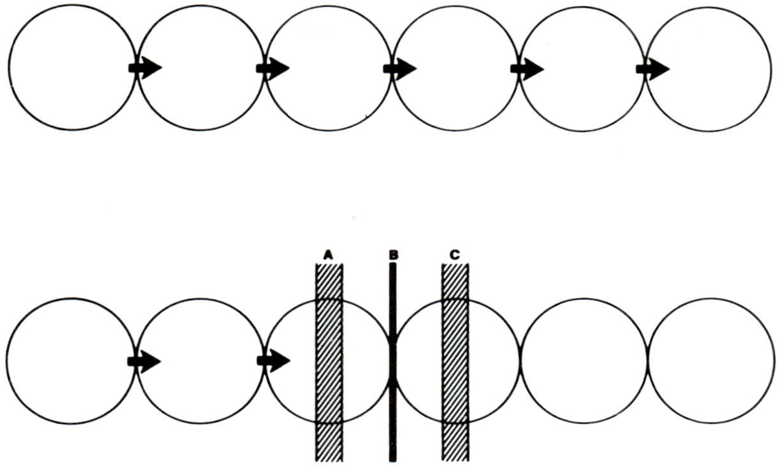

Fig. 1. Sites at which leukocytes act to prevent cell-to-cell spread of virus. Arrows indicate direction of viral spread. The spread of the infection can be stopped by destroying infected cells (A); breaking the contact between infected and uninfected cells (B), preventing viral replication in the adjacent uninfected cells (C).

the host could stop the cell-to-cell spread of HSV by acting at one or more of three different sites (Fig. 1). First, it could destroy the infected cell (site A). Second, it could break cell-to-cell contact or prevent cell fusion (site B); if this happens the virus would have to travel by the extracellular route to infect adjacent cells and would be readily neutralized by antiviral antibody. Third, if the host's immune response either damages the surrounding uninfected cells (site C) or inhibits viral replication in those cells, the spread of the virus also would be halted. I will now present evidence showing that the host's immune response can act on each of these sites.

A. Direct Immunological Attack on Infected Cells

If the host's immune defense mechanisms are to act at site A, the virus must induce new antigens on the surface of the infected cell so that antibody and/or immune lymphocytes can recognize these cells as foreign. Over the last few years it has become quite clear that many nononcogenic viruses induce new antigens on the surface of infected cells (5). These antigens can be detected by a variety of techniques, but one of the most sensitive and quantitative is the binding of radiolabeled antibody to the surface of infected cells (Table 2). For example, within 8 hours after infection of cells with HSV over 9 times as much ^{125}I-labeled anti-HSV antibody is bound to the surface of infected cells

TABLE 2. SPECIFICITY OF ATTACHMENT OF RADIOLABELED ANTIVIRAL ANTIBODY TO RABBIT KIDNEY CELLS INFECTED WITH DIFFERENT VIRUSES[a]

Virus used to infect monolayers	[^{125}I]-Labeled antiviral antibody (counts/min/monolayer of 10^6 cells)		
	Anti-HSV	Antivaccinia	Anti-influenza
Herpes simplex	2,840	230	175
Vaccinia	312	12,600	217
Influenza	264	241	12,200
Uninfected	320	253	126

[a] Monolayers of about 10^6 rabbit kidney cells were infected with herpes simplex, vaccinia, or influenza virus. Uninfected cells served as controls. After 8 hours each monolayer was incubated with [^{125}I]-labeled antibody against HSV, vaccinia, or influenza and the amount of [^{125}I] bound to the monolayers determined. From Rosenthal *et al.* (*19*).

as to uninfected cells (*19*). In the case of vaccinia, almost 50 times as much ^{125}I-labeled antivaccinia antibody is bound to vaccinia-infected as to uninfected cells. By this technique the actual number of antibody molecules bound per cell can be estimated. For example, close to $1.0 \times 10^{6.0}$ antivaccinia antibody molecules were bound to each vaccinia infected cell (*10*). When complement is now added to cells that have antiviral antibody bound to their surface, cell lysis occurs (*4, 5*). This has been demonstrated *in vitro* with a number of DNA and RNA viruses, including members of the herpes, pox, papova-, oncorna-, myxo-, paramyxo-, arbo-, rhabdo-, and arenavirus groups (*5, 16*).

Immune lymphocytes also can react with viral antigens on the surface of infected cells and lead to cell destruction. This has been demonstrated with mumps, measles, rubella, lymphocytic choriomeningitis, and murine leukemia virus (*5*).

It is clear that, at least *in vitro,* antibody plus complement, as well as immune lymphocytes can destroy virus-infected cells. The critical question is whether the destruction of the infected cell will stop the spread of the infection. As schematically illustrated in Fig. 2, this will occur only if immunological destruction of infected cells (A) occur before the virus is transmitted to contiguous uninfected cells (B). If, however, infected cell A is immunologically destroyed after the virus is transferred to cell B, the infection would not be stopped. This latter situation is what appears to happen in the case of HSV (Fig. 3). Although HSV can induce new antigens on the surface of infected cells and these cells can be destroyed by antiviral antibody and complement,

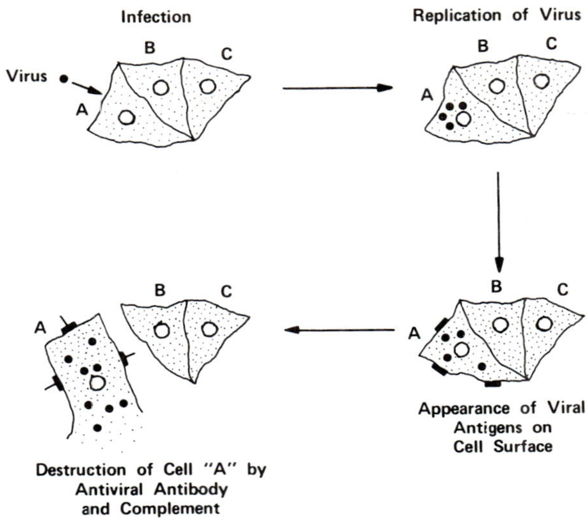

Fig. 2. Destruction of infected cell *before* virus spreads to contiguous cell.

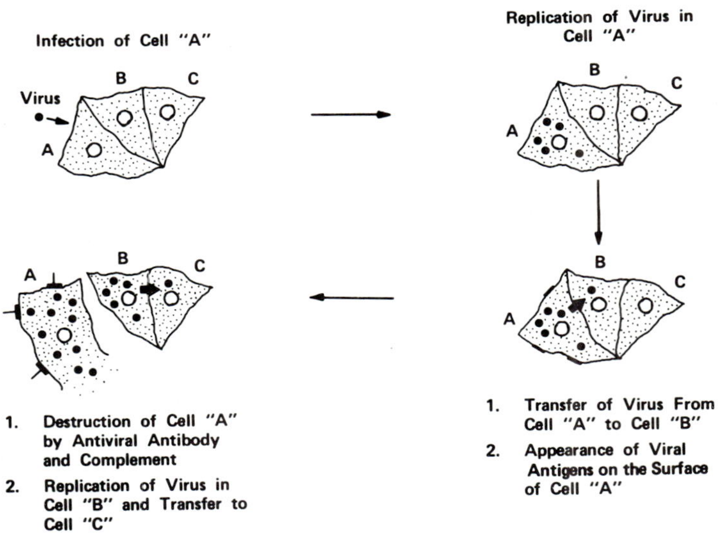

Fig. 3. Destruction of infected cell *after* virus spreads to contiguous cell.

this treatment does not stop the development of plaques (*4, 11*). Evidence that the virus spreads from cell A to cell B, before cell A is immunologically destroyed comes from the following type of experiment. We attempted to determine the time at which viral antigens appeared on the surface of infected cells, the time at which the infected cells became susceptible to immune lysis by antiviral antibody and comple-

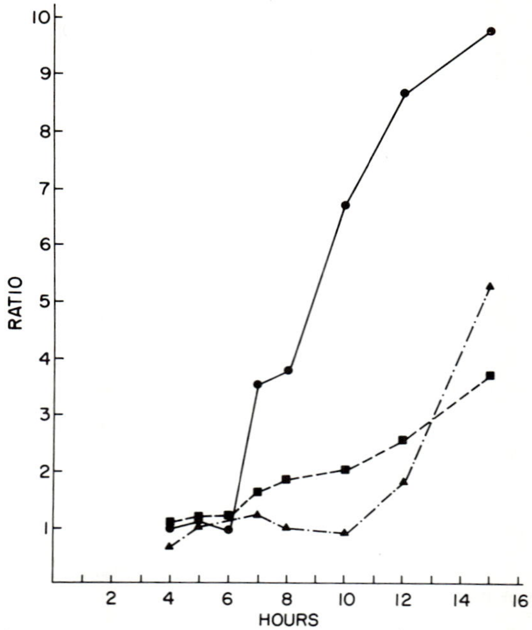

Fig. 4. Relationship of time of appearance of virus-induced cell surface antigens, immune lysis, and transfer of virus to adjacent cells. Ratio of ^{125}I-labeled anti-HSV antibody (cpm) bound to infected cells to ^{125}I-labeled anti-HSV antibody (cpm) bound to uninfected cells (■); ratio of ^{51}Cr (cpm) released from infected cells by anti-HSV antibody plus complement to ^{51}Cr (cpm) released from infected cells by anti-HSV antibody plus heat-inactivated complement (▲); ratio of number of infected cells at each time interval to number of infected cells at 4 hours (●). [From Lodmell *et al.* (*11*).]

ment, and the time at which the virus was actually transferred to adjacent cells (*11*) (Fig. 4). By the binding of ^{125}I-labeled antiviral antibody we found that viral antigens appeared on the surface of infected cells at about 6 hours after infection. By the ^{51}Cr-release assay we found that the infected cells were not susceptible to immune lysis until after the antigens had appeared on the cell surface. By the infectious center assay, we found that virus was transferred from cell A to cell B at 6 to 7 hours after infection. These experiments showed that the virus was transferred to uninfected cells slightly before the infected cells were destroyed by antiviral antibody and complement. Thus, viral spread was one step ahead of the immune attack; the virus was protected from neutralization and the infection persisted.

The rapidity of the spread of the virus suggested that immune lymphocytes would have no more opportunity to recognize and destroy infected

TABLE 3. EFFECT OF LEUKOCYTES ON THE DEVELOPMENT OF VIRAL PLAQUES[a]

Ratio of leukocytes to rabbit kidney cells	No. of plaques	Virus titer (PFU, $\log_{10}$)
0	18	5.0
10	6	3.8
25	0	3.0
50	0	2.8

[a] Monolayers were infected with 15–20 PFU of HSV and then incubated with different concentrations of leukocytes. At the end of 72 hours the leukocytes were removed, the monolayers were washed, and the number of plaques was determined. The cells then were removed and titrated for infectious virus. From Lodmell *et al.* (*11*).

cells, before the virus spread to contiguous cells, than did antiviral antibody and complement. The apparent inability to stop the spread of the infection by an immunological attack directed specifically at the infected cells led us to look at the possibility that factors able to disrupt the connection between contiguous cells or to inhibit viral replication in adjacent cells might stop the viral spread.

B. Nonspecific Inhibition of Viral Infection

Since inflammatory cells are often seen at the site of a number of viral infections, we asked whether these cells could inhibit the spread of the infection. To study this possibility we infected monolayers of rabbit kidney cells with 15–20 PFU of HSV. Then we added peritoneal exudate cells (leukocytes), containing about 80% macrophages, obtained from animals that had never been exposed to HSV. We found that high concentrations of leukocytes prevented the appearance of plaques and markedly reduced the virus titer when measured 72 hours later (Table 3). Several lines of evidence showed that the leukocytes were actually preventing the virus from spreading from one cell to another. First, by immunofluorescence, we found that in the presence of leukocytes the number and size of the fluorescent foci were markedly reduced (*11*). Second, by infectious center assay, we found that at 18 hours after infection there were 6 times more infected rabbit kidney cells in the absence of leukocytes than in the presence of leukocytes (Table 4).

It soon became apparent, however, that the leukocytes were not always producing a permanent cure. Table 5 shows that when leukocytes were removed at the end of 72 hours and the monolayers examined 3 days later (144 hours), normal plaques reappeared, although there was a reduction in the number of plaques as compared to the 20 plaques

TABLE 4. DEVELOPMENT OF INFECTIOUS CENTERS IN PRESENCE AND ABSENCE OF LEUKOCYTES[a]

Hours after infection	Leukocytes absent	Leukocytes present
	(Percentage of cells infected)	
6	0.5	0.3
8	0.4	0.1
10	2.1	0.6
12	2.1	0.6
14	4.5	1.7
16	3.7	1.1
18	9.5	1.5

[a] Monolayers were infected with HSV, washed, and then incubated with peritoneal exudate leukocytes at a ratio of leukocytes to rabbit kidney cells of 50. At different times thereafter the leukocytes were removed, the monolayers trypsinized, and the number of cells infected with HSV determined as previously described (*11*). Unpublished data from Niwa, A., Hayashi, K., and Notkins, A. L.

TABLE 5. APPEARANCE OF VIRAL PLAQUES AFTER REMOVAL OF LEUKOCYTES[a]

Ratio of leukocytes to rabbit kidney cells	No. of plaques	
	72 hours	144 hours
0	20	Confluent
10	4	16
25	0	12
50	0	8

[a] Monolayers were infected with 15–20 PFU of HSV and then incubated with different concentrations of leukocytes. At the end of 72 hours the leukocytes were removed and the number of plaques was determined. The monolayers then were refed, and at 144 hours the number of plaques was again determined. From Lodmell *et al.* (*11*).

in the control group. Thus, leukocytes only temporarily suppressed the development of plaques. Our system, however, was not truly mimicking *in vivo* conditions. *In vivo,* all the immunological components are present (antibody, complement, immune lymphocytes, macrophages, etc.), whereas in our *in vitro* system, one or more of these components were missing. When in subsequent experiments we added in combination, antibody, complment, and leukocytes, we were able to cure many of the infected cultures (*11*).

Precisely, how do the leukocytes exert their effects? When we examined the monolayers microscopically, we found that leukocytes, especially at high concentrations, were slightly toxic to the monolayers. The cytoplasm of the rabbit kidney cells appeared thin and granular and in some cases there was rounding of cells and spaces could actually be seen between adjacent cells. To exert their effects, we found that the leukocytes must be viable and in contact with the monolayers. Under the conditions of our experiments, supernatant fluids from the leukocyte cultures did not inhibit the formation of HSV plaques, and there was no evidence for the presence of interferon. We concluded from these and other experiments that the leukocytes were acting by (*a*) nonspecifically damaging the surrounding uninfected cells and (*b*) interrupting the connection between adjacent cells thereby interfering with cell fusion (*11*). Thus, in the presence of leukocytes, antibody and complement could destroy the infected cells before cell-to-cell contact was reestablished. In this respect, the leukocytes were converting a Type II spread into a Type I spread.

In the experiments I have described thus far, the leukocytes came from normal rabbits that had not been experimentally sensitized to a specific antigen. To see if leukocytes specifically sensitized and stimulated might be more effective in inhibiting viral replication, rabbits were immunized with Freund's complete adjuvant, containing tuberculin. Peritoneal exudate leukocytes obtained from these animals were added to HSV-infected monolayers; purified protein derivative of tuberculin (PPD) was then added to stimulate the sensitized cells (*12*). The data in Table 6 show that in the absence of leukocytes, PPD had no inhibitory effect on viral replication. Similarly, PPD added to leukocytes from animals

TABLE 6. INHIBITION OF HSV REPLICATION BY LEUKOCYTES FROM ANIMALS SENSITIZED WITH FREUND'S COMPLETE ADJUVANT[a]

		Leukocytes present	
Treatment	Leukocytes absent (PFU, $\log_{10}$, HSV)	Freund's incomplete adjuvant (PFU, $\log_{10}$, HSV)	Freund's complete adjuvant (PFU, $\log_{10}$, HSV)
No PPD	5.5	4.3	4.2
PPD	5.5	4.3	2.8

[a] Monolayers were infected with 15–20 PFU of (HSV) and then incubated with 250 tuberculin units of purified protein derivative (PPD) and with leukocytes at a leukocyte-to-rabbit kidney cell ratio of 10. At the end of 48 hours the leukocytes were removed and the monolayers washed and titrated for infectious virus. From Lodmell and Notkins (*12*).

TABLE 7. EFFECT OF DIFFERENT CONCENTRATIONS OF PPD ON HSV REPLICATION[a]

PPD[b]	Leukocytes absent (PFU, $\log_{10}$, HSV)	Leukocytes present (PFU, $\log_{10}$, HSV)
None	5.7	3.7
50	5.4	1.3
500	5.5	0
5000	5.5	0

[a] Leukocytes were obtained from animal sensitized with Freund's complete adjuvant. See footnote to Table 6. From Lodmell and Notkins (*12*).
[b] Purified protein derivative concentration in tuberculin units.

immunized with Freund's incomplete adjuvant (which does not contain tuberculin) failed to inhibit viral replication above and beyond that produced by the presence of leukocytes alone. However, when PPD was added to leukocytes from animals immunized with Freund's complete adjuvant, the virus titer was reduced from $10^{4.2}$ to $10^{2.8}$ PFU.

The data in Table 7 show that the ability of leukocytes from animals immunized with Freund's complete adjuvant to inhibit viral replication is directly related to the concentration of PPD in the medium. In the absence of PPD the virus titer was $10^{3.7}$ PFU; in the presence of 500 tuberculin units (TU), no virus could be recovered.

The effect of different concentrations of sensitized leukocytes on viral replication is illustrated in Table 8. It can be seen that low concentrations of leukocytes, in the absence of PPD, did not inhibit viral replication. Only when the ratio of leukocytes to rabbit kidney cells exceeded 2.5

TABLE 8. INHIBITION OF VIRAL REPLICATION BY LEUKOCYTES FROM ANIMALS IMMUNIZED WITH FREUND'S COMPLETE ADJUVANT AND STIMULATED *in Vitro* WITH PPD[a]

Ratio of leukocytes to rabbit kidney cells	No PPD (PFU, $\log_{10}$, HSV)	PPD 50[b] (PFU, $\log_{10}$, HSV)	PPD 500[b] (PFU, $\log_{10}$, HSV)
Control	6.2	6.3	6.2
0.1	6.3	6.1	5.0
0.25	6.3	5.4	4.2
0.5	6.0	5.2	4.2
1.0	6.2	4.8	3.2
2.5	5.7	3.0	3.2
5	4.7	2.9	2.0
10	4.2	0.9	0.3

[a] See footnote to Table 6. From Lodmell and Notkins (*12*).
[b] Purified protein derivative concentration in tuberculin units.

was viral replication inhibited. However, when both sensitized leukocytes (ratio of 10) and PPD were added, viral replication was inhibited an additional 3–4 logs. Even more important is the fact that PPD-stimulated leukocytes inhibited viral replication by over 90% when the ratio of leukocytes to kidney cells was as low as 0.1.

Examination of the monolayers at these low concentrations of leukocytes failed to reveal any evidence of cytotoxicity as compared to the higher concentrations of leukocytes. The fact that low concentrations of sensitized leukocytes exposed to PPD inhibited viral replication in the absence of cytotoxicity, whereas high concentrations of leukocytes inhibited viral replication by producing cytotoxicity, suggested that different mechanisms must be involved in these two situations. For this reason we began to suspect that an immunologically induced mediator was being produced by the stimulated cells. When supernatant fluids from the stimulated cultures were added to uninfected monolayers for 12 hours prior to infection with HSV, viral replication was markedly inhibited (Table 9). Supernatant fluids from unstimulated cultures produced little if any inhibition. Further studies showed that this mediator was nondialyzable, did not sediment at 100,000 *g*, was resistant to pH 2.0, was species-specific, but was not HSV-specific. From these studies we concluded that the mediator was immunologically induced interferon. Earlier studies from several other laboratories have similarly shown that immunologically stimulated leukocytes produce interferon (*6, 8, 9, 20, 22*).

The experiments I have discussed so far show that (1) normal nonimmunologically stimulated leukocytes (presumably macrophages) and (2) immune leukocytes (presumably lymphocytes) specifically stimulated by an antigen unrelated to HSV can inhibit HSV replication. In one case

TABLE 9. INHIBITION OF VIRAL REPLICATION BY SUPERNATANT FLUIDS PREPARED FROM SENSITIZED LEUKOCYTE CULTURES STIMULATED WITH PPD[a]

Supernatant fluid from leukocyte culture (treatment)	PFU ($\log_{10}$)
None	4.0
PPD (500 tuberculin units)	1.5
PPD (5000 tuberculin units)	0

[a] Leukocyte cultures from animals sensitized with Freund's complete adjuvant were stimulated with purified protein derivative (PPD) and the supernatant fluids were added to monolayers of rabbit kidney cells. At the end of 12 hours the monolayers were washed and infected with 15–20 PFU of herpes simplex virus. The virus titer was determined at the end of 48 hours. From Lodmell and Notkins (*12*).

TABLE 10. EFFECT OF (HSV) SENSITIZED LEUKOCYTES AND DIFFERENT CONCENTRATIONS OF UV-INACTIVATED HSV ON VIRAL REPLICATION[a]

UV-inactivated HSV (PFU)	Leukocytes absent (PFU, $\log_{10}$, HSV)	Leukocytes present (PFU, $\log_{10}$, HSV)
None	7.3	5.9
1×10^6	7.6	4.6
1×10^7	7.4	4.0
1×10^8	7.6	3.2

[a] Monolayers were infected with 15–20 PFU of HSV and then incubated with UV-inactivated virus and with HSV-sensitized leukocytes at a leukocyte-to-rabbit kidney cell ratio of 10. At the end of 48 hours the leukocytes were removed and the monolayers washed and titrated for infectious virus. From Lodmell and Notkins (*12*).

the leukocyte effect was due largely to cytotoxicity preventing cell-to-cell contact, whereas in the other case it was due to the added inhibitory effect of interferon. The specific cell type(s) involved in each situation requires further study.

The next set of experiments shows that leukocytes from animals specifically immunized with HSV can exert these same effects. The data in Table 10 show that in the presence of HSV-sensitized leukocytes, UV-inactivated virus inhibited HSV replication in rabbit kidney cells by over 99%. In the absence of leukocytes, UV-inactivated virus did not inhibit HSV replication. As in the case of PPD, the supernatant fluids from the stimulated leukocyte cultures contained a mediator that markedly inhibited viral replication. This mediator is still in the process of being characterized but all indications suggest that it is interferon.

Since leukocytes seem to be so important in halting viral infections, how do they get to the site of the infection? In the case of HSV, we have shown that at least two different factors may be involved. First, the cleavage product of the fifth component of complement, C5a, is chemotactic for polymorphonuclear and mononuclear leukocytes. Complement (C5) can be activated directly by lysomal enzymes released from injured cells (*3*) or immunologically by the fixation of complement by virus–antibody complexes on the surface of or in the vicinity of infected cells (*21*) (Table 11). Second, a chemotactic factor can be released from sensitized lymphocytes stimulated by viral antigens (*18*). When lymphocytes from patients with a history of HSV were exposed to HSV antigens, they underwent blastogenesis. The supernatant fluids from these stimulated cultures contained a factor that was chemotactic for mononuclear leukocytes (Table 12). By contrast, lymphocytes from antibody-negative patients were not stimulated by viral antigens and pro-

TABLE 11. CHEMOTACTIC ACTIVITY FOR POLYMORPHONUCLEAR LEUKOCYTES AND MACROPHAGES PRODUCED BY THE INTERACTION OF ANTI-HSV ANTIBODY AND COMPLEMENT WITH HSV-INFECTED CELLS[a]

Material tested[b]	Chemotaxis (cells per high-power field)	
	Polymorphonuclear leukocytes	Macrophages
Infected cells		
Anti-HSV + complement	1,018	114
Anti-HSV + heat-inactivated complement	163	23
Nonimmune serum + complement	228	14
Uninfected cells		
Anti-HSV + complement	312	17
Anti-HSV + heat-inactivated complement	286	7
Media alone	0	0

[a] Chemotaxis was evaluated by counting the number of cells that migrated through a micropore filter. From Snyderman *et al.* (*21*).

[b] Supernatant fluids from infected and uninfected cultures treated as indicated.

TABLE 12. PRODUCTION OF LYMPHOCYTE-DERIVED CHEMOTACTIC FACTOR BY HUMAN PERIPHERAL BLOOD LEUKOCYTES[a]

Clinical status of patient	Reagent incubated with peripheral blood leukocytes	
	UV-Inactivated HSV (chemotactic index)[b]	Phytohemagglutinin (chemotactic index)[b]
Antibody positive	9.9	3.9
Antibody negative	2.0	6.7

[a] From Rosenberg *et al.* (*18*).

[b] The chemotatic index represents the ratio of mononuclear leukocytes that migrated through a micropore filter in the presence of supernatant fluid obtained from human, peripheral blood, leukocyte cultures incubated with UV-inactivated HSV or phytohemagglutinin as compared to supernatant fluid from peripheral blood leukocyte cultures incubated with appropriate control antigens.

duced only small amounts of chemotactic factor. *In vivo,* C5a and the lymphocyte-derived chemotactic factor are thought to attract inflammatory cells to the site of the infection where they exert their various effects.

TABLE 13. PRODUCTION OF LYMPHOTOXIN BY HUMAN PERIPHERAL BLOOD LEUKOCYTES

Clinical status of patient	% Inhibition of L-leucine-^{14}C incorporation[b]	
	Leukocytes incubated with UV-inactivated HSV	Leukocytes incubated with phytohemagglutinin
Antibody positive	29.2	50.2
Antibody negative	4.1	38.7

[a] From Rosenberg *et al.* (*18*).

[b] Mouse L-cell monolayers were incubated with supernatant fluid from peripheral blood leukocytes that had been incubated with UV-inactivated herpes simplex virus (HSV), phytohemagglutinin, or phosphate-buffered saline (PBS). L-leucine-^{14}C incorporation into mouse L cells was then measured. Lymphotoxin activity is expressed as percent inhibition of L-leucine-^{14}C incorporation produced by experimental (HSV or phytohemagglutinin) as compared to control (PBS) supernatant fluids.

In addition to chemotactic factors and interferon, a number of other mediators can be released from lymphocytes stimulated by viral antigens. One of these, lymphotoxin, may be particularly important, because it can damage both infected and uninfected cells. The data in Table 13 show that lymphocytes from patients with a history of HSV, when exposed to inactivated viral antigens, produce lymphotoxin; lymphocytes from patients without a history of HSV do not produce lymphotoxin (*18*). Lymphocytes from both antibody-positive and antibody-negative subjects, however, produce lymphotoxin when stimulated with phytohemagglutinin.

IV. CONCLUSION

From the studies that I have attempted to summarize, the following picture is beginning to emerge as to how the host defends itself against Type II spread (Fig. 5). The immunological defense can be thought of as consisting of two phases. The first phase is an immunologically specific antigen recognition phase. Antiviral antibody, complement, and immune lymphocytes react with virus or virus-infected cells. This leads to the release of mediators (e.g., C5a and lymphocyte-derived chemotactic factor) which can attract inflammatory cells to the site of infection. In addition, other mediators, such as lymphotoxin and interferon are released. The nonspecific phase of the defense consists of the action of these inflammatory cells and mediators on infected and adjacent uninfected cells, thereby breaking cell-to-cell contact and inhibiting viral replication. Because the virus spreads so rapidly to contiguous cells, the specific phase of the defense that destroys virus-infected cells appears

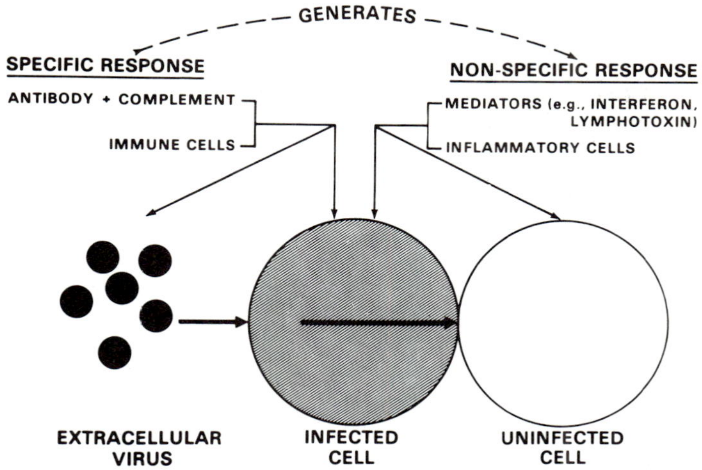

Fig. 5. Sites of action of the specific and nonspecific immune response to viral infections. From Lodmell and Notkins (*12*).

to have little effect on halting the infection. It is the nonspecific phase of the defense, that is, nonspecific cytotoxicity produced by inflammatory cells and lymphotoxin, and the inhibition of viral replication produced by immunologically induced interferon, which appear to be responsible for finally stopping the cell-to-cell spread.

Although the host can recognize and respond to minor antigenic differences in closely related viruses, the host does not seem to have developed a mechanism for recognizing the different routes by which viruses spread. It may be for this reason that the host has evolved a variety of specific and nonspecific immunological responses. Thus, when the host is exposed to a particular virus, the entire armament of immunological weapons may be activated; this would make available the proper combination of responses to cope with any of the different modes of spread. For example, with Type I spread, only antibody may be needed to combat the infection; the rest of the immune system, although activated, may not be used. By contrast, to combat Type II spread, antibody, immune lymphocytes, inflammatory cells, and lymphokines acting in conjunction may be required.

Recognizing that how the host stops a particular viral infection depends, at least in part, on the way in which that virus spreads, we might now ask why do patients with intact humoral immunity, but deficient cellular immunity, have problems in defending themselves against HSV, cytomegalovirus, varicella zoster, vaccinia and measles, but not polio? It appears that these viruses have certain properties in common. First, they induce new antigens on the surface of infected cells; second, they cause cell fusion; and third, although our studies are still incomplete, these viruses seem to spread *in vitro* in the presence of high-titered neutralizing antibody. Thus, they appear to be characterized by a Type II spread. If, as in the case of HSV, these viruses spread to contiguous cells before the infected cells are immunologically destroyed, then the nonspecific phase of the cell-mediated immune response may also be required to stop their spread. This problem is presently under investigation.

REFERENCES

1. Allison, A. C., *Sci. Basis Med.* p. 49 (1972).
2. Blanden, R. V., *J. Exp. Med.* **133,** 1074 (1971).
3. Brier, A. M., Snyderman, R., Mergenhagen, S. E., and Notkins, A. L., *Science* **170,** 1104 (1970).
4. Brier, A. M., Wohlenberg, C., Rosenthal, J. Mage, M., and Notkins, A. L., *Proc. Nat. Acad. Sci. U. S.* **68,** 3073 (1971).
5. Burns, W., and Allison, A. C., *in* "Antigens" (M. Sela, ed.), Vol. 3. Academic Press, New York, 1975 (in press).
6. Epstein, L. B., Cline, M. J., and Merigan, T. C., *Cell. Immunol.* **2,** 602 (1971).
7. Fulginiti, V. A., Kempe, C. H., Hathaway, W. E., Pearlman, D. S., Sieber, O. F., Eller, J. J., Joyner, J. W., and Robinson, A., *Birth Defects, Orig. Art. Ser.* **4,** No. 1, 128 (1968).
8. Glasgow, L., *J. Bacterol.* **91,** 2185 (1966).
9. Green, J. A., Cooperband, S. R., and Kibrick, S., *Science* **164,** 1415 (1969).
10. Hayashi, K., Rosenthal, J., and Notkins, A. L., *Science* **176,** 516 (1972).
11. Lodmell, D. L., Niwa, A., Hayashi, K., and Notkins, A. L., *J. Exp. Med.* **137,** 706 (1973).
12. Lodmell, D. L., and Notkins, A. L., *J. Exp. Med.* **140,** 764 (1974).
13. Murphy, B. R., and Glasgow, L. A., *J. Exp. Med.* **127,** 1035 (1968).
14. Nahmias, A. J., and Roizman, B., *N. Engl. J. Med.* **289,** 667 (1973).
15. Notkins, A. L., *Cell Immunol.* **11,** 478 (1974).
16. Porter, D. D., *Annu. Rev. Microbiol.* **25,** 283 (1971).
17. Rager-Zisman, B., and Allison, A. C., *J. Gen. Virol.* **19,** 339 (1973).
18. Rosenberg, G. L., Snyderman, R., and Notkins, A. L., *Infec. Immunity* **10,** 111 (1974).
19. Rosenthal, J. D., Hayashi, K., and Notkins, A. L., *J. Gen. Virol.* **18,** 195 (1973).
20. Salvin, S. B., Youngner, J. S., and Lederer, W. H., *Infec. Immunity* **7,** 68 (1973).
21. Snyderman, R., Wohlenberg, C., and Notkins, A. L., *J. Infec. Dis.* **126,** 207 (1972).
22. Stinebring, W. R., and Absher, P. M., *Ann. N. Y. Acad. Sci.* **173,** 718 (1970).
23. Wheelock, F. E., and Toy, S. T., *Advan. Immunol.* **16,** 123 (1973).

DISCUSSION

DR. NAHMIAS: As viruses are becoming related to various cancers, I believe we should look at immune responses not just to viruses alone, or to cancer alone, but to both—and also to learn from one field and to apply the knowledge to the other.

I would like to mention two examples of this principle as obtained from studies in our laboratory carried out with Drs. C. Chang, S. Shore, and S. Starr.

In tumor immunology, it has been demonstrated that there are specific tumor antibodies that can lyse tumor cells when acting synergistically with immune or nonimmune lymphocytes. We have also demonstrated such antibodies, called by some "arming antibodies," with dilutions as high as 1 in 1 million, in the serum of herpes simplex virus (HSV)-immune individuals and not in nonimmune ones. Here then is another variable to add to Dr. Notkin's scheme which ties up specific antibody with nonspecific lymphocytes.

A second example of the need to consider virus–cancer immunology as a unit is related to the HSV-2 cervical cancer complex. We have adapted the anticomplement immunofluorescent assay, and, by George Klein's group, with the Epstein-Barr herpes virus to demonstrate nuclear antigens (EBNA). Our assay can detect herpes simplex nucleoantigens (HESNA) in 2–54% of eleven cervical cancers obtained by biopsy. The normal cervical cells, differentiable in the same preparations by their morphology, are negative, as are normal cervical cells from 8 women without neoplasia. The HESNA was also negative with one Epstein-Barr-producing cell line and six human cancers from various sites.

Of particular interest was the finding of HESNA in two of the hamster tumors we reported at these meetings 4 years ago. These tumors had been demonstrated after inoculation of noninactivated or UV-inactivated HSV-2; however, we were not heretofore able to show to our satisfaction HSV antigens in the tumors or HSV antibodies in tumor-bearing animals. The HESNA was also positive in hamster tumors produced with hamster cells transformed *in vitro* with uninactivated HSV-1 or HSV-2 by Duff and Rapp. The tumor associated with HSV-1 yielded negative HESNA with HSV-2 antiserum, but was positive with HSV-1 antiserum.

Much remains to be done with this assay, including attempts to detect which specific HSV antigens may be involved. Once these antigens are indentified, then one can use them in the various immunological assays to determine immune responses to these HSV antigens shared with certain human cancers.

As a pediatrician, I am interested in the effects of transfer factor (TF) in another parasite—the fetus. If I understood you correctly, TF may be circulating in free form in the blood; if so, would TF pass from the mother to the fetus transplacentally? Is there any direct evidence of such transplacental transmission of TF in humans?

DR. TYRRELL: Certain lymphocytes from immune or nonimmune animals and man directly destroy the infectivity of various viruses, including, for example, influenza and parainfluenza viruses. This "nonspecific" mechanism is intracellular and quite distinct from the nonspecific mechanism which you described [Zisman, B., and Denman, A. M., *J. Gen. Virol.* **20**, 211 (1973); Denman, A. M., Rager-Zisman, B., Tyrrell, D. A. J., and Merigan, T. C., *Infec. Immunity* (in press)].

In Type I diseases, delayed hypersensitivity may, nevertheless, occur. Habershon and others [Habershon, R. B., Molyneux, M. E., Slavin, G., Loewi, G., and Tyrrell, D. A. J. *J. Hyg.* **71**, 755 (1973)] and Molyneux and Cole (in preparation), also at the Clinical Research Centre, have shown that influenza virus produces a typical skin reaction and that the size of the lesion is related to the dose of antigen and correlated with the degree to which lymphocytes of the individual are transformed by the same antigen. The skin reaction is also produced by highly purified virus hemagglutin. The degree of hypersensitivity is not closely related to circulating HI antibody titers. Further, vaccination with saline antigens, which increases immunity, does not increase cell hypersensitivity, and there is no close correlation between skin reactions and resistance to infection by experimental challenge. Thus, cellular hypersensitivity occurs but may well be an epiphenomenon.

DR. NOTKINS: I think that any time the host is exposed to a foreign antigen, most of the immunological responses of the host are going to be elicited. The critical question is whether or not the delayed hypersensitivity response plays any real role in stopping Type I spread.

DR. KAUFMAN: We've been looking at similar phenomena from a slighty different point of view. It's clear that what you find in tissue culture systems depends a great deal on exactly how you set up the system. Our group has been fascinated by the work of Ennis. He found that washed, mouse, immune spleen cells would markedly reduce plaque size although they did relatively little to plaque numbers, and immune serum did relatively little to plaque size.

We've carried on this system and adapted it both to Hep 2 cells and conjunctival cells. Sometimes the previously used more indirect measures of cell-mediated immunity correlate with each other and correlate with what seems to happen clinically, and sometimes there seems to be a very poor correlation among blast transformation, macrophage inhibition, lymphokine production, chemotactic factors, and other cellular responses. In a direct antiviral assay, we saw the possibility of using immune cells and getting a direct measure of antiviral cellular activity. We've done this and can easily show that nonimmune cells in our systems do absolutely nothing to plaque size and are a good control. Immune cells dramatically reduce the plaque size in a very easily quantitated manner offering us perhaps not only an additional tool to study mechanisms but a very simple reproducible straightforward way to measure direct antiviral cellular activity using washed leukocytes. The active cell seems to be the lymphocyte. We hope this method can be applied to humans as well as to animals. Certainly we need to find simpler and more direct measures of cellular immunity and correlate these with the indirect measures of cellular immunity.

DR. FUDENBERG: You showed some very nice data on leukocytes having an effect. You're equating this with cell immunity. As far as I can see there were no cell fractionation experiments to determine whether these were really T or B lymphocytes, macrophages, or even, perhaps remotely, granulocytes. I wonder if macrophages alone did the job, if lymphocytes alone will do a good job; if a combination of lymphocytes and macrophages is necessary; or if lymphocytes, are these T or B cells? Without answering these questions I don't think you can extrapolate from your data the cellular immunity.

DR. NOTKINS: It would certainly be nice to have this information and some of these studies are under way. At this stage, however, I don't think it is critical to the main point that I was trying to make, namely, that the spread of HSV is not stopped by a direct immunological attack on the infected cells but rather by the action of inflammatory cells and immunologically induced interferon on the surrounding uninfected cells.

DR. TEVETHIA: I would like to point out the elegant studies by Barry Bloom and by Marcus, that viruses do replicate preferentially in T lymphocytes that are activated either by PHA, concanavalin A, or even by PPD. I think, when we talk about cell immunity to viruses, we have to keep in mind that these activated lymphocytes may actually provide a much better host cell for these viruses in which to replicate.

Dr. Notkins: I think this is an interesting point. The critical issue, however, is whether the interaction of virus with immune leukocytes results in rapid production and high levels of interferon at the local site of the infection, thereby stopping the cell-to-cell spread of the virus. In my opinion this is more important then the question of whether a few activated lymphocytes are destroyed.

Dr. Tevethia: In experiments carried out in my laboratory, I can include along the list of the vesicular stomatitis virus (VSV), the herpes virus. If you activate the lymphocytes with concanavalin A, the herpes virus is replicated much more preferentially, as you can test by the infectious center assays or by doing a total titration.

Dr. Notkins: This certainly could be important if the replication of the virus destroys leukocytes before interferon is produced.

Dr. Lehmann-Grube: If I understood Dr. Notkins correctly, he concludes from his observations that elimination of virus from within an infected cell requires its exposition to the defense forces of the organism by lysis of the cell. I find it difficult to accept this notion in the general sense it was presented. We know of numerous examples where removal of virus even from heavily infected tissues is accomplished without apparent destruction of cells. A striking example is the disappearance of lymphocytic choriomeningitis virus from intestinal organs of the mouse after intraperitoneal infection that precedes, rather than follows, immunopathological destruction of infected cells. The question may be raised whether experiments performed *in vitro* tell us much about the mechanism(s) by which a host disposes of virus from sites where immune processes cannot be expected to play a relevant role. Interferon may be one way but most probably not the only one. After all, in the intact host, virus is not restricted to monolayers, allowing free access to virtually every cell of anything we may add. A solid tissue is a different matter. I have no good hypothesis to offer for the disappearance of virus from within cells, but I think every explanation must take into account the fact that this can occur without apparent damage to the cellular host.

Dr. Chany: I should like to add another nonspecific defense mechanism against viral replication to the one you reported. In brief, some years ago, Dr. Lwoff *et al.* [Lwoff, A., and Lwoff, M., *Ann. Inst. Pasteur, Paris* 98, 173 (1960)] showed that a decrease of pH in the inflammatory area might affect the replication of viruses. In tissue cultures, infected at low multiplicities or with slowly replicating viruses, the pH decreases very fast in the medium. In our laboratory, Drs. Fiszman and Léauté studied the replication of VSV at different pH levels. A decrease from pH 7.4 to only pH 7 inhibited VSV replication by about 90%. At such a pH, viral RNA and protein synthesis are not affected and the number of virions produced is about the same as at permissive pH levels. However, the majority of these particles is noninfectious. As they have shown, lack of infectivity is due to disfunctioning of the transcriptase which is either abnormally inserted in the virion or altered [Léauté, . B., Fiszman, M., Chany, C., and Girard, M., *C. R. Acad. Sci.* 273, 2176 (1971); Fiszman, M., Léauté, J. B., Chany, C., and Girard, M., *J. Virol.* (in press)]. This adds another noninfectious particle to those previously described by Dr. Huang. I think, therefore, that pH effect could interfere with viral replication and I wonder if you have taken this into consideration.

DR. NOTKINS: Yes. We looked carefully for such changes in pH in our medium but there were none under the conditions of our experiments.

DR. MERIGAN: In regard to Dr. Fudenberg's question as to the cellular source of the interferon in the leukocytes, I think we have some very clear-cut information in the human system with HSV antigen. We've studied people after HSV lesions developed and find that the T lymphocyte seems to be the source of the immune specific interferon that is produced.

CHAPTER 8

Transfer Factor and Cellular Immunity to Viral Infection

H. Sherwood Lawrence*

I. INTRODUCTION

There has been a tendency to equate the onset and presence of humoral antibody to viral antigens with mechanisms of host resistance and ultimate recovery from such infections. This thinking, until recently, has dominated experimental approaches to the assessment of host immune responses and has influenced the strategy of prophylactic vaccination. There is, of course, no doubt that humoral antibody of the appropriate specificity exerts a critical influence in determining the host's initial response to viral infection from point of entry of virions in the gut or respiratory tract to the emergence of the blood-borne viremic phase. Nevertheless, once the virus has moved to its privileged intracellular locus, it can settle down and successfully evade the deleterious effects of antibody since the latter does not usually penetrate the cell (*54*).

Examples of this state of affairs may be had in the precise time schedules that must be met if passively administered IgG is to abort an attack of measles, infectious hepatitis, or poliomyelitis that is already in motion. Other examples abound, where agammaglobulinemic patients can and do recover from viral infections with little or no intercession of antibody. By contrast, patients with congenital cellular immunodeficiency and rela-

* Work from the author's laboratories has been supported by the U. S. Public Health Service Research Grant AI-01254-18 and Training Grant AI-00005-15 and in part by the Streptococcal and Staphylococcal Commission of the Armed Forces Epidemiological Board.

tively intact immunoglobulin synthesis, such as Wiskott-Aldrich syndrome, can succumb to disseminated viral illness resulting from smallpox vaccination or natural infection. Additionally, the plight of patients with acquired deficiencies of T-cell function, in consequence of disease states such as Hodgkin's or of immunosuppressive therapy, are particularly susceptible to disseminated viral infections despite B-cell function and antibody production that is relatively intact.

Much of the recent evidence derived from clinical and experimental viral infections has reemphasized the critical role of cell-mediated immunity (CMI) in eradication of such intracellular parasitism once it has been established (*54*). Moreover, exuberant antibody production in intracellular parasitism has been shown to depress the CMI response to the infectious agent upon which the patient's recovery, if not his life, depends. This type of ineffective immune response has been documented in leprosy, candidiasis, coccidioidomycosis, tuberculosis, and syphilis. In each instance, afflicted patients have blocking factors detected in their serum that coat reactive lymphocytes and prevent activation of such cells by the appropriate antigen *in vitro*. If such cells are washed and resuspended in normal sera or tissue culture medium, they will now respond with transformation, proliferation, and lymphokine production when confronted with the appropriate antigen.

Analogous evidence has been secured in tumor-bearing animals and in humans demonstrating the presence of "blocking factors" (antigen–antibody complexes) that serve to blindfold reactive lymphocytes and camouflage tumor target cells *in vitro* (*13*). Again washing of lymphocytes and resuspension restores their capacity to destroy target cells. It is of interest, of course, to virologists whether this state of affairs could be detected either as a cause or in consequence of disseminated viral infections as well. The question naturally arises, then, how could one intervene and deflect the virus-infected host from such ineffective immune responses that fail to eradicate virus and further depress CMI. This brings us to a consideration of transfer factor—an immunological reagent that has been found particularly effective in selectively elevating CMI against intracellular parasites.

II. TRANSFER FACTOR IN THE INITIATION AND AUGMENTATION OF CMI

Transfer Factor (TF) is a convenient shorthand description of the material(s) derived from circulating leukocytes of immune donors that confers a state of specific cellular immunity on nonimmune recipients.

The donor cell is not glass-adherent and is a mononuclear cell of the lymphoid series (*19, 21, 22*).

It has been found that extracts of such cell populations are as effective as viable cells in the transfer of cutaneous delayed-type hypersensitivity (DTH) and CMI (*20, 21*). The transfer of DTH was found susceptible to serial passage from individual A to B to C using leukocyte extracts obtained from B as early as the third day or as late as the third week following initial transfer of DTH from A (*20*). This result suggested that the initial activity may be replicated and the recipient produces more TF in the process; it tended to exclude the notion that TF is a "superantigen" or a unique type of immunoglobulin. However, the search for a replicating macromolecule proved disappointing when it was found that the activity resisted enzymic digestion with DNase, pancreatic RNase, and trypsin (*20, 21*).

The biological properties of TF in leukocyte extracts include: (*a*) the initiation and prompt onset of the transferred DTH (hours) and the prolonged duration of this state of altered reactivity (1–2 years); (*b*) the capacity to transfer is governed by the pattern and intensity of the immunological experience of the donor and the quantity of cells used; (*c*) blood leukocyte preparations can transfer only DTH and CMI and not a primary or secondary antibody response to the same antigenic determinants so that recipients express pure DTH without the intercession of antibody (*27*); and (*d*) incubation of mixed cell populations of differing immunological specificities with specific antigen, (PPD), results in the release of the related TF (tuberculin) into the cell-free supernatant solution whereas TF's residing in other cells of differing specificity (toxoid) remains in the cell pellet. Moreover, TF activity was not neutralized or blocked by interaction with specific antigen and the cell populations from which the specific TF was liberated were desensitized and lost their capacity to transfer cutaneous DTH *in vivo* (*28*).

Additional evidence for the immunological specificity of TF was achieved *in vivo* in studies of mechanisms of allograft rejection (*29*). Here it was demonstrated that one individual does not possess a preexistent TF versus another individual's HL-A antigens. It, therefore, became necessary to raise a TF in prospective donors by repeated exposure to the antigenic stimulus supplied by another individual's skin allograft. Transfer factor was raised *de novo* in this fashion not after exposure to one allograft, but only after immunization by two to four allografts from the same individual. When TF prepared in this fashion versus A's skin was injected into another indifferent individual, it caused him to react to his first exposure to A's skin as if he had been previously sensitized (i.e., accelerated rejection), whereas control allografts from individuals B, C, or D were accorded a first-set response characteristic of the unsensitized individual. These results indicate that one does not possess a TF directed against other individuals HL-A antigens; that an allograft-specific TF can be raised *de novo* following adequate immunization; and that the new TF confers specific immunological reactivity versus the immunizing tissues of the allograft obtained from individual A only and not against allografts obtained from other individuals. Moreover, the results established that prior testing of TF recipients with antigen was not the cause of the transferred reactivity. Additional studies demonstrating the transfer of coccidioidin sensitivity with leukocyte extracts from immune donors into nonimmune recipients extended these observations to DTH and CMI induced by fungal antigens (*42*). The broad scope of antigenic specificities for DTH and CMI that TF has transferred from immune donors to nonimmune recipients is summarized in Table 1.

TABLE 1. CUTANEOUS DELAYED-TYPE HYPERSENSITIVITY OR CELL-MEDIATED IMMUNITY CONFERRED BY TRANSFER FACTOR FROM IMMUNE DONORS[a,b]

Microbial antigens:			
Bacterial	Fungal	Viral	Parasitic
Tuberculin	Coccidioidin	Vaccinia	Schistosomal (monkey)
Streptococcal	Histoplasmin	Herpes	Coccidioides (rat)
Diphtheria toxoid	Candidin	Rubeola	
Lepromin	*Trichophyton*	Mumps	
Tularin	*Trichophyton rubrum*		
	Blastomycin		
Neoantigens:			
Ethylene-oxide human serum protein			
Keyhole limpet hemocyanin			
Histocompatibility antigens:			
HL-A skin allograft specificities			
Tumor antigens:			
Melanoma			
Sarcoma			
Other antigens:			
Kviem			
"Cat-scratch"			

[a] From Lawrence (*24*); reprinted by permission.

[b] Recipients express pure delayed-type hypersensitivity or cell-mediated immunity in absence of antibody to same determinants.

In pursuit of the identity and mechanism of action of this potent molecule(s), it was observed that, following dialysis of leukocyte extracts in a visking cellulose tubing, most of the TF activity of the parent leukocyte extract appeared in the dialyzate. This preparation has been designated dialyzable transfer factor (TF_D). The immunological effects of TF_D *in vivo* are indistinguishable from those produced by administration of the viable leukocytes or the leukocyte extracts from which it is prepared. The properties of TF_D are summarized in Table 2, and are schematically diagrammed in Figs. 1 and 2 (*22–26*).

This carefully laid groundwork plus the availability of a highly purified preparation of TF as a nonantigenic moiety of <10,000 mol wt with the property of selective initiation and/or augmentation of CMI without concomitant humoral immunity, set the stage for the application of TF to the immunotherapy of disease.

III. IMMUNOTHERAPY WITH TRANSFER FACTOR (TF_D)

The general principle that has emerged is that TF_D administration may be expected to benefit those patients afflicted with disease states that are caused by or result in the abrogation or suspension of CMI responses and the attendant consequences. Administration of TF_D and

TABLE 2. SOME PROPERTIES OF DIALYZABLE TRANSFER FACTOR[a]

Biological	Biochemical	Immunological
Properties of WBC extract: Prompt onset (hr) Long duration (>1 yr)	Soluble dialyzable, lyophilizable	Not immunoglobulin
Equal intensity	<10,000 mol wt	Not immunogenic
	No protein, albumin, α- or γ-globulin	Immunologically specific
Dissociable from transplantation antigens	Orcinol-positive	Converts normal lymphocytes *in vitro* and *in vivo* to antigen-responsive state
	Biuret-positive	
	Polypeptide/polynucleotide composition	Transformation and clonal proliferation of converted lymphocytes exposed to antigen
Small quantities → magnified effects	Inactivated 56°C—30 min	
	Resists pancreatic RNase	Informational molecule/depressor/receptor site?
	Retains potency—5 yr	

[a] From Lawrence (*22*); reprinted by permission.

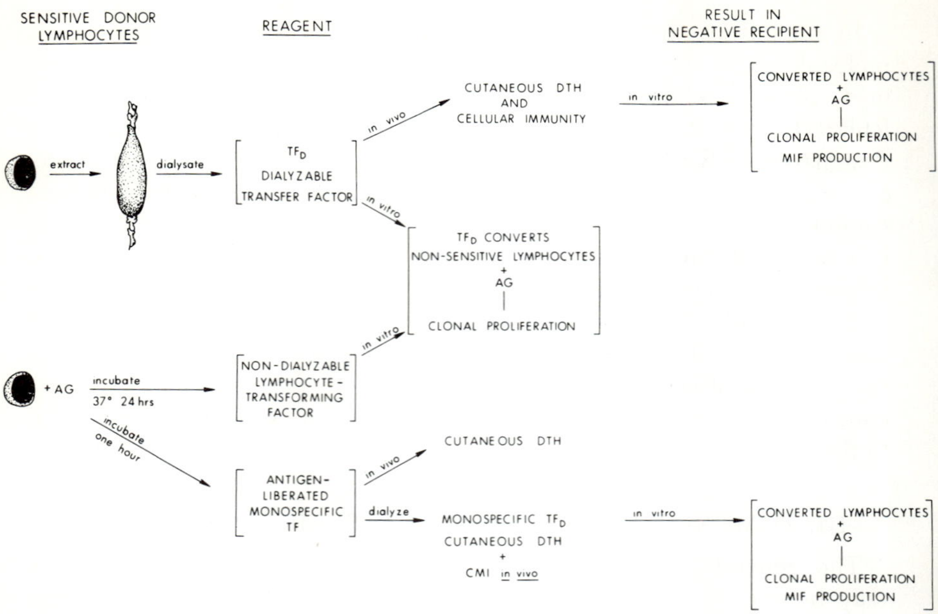

Fig. 1. Diagrammatic comparison of *in vivo* and *in vitro* activities of dialyzable transfer factor; antigen-liberated transfer factor; and lymphocyte-transforming factor. DTH, delayed-type hypersensitivity; AG, antigen; MIF, migration-inhibitory factor; CMI, cell-mediated immunity. [From Lawrence (*22*) reprinted by permission.]

initiation, restoration, or augmentation of CMI have exerted a beneficial effect on the clinical course of congenital and acquired cellular immunodeficiency syndromes; of metastatic malignancies; and of intracellular infections caused by viruses, mycobacteria, and fungi.

Following administration of TF_D, both normal and diseased nonimmune recipients acquire promptly the specific cutaneous DTH and cellular immunities usually concordant with those expressed by the immune TF donor. In addition, there appear in the recipient's circulation a new clone of lymphocytes that respond *in vitro* to the appropriate antigen by transformation and proliferation and/or production of lymphokines, such as migration-inhibitory factor (MIF); lymphotoxin (LT); and interferon (IF). Thus the recipient and his lymphocytes have somehow received information to respond *in vivo* and *in vitro* to antigen with the naturally acquired cellular immunity of the donor.

In addition to the acquisition of such immunological reactivity, certain nonspecific consequences, indicating a general augmentation of CMI, have been detected in immunodeficient children who have been elevated from "null" to almost completely normal cellular immunity following administration of TF_D. The latter include acquisition of lymphocytic

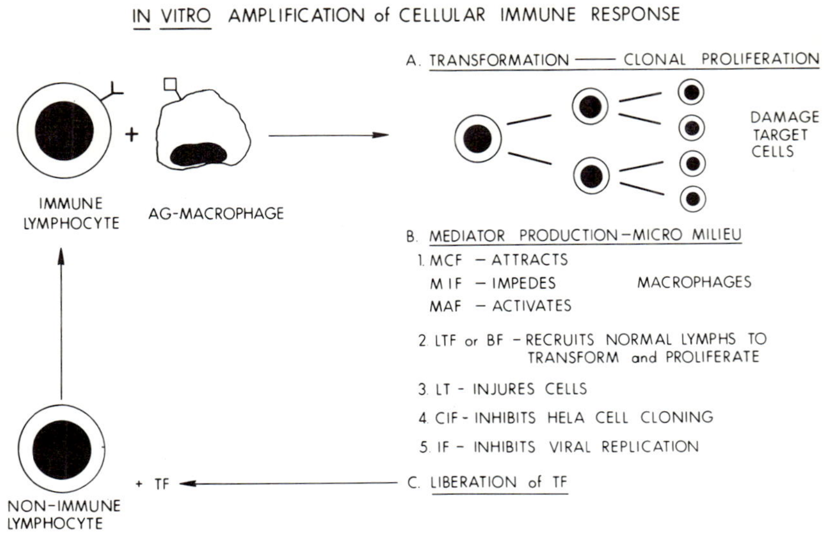

Fig. 2. Diagrammatic concept of the transfer factor (TF) shunt: Antigen-liberated TF engages non-immune lymphocytes and proliferation plus mediator production ensues MCF, macrophage chemotactic factor; MIF, migration-inhibitory factor; MAF, macrophage activation factor; LTF, lymphocyte transforming factor; BF, blastogenic factor; LT, lymphotoxin; CIF, cloning inhibitory factor; IF, interferon. [From Lawrence (*24*); reprinted by permission.]

receptor sites for PHA, for E-rosettes with sheep RBC, and for histocompatibility antigens to respond in mixed-lymphocyte culture (MLC) (*24*). These effects of TF_D are summarized in Table 3.

TABLE 3. SPECIFIC ACTIVITIES AND NONSPECIFIC CONSEQUENCES OF TF[a]

Specific activities[b]	Nonspecific consequences[b]
1. Cutaneous DTH and CMI	1. General stimulus CMI
2. Homograft rejection	2. Transfer PHA response
3. Lymphocyte transformation	3. Transfer E-rosette response
4. Mediator production	4. Transfer MLC response
5. New TF's raised *de novo*	5. Transfer capacity for DNCB
6. No transfer secondary AB response	6. Transfer polychemotaxis

Macrophage activation—probable
T–B Cell facilitation—possible

Are nonspecific consequences of TF sufficient for effective immunotherapy or what additional beneficial increment is conferred by TF specific for related infection or tumor?

[a] From Lawrence (*24*); reprinted by permission.
[b] DTH, delayed-type hypersensitivity; CMI, cell-mediated immunity; TF, transfer factor; AB, antibody; PHA, phytohemagglutinin; MLC, mixed-lymphocyte culture.

I will limit this discussion to the experience with TF_D immunotherapy of infectious disease with emphasis on viral infections. For those interested, the broader implications of this principle and the impact of TF_D on the management of immunodeficiency syndromes and cancer are dealt with elsewhere (*22–25*). Of interest to students of infectious disease the restoration of DTH and CMI by TF_D to immunodeficient children has resulted in their ability to resist the multiple infections that had formerly been their lot (*8, 50*).

Dialyzable transfer factor has been found effective in ameliorating and/or eradicating disseminated infections resistant to antibiotic therapy. This category includes leprosy and tuberculosis, mucocutaneous candidiasis, and coccidioidomycosis. Among the viral infections that have responded to TF_D administration are measles giant-cell pneumonia, congenital herpes, subacute sclerosing panencephalitis (SSPE), and herpes zoster. The immunological and clinical effects of TF_D administration in these diseases are summarized in Table 4. We have outlined in Table 4 the cumulative data showing restitution of specific cellular immunity with conversion of reactivity to the related antigen (cutaneous DTH, MIF production, lymphocyte transformation) and clinical improvement (curtailment or eradication of the infective agent) following administration of TF_D to patients with refractory infections caused by obligatory or facultative intracellular microorganisms.

A. Immunotherapy of Viral Infections

1. *Disseminated Vaccinia*

Not shown in Table 4 are the earlier efforts at immunotherapy of established viral disease employing viable leukocytes. The leukocytes were obtained vaccinia-immune donors and were administered to 3 patients with disseminated vaccinia (*12, 16, 38*). The patients responded promptly following cell transfer with acquisition of cutaneous DTH reaction to vaccinia antigen and concomitant recovery from the disseminated vaccinia. Two of the patients experienced a control period in that they had failed to respond to specific antivaccinia γ-globulin administration prior to cell transfer.

Dialyzable transfer factor has not been used to treat disseminated vaccinia as yet but the above results, taken in conjunction with subsequent experience in other viral diseases, suggest that it would be effective in initiating cellular immunity to vaccinia virus.

2. *Subacute Sclerosing Panencephalitis*

Subacute sclerosing panencephalitis (SSPE) is considered a degenerative disease of the brain arising in consequence of a "slow-virus" infection. Recent evidence from a number of investigations has implicated

TABLE 4. TF_D IMMUNOTHERAPY—INFECTIOUS DISEASE[a,b]

Disease	TF_D source	Responders	Immune conversion			Clinical improvement	Reference
			DTH	MIF	Lymphocyte		
Viral							
Measles	Rubeola +	2/2		+	0	2/2	(*35*)
SSPE	Rubeola +	5/7		+	0	2/7	(*35*)
Neonatal herpes	Herpes +	1/1				1/1	(*35*)
Herpes zoster	Herpes +	1/1				1/1	(*6*)
SSPE	Rubeola +	2/2				2/2	(*55*)
		11/13				9/13	
Mycobacterial							
Leprosy	Lep +	6/9	+		0	6/9 Reversal	(*3*)
Tuberculosis	Tub +	1/1	+	+	+	1/1	(*57*)
Fungal							
Coccidioidomycosis	Cocci +	3/3	+	+	+	2/3	(*10*)
Candidiasis	*Candida* +	1/2	+	+		0/2	(*45*)
		1/1	+	+		1/1	(*46*)
		1/1	+		+	1/1	(*41*)
		4/5	+	+	+	2/5	(*17, 18*)
		5/7	+	+		1/7	(*48*)
		1/1	+	+	+	1/1	(*56*)
		5/5	+	+	0	5/5	(*14*)
		1/1	+		+	1/1	(*11*)
		2/2	+			2/2	(*1*)
		21/25				14/25	
Total		39/48 (80%)				30/50 (60%)	

[a] From Lawrence (*24*); reprinted by permission.

[b] TF_D, Dialyzable transfer factor; DTH, delayed-type hypersensitivity; MIF, migration-inhibitory factor; SSPE, subacute sclerosing panencephalitis.

measles virus as an etiological agent. In a series of studies of normal and diseased children, Reinert *et al.* (*44*) reported evidence to show that patients with SSPE exhibited depressed or absent lymphocyte transformation responses and leukocyte migration test (LMT) responses when their cells were exposed to measles antigen *in vitro*. Children with SSPE responded like the nonimmune infants and unlike the immune children who served as controls. When the children with SSPE were given either lymphocyte infusions or TF_D prepared from such lymphocytes of immune donors, the negative *in vitro* tests of CMI immunity to measles antigen (lymphocyte transformation; leukocyte migration) became positive.

The results of this study led to a trial of immunotherapy by Moulias *et al.* (35) of 7 children with SSPE using measles-immune TF_D. Of this group, 5 children responded immunologically with conversion of their LMT test to positive when exposed to measles antigen *in vitro;* only 2 children responded clinically with improvement in their neurological, electroencephalographic, and cerebrospinal fluid findings. One child who failed to respond to TF_D administration with immunological conversion experienced rapid clinical deterioration.

Vandvik *et al.* (55) have also reported similar clinical and immunological improvement in a 16-year-old boy with rapidly progressive SSPE following administration of measles-immune TF_D which was given on three separate occasions over an 8-month period of observation. Of considerable interest is the finding of a marked fall in the IgG content of the cerebrospinal fluid and a marked fall in titer of measles virus HI[a] antibodies in the cerebrospinal fluid and serum following TF_D administration. Additionally, the patient's serum interferon levels rose from 30 to 220 following the first injection of TF_D. The patient experienced marked mental and neurological improvement and suffered no untoward effects from TF_D injections.

The authors refer to a second patient with SSPE who also responded favorably to a single dose of TF_D with clinical remission but who was lost to follow-up immunological and clinical studies and returned in relapse 10 months later. Vandvik *et al.* (55) are properly circumspect in their interpretation of the events observed, yet feel the results achieved warrant further attention.

Emodi *et al.* (7) also report on the prompt appearance of interferon in the serum of another patient with SSPE following treatment with TF_D. On each of three occasions, TF_D administration was followed by a rise in titer from 0 to 30 units/ml. There is no comment on the clinical response of the patient.

3. *Measles Pneumonia*

Moulias *et al.* (35) have treated 2 children afflicted with giant-cell measles pneumonia with TF_D obtained from immune donors. The prompt acquisition of CMI to the virus following TF_D caused an initial inflammatory response in the lungs of each patient. This inflammation subsided and both patients recovered uneventfully. In each patient, following TF_D administration, Soberg's LMT converted from negative (no response to measles antigen) to positive (inhibition of migration), concomitant with the clinical improvement noted.

[a] Hemagglutination inhibition.

4. *Herpes Virus Infections*

Moulias *et al.* (*35*) have treated a child with neonatal herpes infection with TF_D from immune donors and report the disappearance of the herpetic eruption in the ensuing days.

More recently Drew *et al.* (*6*) have reported on the recovery of an immunosuppressed patient with Hodgkin's disease from disseminated herpes zoster following administration of TF_D from immune donors. The initiation of CMI followed promptly the injection of TF_D and was associated with fever and the development of intense erythema around the vesicles. Additionally, the patient's lymphocytes acquired immune responsiveness to varicella zoster antigen and to PHA. The viral cultures taken after TF_D administration had also become negative.

5. *Prospects for Multiple Sclerosis*

There are recent reports on the inability of the blood lymphocytes of patients with multiple sclerosis to respond to certain viral antigens in the *in vitro* LMT. Ciongoli *et al.* (*5*) have compared this immune deficiency in 18 multiple sclerosis patients with 18 normal individuals and find it selectively expressed against myxoviruses (measles, parainfluenza), whereas the response to tuberculin is unimpaired in the patient group. In the same issue of *Lancet,* Utermohlen and Zabriskie (*52*) report on the inability of the leukocytes of 15 patients with multiple sclerosis to respond to measles antigen with inhibition of leukocyte migration compared to 15 controls. Additionally, in extension of earlier observations where tuberculin PPD was used as antigen (*43*), they report that measles-immune TF_D converts the leukocyte populations of the multiple sclerosis patients to an antigen-responsive state as detected by inhibition of leukocyte migration when measles viral antigen is added.

Utermohlen and Zabriskie (*52*) suggest that this *in vitro* endowment of CMI to measles virus by immune TF_D should also occur *in vivo* and provide a rationale for therapy of multiple sclerosis. Preliminary reference for such *in vivo* transfer of CMI to measles antigen is cited in another paper (*53*). Jersild *et al.* (*15*) also cite the beginning of a trial of TF_D therapy in 5 patients with progressive multiple sclerosis and report that, following administration of TF_D 4 patients converted their MIF test to measles antigen as well as their MLC test. The principle involved in this approach to multiple sclerosis is well founded and at this writing, carefully controlled clinical trials are being conducted by Dr. Zabriskie and his associates at Rockefeller University and Dr. Jersild and his associates at the Rigshospitalet in Copenhagen. The results of these studies are awaited with interest.

B. Potential of TF_D As an Immunizing Agent

There is thus cumulative evidence that TF_D can initiate or augment sustained CMI responses in patients afflicted with disseminated microbial fungal and viral infections and result in eradication of the infectious agent. Such effects in a compromised host give promise for the possible application of TF_D as an immunizing agent for potentially lethal infectious processes for which no vaccine is available. There is no great demand for such usage, except that the need may exist among scientists working with highly infective microbes or viruses or among patients who are immunosuppressed as a result of lymphomatous disease and/or its treatment.

C. Early Applications of TF_D to Cancer Immunotherapy

There is increasing awareness that TF_D selectively augments CMI and not Ig production to the same antigenic determinants (*22, 27*). Additionally, the work of Hellström and Hellström (*13*) has called attention to the possibility that immunological plight of the cancer patient may arise from the production of serum "blocking factors" (tumor antigen–antibody complexes) that blindfold potentially reactive lymphocytes and camouflage the tumor target cells. In consequence the tumor-bearing host tolerates an allograft with antigens that differ from his own in the face of such "natural" immunosuppression.

The availability of TF_D as an immunological reagent that selectively augments CMI and has proven effective in the reconstitution of congenital immunodeficiency syndromes and in the recovery from infection with intracellular microbes, has led to its recent trials in the immunotherapy of cancer (*24, 25*) as outlined in Table 5.

In an early study with my colleagues at the Sloan-Kettering Institute, we evaluated the effects of pooled nonspecific TF_D in 5 patients with inflammatory cancer of the breast (*39, 40*). Here the basic assumption arose from the early suggestion of Thomas (*51*) extended by Burnet (*4*) and termed "immunologic surveillance." We reasoned that a proportion of healthy women over 40 years of age would have met and responded with CMI to mutant breast cells from puberty onward and should have in their leukocytes a corresponding tumor-specific TF_D. A group of such normal women served as blood donors and the TF_D prepared from their leukocytes was pooled and administered repeatedly to 5 patients with inflammatory breast cancer. Only 1 of the 5 patients experienced marked partial regression of the breast cancer following extended administration of TF_D which lasted for 6 months and was followed by relapse despite continued TF_D administration.

TABLE 5. TF_D IMMUNOTHERAPY OF CANCER[a,b]

Disease	TF_D source	Responders	Immune conversion	Clinical response	Reference[c]
Breast carcinoma	Pooled nonspecific	3/5	+ DTH to PPD, SK-SD	Remission, 1/5	(*39, 40*)
Melanoma	Sens. donor-specific WBC extract	1/1	N.D.	Reject. metastatic nodules, 1/1	(*2*)
	Family-specific	2/9	+ Lymph. trans to TSTA	Regress. lesions, 2/9	(*49*)
	Sens. donor-specific	4/10	N.D.	Regress. lesions, 4/10	(*34*)
Sarcoma					
Alveolar	Twin-specific	1/1	+ MIF to TSTA	0/1	(*33, 37*)
Osteogenic	Family-specific	8/12	+ RFC; + TSTA	8/12	(*8, 32*)
Nasopharyngeal carcinoma	Infection mononucleosis patients	2/2	Fall AB titer to EB-viral AG	Regress., 2/2	(*9*)
Total		21/40 (52%)		18/40 (45%)	

[a] From Lawrence (*24*); reprinted by permission.

[b] TF_D, dialyzable transfer factor; DTH, delayed skin test; Lymph. trans., lymphocyte transformation; PPD, purified protein derivative; MIF, migration-inhibitory factor; RFC, rosette-forming cells; AB, antibody; AG, antigen; N.D., not determined.

In this study TF_D was administered repeatedly in massive doses over a prolonged period, and it was evaluated as if it were a new drug. In such large dosage TF_D did not cause any inflammatory or hypersensitivity reactions to itself in the 5 patients studied. There were no harmful side effects observed and no hematological, immunological, biochemical, or enzymic abnormalities detected. Whether "pooled, nonspecific" TF_D holds promise for cancer immunotherapy cannot be decided from these limited observations. The conclusion one can derive from our results is that TF_D can be given safely in large dosage (257 ml TF_D equivalent to 217 billion cells), repeatedly (125 injections), and over a prolonged period (310 days). This dose magnitude, repetitive administration, and total duration of therapy far exceeds the requirements that have been found effective in the reconstitution of immunodeficiency syndromes or the eradication of infectious diseases. Hence this study provides a clue to the margin of safety that obtains under the ordinary circumstances of TF_D usage (*40*).

Of considerable interest to students of virology as well as to tumor immunobiologists are subsequent findings that the domiciliary relatives of certain cancer patients will possess tumor-specific CMI versus the antigens of the patient's tumor. This has been described for malignant melanoma (*47, 50*), for alveolar carcinoma (*33*), and for osteogenic sarcoma (*8, 32*).

This is usually revealed by testing blood lymphocyte reactivity of the family contact to the patient's tumor antigens and preparing a tumor-specific TF_D from such immune lymphocytes. The administration of tumor-specific TF_D to the respective patients results in the acquisition of CMI by the patient to his own tumor antigens as measured by lymphocyte reactivity *in vitro* and occasional regression of tumor metastases *in vivo*. This is a challenging observation which suggests, at least for the types of cancer studied so far, that neoplasia may be more of an infective disease than has been realized. The finding also suggests a source of tumor-specific TF_D for immunotherapy.

Another topic of interest to virologists in particular is the finding of Goldenberg and Brandes (*9*) that TF_D obtained from EB-positive donors who have recovered from infectious mononucleosis resulted in tumor regression in 2 patients with nasopharyngeal carcinoma. The clinical improvement in 1 patient was associated with lymphocytic infiltration of the tumor and a fall in antibody titer to EB viral capsid antigen. This observation may provide another link in the possible relationship of EB viral infection, mononucleosis, and nasopharyngeal carcinoma. The Canadian National Research Council has since launched a controlled field trial of the effects of EB-positive TF_D in nasopharyngeal carcinoma in Hong King where the disease is endemic, that is being conducted by Goldenberg and Brandes.

Fudenberg *et al.* report on the experience of his group in the treatment of osteogenic sarcoma with tumor-specific TF_D, obtained from contacts of the index case, in Chapter 9 of this volume, and the subject will not be covered here (cf. *8, 32*).

Thus the application of the principle of TF_D to the cancer patient's plight is just beginning and any evaluation would be premature at this time. The promise that TF_D affords rests on its capacity to elevate selectively the specific CMI responses of the host to his tumor and nonspecifically augment CMI in general without concomitant antibody production. That TF_D can be administered with safety over a wide range of dosage is not a disadvantage (*39, 40*).

D. Prospects for TF_D Immunotherapy

Since most of the clinical experience with TF_D is still at the anecdotal stage this is not the time for pronouncements or generalizations. It is hoped that the several clinical trials currently in motion will supply data that will enable a critical assessment of the role of TF_D as primary or as adjunctive therapy of the various diseases for which the initial results appear either suggestive or promising.

REFERENCES

1. Bläker, F., *et al., Deut. Med. Wochenschr.* **98,** 415 (1973).
2. Brandes, L., Galton, D. A. G., and Wiltshaw, E., *Lancet* **1,** 293 (1971).
3. Bullock, W. E., Fields, J. P., and Brandriss, M. W., *N. Engl. J. Med.* **287,** 1053 (1972).
4. Burnet, F. M., *Lancet* **1,** 1171 (1967).
5. Ciongoli, A. K., Platz, P., Dupont, B., Svejgaard, A., Fog, T., and Jersild, C., *Lancet* **2,** 1147 (1973).
6. Drew, W. L., Blume, M. R., Miner, R., Silverberg, I., and Rosenbaum, E. H., *Ann. Intern. Med.* **79,** 747 (1973).
7. Emodi, G., Just, M., and Grob, P., *Lancet* **2,** 1382 (1973).
8. Fudenberg, H. H., Levin, A. S., Spitler, L. E., Wybran, J., and Byers, V., *Hosp. Pract.* **9,** 95 (1974).
9. Goldenberg, G., and Brandes, L., *Clin. Res.* **20,** 947 (1972).
10. Graybill, J. R., Silva, J., Alford, R. H., and Thor, D. E., *Cell. Immunol.* **8,** 120 (1973).
11. Griscelli, C., *et al., Biomedicine* **18,** 120 (1973).
12. Hathaway, W. E., Githens, J. N., Blackburn, W. R., Fulginiti, V., and Kempe, C. H. *N. Engl. J. Med.* **273,** 953 (1965).
13. Hellström, K. E., and Hellström, I., *Annu. Rev. Microbiol.* **24,** 373 (1970).
14. Hitzig, W. H., *Schweiz. Med. Wochenschr.* **102,** 1237 (1972).
15. Jersild, C., Platz, P., Thomsen, M., Hansen, G. S., Svenjgaard, A., Dupont, B., Fog, T., Ciongoli, A. K., and Grob, P., *Lancet* **2,** 1381 (1973).
16. Kempe, C. H., *Pediatrics* **26,** 176 (1960).

17. Kirkpatrick, C. H., Chandler, J. W., and Schimke, R. N., *Clin. Exp. Immunol.* **6**, 375 (1970).
18. Kirkpatrick, C. H., Rich, R. R., and Smith, T. K., *J. Clin. Invest.* **51**, 2948 (1972).
19. Lawrence, H. S., *Proc. Soc. Exp. Biol. Med.* **71**, 516 (1949).
20. Lawrence, H. S., *J. Clin. Invest.* **34**, 219 (1955).
21. Lawrence, H. S., *in* "Cellular and Humoral Aspects of the Hypersensitive States" (H. S. Lawrence, ed.), p. 279. Harper (Hoeber), New York, 1959.
22. Lawrence, H. S., *Advan. Immunol.* **11**, 195 (1969).
23. Lawrence, H. S., *in* "Mediators of Cellular Immunity" (H. S. Lawrence and M. Landy eds.), p. 143. Academic Press, New York, 1969.
24. Lawrence, H. S., *Harvey Lect.* **78**, 1972–1973 (1974).
25. Lawrence, H. S., *Clin. Immunobiol.* **2**, 116 (1974).
26. Lawrence, H. S., and Al-Askari, S., *in "In Vitro* Methods in Cell-Mediated Immunity" (B. R. Bloom and P. Glade, eds.), p. 531. Academic Press, New York, 1971.
27. Lawrence, H. S., and Pappenheimer, A. M., Jr., *J. Exp. Med.* **104**, 321 (1956).
28. Lawrence, H. S., and Pappenheimer, A. M., Jr., *J. Clin. Invest.* **36**, 908 (1957).
29. Lawrence, H. S., Rapaport, F. T., Converse, J. M., and Tillett, W. S., *J. Clin. Invest.* **39**, 185 (1960).
30. Lawrence, H. S., Rapaport, F. T., Converse, J. M., and Tillett, W. S., *Transplant., Ciba Found. Symp., 1961* p. 271 (1962).
31. Levin, A. S., Spitler, L. E., and Fudenberg, H. H., *Annu. Rev. Med.* **24**, 175 (1973).
32. Levin, A. S., Spitler, L. E., Wybran, J., Fudenberg, H. H., Hellström, I., and Hellström, K. E., *Clin. Res.* **20**, 568 (1972).
33. LoBuglio, A. F., Neidhart, J. A., Wilberg, R. W., Metz, E. N., and Balcerzak, S. P., *Cell. Immunol.* **7**, 159 (1973).
34. Morse, P. A., *et al., Clin. Res.* **21**, 71 (1973).
35. Moulias, R., Goust, J. M., Reinert, P., Fournel, J. J., Deville-Chabrolle, A., Duong, N., Muller-Berat, C. N., and Berthaux, P. (1973). *Nouv. Presse Med.* **2**, 1341.
36. Moulias, R. *in "Progress in Immunology"* (L. Brent and J. Holborow, eds.), Vol. II, North-Holland Publ., Amsterdam, 1974.
37. Neidhart, J. A. *et al., Clin. Res.* **20**, 748 (1972).
38. O'Connell, C. J., Karzon, D. T., Barron, A. L., Plaut, M. E., and Ali, V. M., *Ann. Intern. Med.* **60**, 282 (1964).
39. Oettgen, H. E., Old, L., Farrow, J., Valentine, F. T., Lawrence, H. S., and Thomas, L., *J. Clin. Invest.* **50**, 71a (abstr) (1971).
40. Oettgen, H. E., Old, L., Farrow, J., Valentine, F. T., Lawrence, H. S., and Thomas, L., *Proc. Nat. Acad. Sci. U. S.* **71**, 2319 (1974).
41. Pabst, H. F., and Swanson, R., *Brit. Med. J.* **2**, 442 (1972).
42. Rapaport, F. T., Lawrence, H. S., Millar, J. W., Pappagianis, D., and Smith, C. E., *J. Immunol.* **84**, 368 (1960).
43. Reed, S. E., and Zabriskie, J., *Transplant. Proc.* **4**, 247 (1972).
44. Reinert, P., Moulias, R., Goust, J. M., Hors, J., and Bussel, A., *Arch. Franc. Med.* **29**, 655 (1972).
45. Rocklin, R. E., *et al., Cell. Immunol.* **1**, 290 (1970).
46. Schulkind, M. L., *et al., Cell. Immunol.* **4**, 606 (1972).
47. Spitler, L. E., Levin, A. S., Blois, M. S., Epstein, W., Fudenberg, H. H., Hellström, I., and Hellström, K. E., *J. Clin. Invest.* **51**, 952 (1972).
48. Spitler, L. E. *et al., Clin. Res.* **20**, 519 (1972).

49. Spitler, L. E. *et al., Clin. Res.* **21,** 221 (1973).
50. Spitler, L. E., Levin, A. S., and Fudenberg, H. H., *Methods Cancer Res.* **8,** 59 (1973).
51. Thomas, L., *in* "Cellular and Humoral Aspects of the Hypersensitive States" (H. S. Lawrence, ed.), p. 529. Harper (Hoeber), New York, 1959.
52. Utermohlen, V., and Zabriskie, J. B., *Lancet* **2,** 1147 (1973).
53. Utermohlen, V., and Zabriskie, J. B., *J. Exp. Med.* **138,** 1591 (1973).
54. Valentine, F. T., and Lawrence, H. S., *Advan. Intern. Med.* **17,** 51 (1971).
55. Vandvik, B., Froland, S. S., Hoyeraal, H. M., Stien, R., and Degre, M., *Scand. J. Immunol.* **2,** 367 (1973).
56. Vladimarsson, H., *et al., Clin. Exp. Immunol.* **11,** 151 (1972).
57. Whitcomb, M. E., and Rocklin, R. E., *Ann. Intern. Med.* **79,** 61 (1973).

DISCUSSION

DR. KATZ: Two questions: (1) In what way was transfer factor response measured in patients with subacute sclerosing panencephalitis (SSPE) and multiple sclerosis? (2) What were the antigens utilized and how was response assayed?

DR. LAWRENCE: In SSPE the immunological response to dialyzable transfer factor (TF_D) was measured by (1) conversion of the leukocyte migration test of Soberg (LMT) from negative (no inhibition of migration) to positive (migration inhibition) in the presence of measles virus antigen; (2) detection of interferon production in the serum of patients following therapy; and (3), in one study, selective fall in HI antibody titer to measles virus in the cerebrospinal fluid (CSF) and serum.

The clinical responses followed were the usual neurological, electroencephalographic, and CSF parameters [Moulias, R. *et al., Nouv. Presse Med* **2,** 1341 (1973); Vandvik, B., Froland, S. S., Hoyeraal, H. M., Stein, R., and Degre, M., *Scand. J. Immunol.* **2,** 367 (1973); Emodi, G., Just, M., and Grob, P., *Lancet* **2,** 1382 (1973)].

In multiple sclerosis the immunological response to TF_D *in vivo* was followed by (1) conversion of migration-inhibitory factor (MIF) production *in vitro* to paramyxoviruses and (2) conversion of reactivity of patients lymphocytes in the mixed-leukocyte culture (MLC) *in vitro* [Jersild, C., Platz, P., Thomsen, M., Hansen, G. S., Svenjgaard, A., Dupont, B., Fog, T., Ciongoli, A. K., and Grob, P., *Lancet* **2,** 1381 (1973)]. The inability of the lymphocytes of patients with multiple sclerosis to respond *in vitro* was detected selectively to paramyxoviruses (measles, parainfluenza) in one study [Ciongli, A. K., Platz, P., Dupont, B., Svejgaard, A., Fog, T., and Jersild, C., *Lancet* **2,** 1147 (1973)] and to measles virus in another study [Utermohlen, V., and Zabriskie, J. B., *Lancet* **2,** 1147 (1973); *J. Exp. Med.* **138,** 1591 (1973)]. Utermohlen and Zabriskie also showed that addition of measles-immune TF_D equipped the patients lymphocytes to respond to measles viral antigen *in vitro.*

DR. NATHANSON: Will transfer factor break tolerance?

DR. LAWRENCE: We do not know for classic tolerance, since breaking tolerance with TF_D is just being attempted in experimental animal models. However, patients with disseiminated microbial or viral infections whose tissues are bathed in a sea of replicating antigen could be viewed as examples of "high-zone" tolerance. In the latter examples, TF_D in elevating cell-mediated immunity (CMI) and eradicating the replicating agent may have broken such a tolerant state (e.g., candidiasis, coccidioidomycosis, lepromatous leprosy). A similar case could be made for the patient with metastatic cancer.

DR. NAHMIAS: There are no direct data on the transfer of transfer factor into babies, transplacentally?

DR. LAWRENCE: None at all.

DR. AARONSON: Can the administration of transfer factor stimulate CMI against a specific antigen in clinical states associated with anergy to that antigen?

DR. LAWRENCE: Yes. There is a growing list of diseases that result in acquired cellular immune deficiency and anergy to a specific antigen where the patients have responded to administration of TF_D with the acquisition of cutaneous delayed-type hypersensitivity (DTH) and CMI to that antigen. A few examples include tuberculin purified protein derivative (PPD) in anergic sarcoid patients [Lawrence, H. S., and Zweiman, B., *Trans. Ass. Amer. Physicians* **81**, 240 (1968)]; SKSD in anergic patients having a variety of metastatic malignancies [Solowey, A. C., Rapaport, F. T., and Lawrence H. S., "Histocompatibility Testing," p. 75. Karger, Basel, 1967]; varicella zoster antigens in an anergic patient with Hodgkin's disease and disseminated herpes zoster [Drew, W. L., Blume, M. R., Miner, R., Silverberg, I., and Rosenbaum, E. H., *Ann. Intern. Med.* **79**, 747 (1973)]; lepromin in lepromatous leprosy [Bullock, W. E., Fields, J. P., and Brandriss, M. W., *N. Engl. J. Med.* **287**, 1053 (1972)]; coccidiodin in disseminated coccidioidomycosis [Graybill, J. R., Silva, J., Alford, R. H., and Thor, D. E., *Cell. Immunol.* **8**, 120 (1973)]; *Candida* in disseminated mucocutaneous candidiasis [Pabst, H. F., and Swanson, R., *Brit. Med. J.* **2**, 442 (1972)]; tumor-specific antigens in sarcoma [LoBuglio, A. F., Neidhart, J. A., Wilberg, R. W., Metz, E. N., and Balcerzak, S. P., *Cell. Immunol.* **1**, 159 (1973)]; and, of course, children with various congenital immunodeficiency diseases who are anergic to a whole array of specific antigens [Levin, A. S., Spitler, L. E., and Fudenberg, H. H., *Annu. Rev. Med.* **24**, 175 (1973)]. A more detailed consideration of this topic may be found in recent reviews of the general subject [Lawrence, H. S., *Advan. Immunol.* **11**, 195 (1968); *Harvey Lect.* **68**, 1972–1973 (1974)].

CHAPTER 9

Immunological Evidence for a Viral Etiology of Certain Human Malignant Tumors*

H. H. Fudenberg
V. Byers
A. S. Levin

I. INTRODUCTION

As has been pointed out by Lawrence (*20*), transfer factor (TF) remained, until recently, primarily a myth among immunologists for several reasons: (*a*) it could not be shown to exist in any other species than humans—not in guinea pigs, rats, mice, etc.; (*b*) its effect could be demonstrated only by skin tests—this was a time when sophisticated immunologists were working with quantitation of serum antibodies (micrograms of antibody nitrogen) or working with chains or fragments of immunoglobulin molecules; and (*c*) TF did not fit into any conventional immunological dogma, since the mechanism whereby such a small molecule could produce such effects could not be conceived.

II. ROLE OF LYMPHOCYTES IN IMMUNOLOGY

Before presenting our evidence for viral etiology of certain human tumors, certain preliminary information is necessary. In addition to mac-

* This work was supported by U. S. Public Health Service grants (HD-05894) and (AI-09145).

rophages (immunologically very important cells), at least two kinds of lymphocytes exist: (*a*) B for bursa in the chicken (*3*) and bone marrow in mammals (*17*) and (*b*) T for thymus-derived or thymus-dependent cells (*12*). The B cells, in response to antigens, make humoral antibodies, IgG, A, M, D, and E, which produce protection against classic microorganisms such as *Streptococcus, Pneumococcus, Meningococcus.* Sensitized T cells *in vitro* respond to antigens in several ways. First, they release a whole host of soluble lymphocyte factors, e.g., chemotactic factors that attract granulocytes, eosinophiles, macrophages, etc., factors that inhibit the normal migration of macrophages, and lymphotoxins that kill cells other than lymphocytes (*5*). These factors are all detectable in soluble supernatants obtained when lymphocytes are incubated with antigens to which the donor is sensitive. None of them have been purified, so we do not really know whether they act *in vivo* the way that they do *in vitro.* I assume they do. There are at least a dozen of these, collectively termed "mediators of cellular immunity." The T cells are also responsible for rejecting homografts. They are presumably responsible for "immunological surveillance" against tumor antigens. They produce protection against most viruses, parasites, and fungi (*12*). In the laboratory they can be quantitated by various parameters; the one most commonly used is incorporation of labeled precursors into DNA under the influence of certain lectins, e.g., phytohemagglutinin (PHA).

For some antigens, T-cell cooperation with B cells is necessary to produce a maximal antibody response. These data are generally accepted, but oversimplified. For example, for some years we have postulated that separate T-cell subpopulations may exist, perhaps one for DNA synthesis, another for mediator production, etc. (*9, 10*).* For many years we also postulated the existence not only of cooperating, but also of suppressor T cells (*9, 10*), although at the time these postulations were put forth most immunologists termed them "heresay." There is now good evidence for these (e.g., *6, 18*). Indeed, I suggested that suppressor T cells are responsible for suppression of both autoimmune disease and certain malignancies [reviewed in Fudenberg (*9*)]. Burnett postulated that autoimmune disease is due to the emergence of "forbidden clones" (*1*). By contrast, I postulated that "autoantibodies" are always being made and that suppressor T cells prevent these autoantibody-forming cells from undue proliferation. (Presumably, autoantibodies are normally produced to remove aged or damaged tissue.) Thus, we believe that "autoimmune" disease results not from mutant B-cell

* We provided evidence to support this concept [e.g., Fudenberg (*9a*) and Spitler *et al.* (*24b*)].

clones producing autoantibodies, but rather from a genetically determined *defect* in thymic-dependent immunity, and we call the concomitant disease states "diseases of immunological aberration" (DIA) rather than "autoimmune disease" (AID) (*7*). The relevance of this to TF will become apparent later.

III. RELATION OF T-CELL DEFICIENCY TO CANCER AND AUTOIMMUNITY

In 1960, based on patient observations, I postulated that immunological deficiency was a necessary but not sufficient condition for development of both cancer and autoimmune disease (*14*). Subsequently, we showed that "adult onset" agammaglobulinemia, where initial onset of the symptoms first occur late in life, is, in reality, a genetically determined condition (*19*), since tests of lymphocyte function of parents of such patients showed them to be obligate heterozygotes. We also demonstrated that patients with pronounced B-cell defects develop defects in cellular immunity (T-cell defects) later in life (*4*). We also showed a high incidence of DIA and also autoantibodies without clinical disease in first-degree relatives of such patients (*11*). We later performed animal experiments to relate T-cell deficiency to diseases of immunological aberration. We neonatally thymectomized mice of strains which do not ordinarily develop the kinds of lesions seen in the New Zealand strains, i.e., nephritis characterized by deposition of complement and IgG, similar or identical to the renal lesions of patients with lupus erythematosus. Normal strains, neonatally thymectomized, developed a renal lesion identical to that of the NZB/W hybrids (*16*). If we gave the thymectomized mice interferon-inducing agents, they did not develop this lesion (*15*), suggesting again to us that diseases of immunological aberration, at least of certain varieties (at the time, we mentioned rheumatoid arthritis, lupus and multiple sclerosis), were due to genetically determined defects (*8*) in cell-mediated immunity, perhaps for one or another virus. We suggested that these diseases were viral diseases, and that restoration of T-cell function would help.

These data taken in a combination led to a series of hypotheses which suggested that genetically determined deficiency in number or function of T-cells, perhaps a selective deficiency for T-cell specific for one or another virus, was responsible for autoimmunity and for cancer. Alternatively, we postulated lack of T-cell suppressor cells as the cause. These hypotheses are not actually exclusive, and, indeed, I felt both of these factors were operative.

IV. CLINICAL TESTS OF TRANSFER FACTOR

Dr. Lawrence has presented his data on TF; until 1969 he had done approximately forty-eight transfers of skin tests using TF extracted from 0.1 cc of white cells of tuberculin-positive donors and converting individuals negative to tuberculin by skin test to tuberculin positive by the same criteria. We first used dialyzable TF for therapy of cellular immune deficiency in 1969; we used about 25 times the dose of TF used for conversion of skin tests (*22*). (In our laboratory a standard dose is now the TF obtained from one billion leukocytes.) We showed that it was possible to transfer specific cellular immunity not only by skin test, but also by the criteria of macrophage migration-inhibitory factor (MIF) production, a much more sophisticated test for cellular immunity than "skin tests"; we used a modification of the MIF developed by Rocklin, *et al.* (*23*).

Our group has treated with TF about 130 patients with various diseases, some of them up to 20 times without complications. Indeed, we have administered more TF than the rest of the world combined (*21*). After concluding about 8 years ago that chronic mucocutaneous candidiasis was a genetically determined cellular immunological deficiency syndrome [on the basis of high incidence of autoantibodies in a patient with this disease and his relatives (*25*)], we treated patients with this syndrome with *Candida*-specific TF with dramatic improvement (*24*). We treated another 30 patients with Wiskott-Aldrich syndrome, a genetically determined defect in cellular immunity, about 60% with dramatic success.

Before initiating therapy of cancer patients with TF, we had to devise a test that would enable us to determine how often each recipient should be treated with tumor-specific transfer factor. This was made possible by a test developed by Wybran and myself (*28*). Human T cells from peripheral blood will bind sheep cells nonspecifically in a formation called a "rosette" (*26*). The total number of rosette-forming cells (RFC) (total T-cell population) in normal individuals is about $65 \pm 5\%$. In another test based on binding of sheep red blood cells, which we term the "active" rosette test, the mean number of RFC for normal subjects, ranging in age from neonate to age 60, is $28.5 \pm 6.5\%$ (*28*). Hence, less than 15% active RFC is subnormal. In patients with cancer, the average number of active RFC in newly diagnosed, local cancer was 12.2%. Newly diagnosed patients with metastases had even lower numbers of active RFC (*28*). A normal individual has the same number of active RFC at all times. However, when a normal individual gets a viral infection, the number drops to subnormal values (*28*). These active RFC are, indeed, T cells, as was proven by studies of various tissues from fetuses of different gestational ages, e.g., thymus, marrow, and spleen. In the thymus of a fetus of 12 weeks gestational age, 85% of the cells in the thymus are active T cells (27). (None are present in the liver, spleen, and bone marrow.) At 19 weeks of gestational age, the RFC in the thymus decline and RFC begin to appear in peripheral blood and other organs (presumably they have left the thymus and disseminated).

In other studies, we have isolated these cells and shown that they are the cells that directly kill human cancer cells, at least cells from tumors of presumably viral etiology, *in vitro* (29). To recapitulate, active T cells, therefore, (*a*) seem to be depressed in cancer, (*b*) kill tumor cells, at least those cells from tumors presumably of viral origin, *in vitro,* and (*c*) protect us from certain virus diseases. These data provide support for the cancer immunologists who believe that cancer is caused by defective immune systems; the cancer virologists, by contrast, believe that cancer is caused by oncogenic viruses. Could these two theories be reconciled? Our data suggest that they can. In an individual with a selective genetic defect for one or another oncogenic virus, a viral illness would further suppress T-cell function so that exposure to the given oncogenic virus will result in cancer (especially to proliferation of oncogenic viruses already present in the host, if the oncogenic theory is correct).

V. COMBINED EFFECT OF IMMUNODEFICIENCY AND ONCOGENIC VIRUSES IN CANCER

In 41 patients with pharyngitis, 100% of the patients with bacterial pharyngitis had normal numbers of active T cells. Of the patients with nonbacterial pharyngitis, 100% had low active RFC; these remained low for up to 3 weeks (28). Thus, we believe that, if the cellular immunity of someone with a selective immunological deficiency for an oncogenic virus (e.g., melanoma) is depressed by other viral infections and the (melanoma) virus is activated, then the subject is far more likely to develop melanoma.

As far as I know, there is no proof of viral etiology of human cancer. The evidence for such etiology in certain animal tumors is compelling, but despite the finding of reverse transcriptase and 70S RNA template similar to RNA tumor virus in human cancer, no direct proof exists that viruses are implicated. In our attempt to add some indirect proof, we did an immunological study to test this possibility. The study was designed to test our hypothesis that tumor viruses induce clinically significant tumors in genetically predisposed humans (*7*), but not in genetically normal individuals. For this purpose, we tested patients with osteogenic sarcoma, their first-degree relatives, and their household contacts for the ability of the lymphocytes of these people to kill osteogenic sarcoma cells. The cells used were free of contaminating fibroblasts by criteria of (*a*) electron microscopy and (*b*) growth of tumor cells only in neonatally immunosuppressed mice inoculated with tumor cells. Control cells were fibroblasts from the same donors, fibrosarcoma cells (a closely related tumor), and hypernephroma cells. We tested contacts and relatives of patients with hypernephroma by the same method. If lymphocytes of a given subject killed hypernephroma cells but not the matching skin, and not cells derived from other tumors, we considered the subjects to have specific immunity to hypernephroma. Killing was measured by chromium-51 release from tumor cells and fibroblasts labeled with ^{51}Cr under standard conditions of time, standard lymphocyte–target cell ratios, etc. To insure absence of fibroblast contamination, we usually used cells from the sixth passage shown to be free of mycoplasma and viruses by electron microscopy. We use only the cell lines where the tumors produced in the injected mice contain 100% tumor cells. (By immunofluorescent criteria, the cells that grow out in the mice, as shown by antisera to osteosarcoma and hypernephroma, are antigenically indistinguishable from the human cells prior to injection into mice.) In the chromium-51 release assay, the amount of chromium released is an indirect measure of the percent of cells killed. Lymphocytes of normals under standard conditions produce less than 7% chromium release; and in people with cellular immunity to osteosarcoma, their lymphocytes produce greater than 35% of the total from the target cells. [Incidentally, as donors for tumor-specific TF, we use only those subjects who have greater than 50% chromium-51 release in our standard assay (*13*).]

Examination of lymphocytes of osteosarcoma patients and their household contacts disclosed T-cell-mediated specific cytotoxicity for osteosarcoma in approximately 20% of first-degree relatives or parents. Other close household contacts (e.g., servants) have significant cytotoxicity. Of the noncontacts, only 1 of 69 had such tumor-specific cytotoxicity, and he was an undertaker. (Although he does not remember having "treated" a patient with osteosarcoma, he had no cytotoxicity against hypernephroma cells, breast tumor cells, etc.)

In terms of breast cancer, we have had a chance to test contacts in only five families, but of the 15 subjects tested, 5 had tumor-specific immunity.* Hence, the incidence of cellular immune-specific cytotoxicity to these tumors is approximately 20%. There is no apparent correlation between cellular immunity in these contacts and relatives and humoral

* Including 2 husbands.

immunity, as measured by indirect immunofluorescence. If we add the people who have had close contact with patients with tumors other than hypernephroma and osteosarcoma (e.g., breast, etc.) to our controls (no known exposure to any kind of tumor), 1 in 200 subjects has developed cellular immunity to one of these tumors. [Normal subjects without known exposure to someone with cancer are very hard to find (*2*).]

Since contacts with tumor-specific immunity are not always blood relatives, these data provide inferential evidence that viruses or other infectious agents infect a given household group at one particular time (bacterial cultures were negative). We hypothesize that the people with genetically intact cellular immunity can deal with the viruses and develop active cellular immunity; those who are predisposed genetically and hence cannot deal with it, will develop that particular kind of tumor (*10*).

REFERENCES

1. Burnet, F. M., "The Clonal Selection Theory of Acquired Immunity." Cambridge Univ. Press, London and New York, 1959.
2. Byers, V. *et al.*, in preparation (1974).
3. Cooper, M. D., Peterson, R. D., and Good, R. A. *J. Exp. Med.* **123,** 75 (1966).
4. Douglas, S. D., Goldberg, L. S., and Fudenberg, H. H., *Amer. J. Med.* **48,** 48 (1970).
5. Dumonde, D. C., Wolstencroft, R. A., Panagi, G. S., Morley, J., and Howson, W. T., *Nature (London)* **224,** 38 (1969).
6. Folch, H., and Waksman, B. H., *Cell. Immunol.* **1,** 12 (1973).
7. Fudenberg, H. H., *Arthritis Rheum.* **9,** 464 (invited review) (1966).
8. Fudenberg, H. H., *Proc. Int. Congr. Hum. Genet., 3rd, 1966* pp. 223–246 (1967).
9. Fudenberg, H. H., *Hosp. Pract.* **3,** 43 (1968).
9a. Fudenberg, H. H., *Immunol. Deficiency Dis. Man* **4,** No. 1, 340–348 (1968).
10. Fudenberg, H. H., *Amer. J. Med.* **51,** 295 (1971).
11. Fudenberg, H. H., German, J. L., III, and Kunkel, H. G., *Arthritis Rheum.* **5,** 565 (1962).
12. Fudenberg, H. H., Good, R. A., Hitzig, W. H., *et al., Bull. W.H.O.* **45,** 927–945 (1971).
13. Fudenberg, H. H., Levin, A. S., Spitler, L. E., Wybran, J., and Byers, V., *Hosp. Pract.* **9,** 95 (requested review) (1974).
14. Fudenberg, H. H., and Solomon, A., *Vox Sang.* **6,** 68 (1961).
15. Guttman, P. H., Davis, W. C., Fudenberg, H. H., and Merigan, T. C., *Vox Sang.* **17,** 278 (1969).
16. Guttman, P. H., Wuepper, K. D., and Fudenberg, H. H., *Vox Sang.* **12,** 329 (1967).
17. Henry, C., Faulk, W. P., Kuhn, L., Yoffey, J. M., and Fudenberg, H. H., *J. Exp. Med.* **131,** 1200–1210 (1970).
18. Kamin, R., Henry, C., and Fudenberg, H. H., in preparation (1974).
19. Kamin, R. M., Fudenberg, H. H., and Douglas, S. D. *Proc. Nat. Acad. Sci. U. S.* **60,** 881 (1968).
20. Lawrence, H. S., *Ann. N. Y. Acad. Sci.* (1974).
21. Levin, A. S., Spitler, L. E., and Fudenberg, H. H., *in* "Primary Immuno-

deficiency Diseases in Man" (Proc. 2nd W.H.O. Int. Workshop-Nat. Found., 1973 (in press).
22. Levin, A. S., Spitler, L. E., Stites, D. P., and Fudenberg, H. H., *Proc. Nat. Acad. Sci. U. S.* **67**, 821 (1970).
23. Rocklin, R. E., Meyers, O. L., and David, J. K., *J. Immunol.* **104**, 95 (1970).
24a. Spitler, L. E., Levin, A. S., and Fudenberg, H. H., 1973. *in* "Primary Immunodeficiency Diseases in Man" (Proc. 2nd W.H.O. Int. Workshop-Nat. Found., 1973) (in press).
24b. Spitler, L. E., *et al., J. Exp. Med.* **131**, 133–148 (1970).
25. Wuepper, K. D., and Fudenberg, H. H., *Clin. Exp. Immunol.* **2**, 71 (1967).
26. Wybran, J., Carr, M. C., and Fudenberg, H. H., *J. Clin. Invest.* **51**, 2537 (1972).
27. Wybran, J., and Fudenberg, H. H., *Trans. Ass. Amer. Physicians* **84**, 239 (1971).
28. Wybran, J., and Fudenberg, H. H., *J. Clin. Invest.* **52**, 1026 (1973).
29. Wybran, J., Hellström, I., Hellström, K. E., and Fudenberg, H. H., *Int. J. Cancer* (in press).

DISCUSSION

Dr. Bang: Since T cells are apparently low in cancer which progresses, does this not confirm the work of others on dinitrochlorobenzene (DNCB) sensitization? In addition, since we have found that T cells are greatly reduced in kwashiorkor, what is state of nutrition of the patients under study?

Dr. Fudenberg: I tried to emphasize that active T cells are far more important than the total number of T cells. We find that in malnourished individuals total T cells are down. A drop of the active, without a drop in the total, is much more significant. Many of our patients with very low values were in excellent nutritional status. In other words, in patients with the primary malignancies removed and no evidence of disease by any radiological technique, there is rarely weight loss. We can detect a drop in active T cells at least 2 months before the appearance of tiny metastases. Metastases require approximately a billion cells to be visible by X-rays. We figure that we are detecting a drop in active T cells in 1 to 5% of that number. It is long before clinical symptoms and these patients are in very good nutritional shape. So were those hypernephroma patients, for that matter.

Dr. Lawrence: I might say further to answer Dr. Bang's question, Brown and Katz have shown that transfer factor (TF) will restore cellular immunity in children with kwashiorkor and marasmus without changing the nutrition. So that if nutrition was a factor it need not be replenished to allow T-cell function to be restored.

Dr. Purchase: Three questions: (*1*) In humans with a congenital lack of a thymus, e.g., ataxia telangiectasia, does TF work? (*2*) Do you test for this? Does this group account for a large proportion of the cases that do not respond to TF therapy? (*3*) Have you tried transplantation of thymus, purified T cells, or whole blood at the time of TF therapy?

Dr. Lawrence: I learn from what I read. It has been shown that dialyzable TF will reconstitute cellular immunity in ataxia telangiectasia patients and improve their clinical state. And to pursue that question further, Hitzig with Fudenberg and Spitler has shown that Swiss-type agammaglobulinemics will be reconstituted for their cellular immunity by TF, an event that would have been predicted as theoretically impossible. But whether a real T-cell deficient individual such as one with DeGeorge syndrome could respond is not known. Some of the evidence coming forth now would suggest that, in addition to engaging mature lymphocytes (perhaps precommitted), TF may cause the differentiation of primitive stem cells that have T-cell function.

DR. FUDENBERG: With regard to using whole blood transfusions or whole T cells, it has been shown that transfusions will help a T-cell deficient patient. So will infusions of purified lymphocytes. However, there are some obstacles. First, Schwartz and Black in Boston have shown that, if you take two nonhistocompatible mice and graft lymphoid cells from one onto the other, and if they will survive a graft-versus-host (GVH) disease, they will eventually come down with malignant tumors in a period corresponding to approximately 20 years of human life. I think that is ethically contraindicated. Second, if you do this, and if the recipient needs a renal transplant some 5 years later (or another kind of organ transplant), he will be sensitized to HLA antigens and reject the transplant. Transfer factors, as Dr. Lawrence has shown, has no HLA antigens. Third, if you give whole blood you always have the risk of hepatitis. This small molecular weight material does not carry that risk. Thus, for a whole variety of reasons, TF is far safer than whole blood or lymphocytes, both on practical and on theoretical grounds.

In terms of people with absence of a thymus, we have treated 8 Swiss-type agammaglobulinemias and 4 have had laboratory responses. These are people who supposedly have severe deficiency in both T and B cells. Four had immunological responses. Of the 4 only 2 were permitted enough clinical time to go on to see if they develop a clinical response. Those 2 did. Two in the other group we tried clinically and failed. So we have 2 out of 8. Even under worse conditions, 2 out of 8 clinical responses using TF for the cellular immune deficiency and giving plasma by a system that we developed some years ago for the humoral response, would imply that severe combined immunodeficiency (SCID) is not a disease but a syndrome with multiple causes. Some people thought this was due to a defect in stem cells which give rise both to T and B cells. Whether there are several varieties of SCID, one of which TF helps, and whether TF works by a mechanism other than the ones that Dr. Lawrence has mentioned, so far is conjectural. In any case, TF does seem to help at least some patients who have no evidence of a thymus. We do test for lack of functioning thymocytes, for example in the active T cells, as I showed you at a normal of 28%. Wiscott-Aldrich (WAS) patients have about 1–3% active T RFC and those responding to TF get a rise in serum T cells to 10–12%; the rise precedes the change in migration-inhibitory factor production and skin tests. We do not try to get the active RFC back to normal levels because you can live a normal life with one-fourth of your kidney mass, one-fourth of your liver mass, and one-fourth of your immunoglobulin level. We try to get the active RFC up to 10%. If they then don't rise, this is the best indication that TF will not work. In about 40% of WAS patients, TF does not work; indeed, by other tests devised in our laboratory, we can actually predict in advance who will not respond; the results suggest that there are two different kinds of WAS disease, both sex linked. In hemophilia, there are two types with the same genetic pattern (Factor 8 and Factor 9 deficiency); further, the two different diseases have the same clinical manifestations and the same severity.

DR. FJELDE: Does the presence of PPLO in cultures from which TF is made make any difference in the production or effect of TF?

DR. FUDENBERG: First of all it wouldn't get through the dialysis tubing. One of the things we do is test a small aliquot of every batch we make that we are going to use therapeutically and test it in a normal human volunteer who is negative for one or more of the specificities present in the donor. But, in addition to the cancer specificity, the preparation also possesses specificity for cocci or another antigen. We don't use it unless a known specificity is transferred to a normal volunteer. An occasional batch doesn't work, and we sometimes wonder, "Is that a batch that's been denatured?."

DR. BRUNELL: (*1*) Are the lymphocytes of recipients of TF endowed with the cytotoxic activity of donors? (*2*) Can you demonstrate specific receptors for tumor cells in lymphocytes similar to those reported in lymphocytes for Burkitt cells by George Klein?

DR. FUDENBERG: The recipients at the time of TF therapy have practically no ability to kill tumor cells, in contrast to their family contacts who have a good ability, suggesting again to us that there is a specific defect in those patients with that antigen. After they are given TF, cytotoxicity rises, but never to the hyperimmune level. It rises to about 40% and we give more TF until it does so.* We use that as an index, too, of when to again administer TF. In answer to, "Do we test for receptors?," we have done some immunofluorescence studies, but not by the Klein method, that show that these lymphocytes do have receptors for specific antigens.

DR. POLLARD: Drs. Lawrence and Fudenberg have discussed immunotherapy of neoplastic and other diseases. In the course of the past 2 years, Dr. Truitt and I have been investigating the prevention and treatment of virus-related spontaneous leukemia in germfree AKR mice. As you know, over 95% of these mice develop lymphatic leukemia at average age of 8.5 months and die within 10 days. At 11 weeks of age, AKR mice were given 1000 R X-rays (whole-body) and 24 hours later they were injected IV with 10^7 bone marrow cells from an allogeneic source (DBA/2). None of the AKR mice developed leukemia over a period of 15 months. In a second stage of investigation, older leukemic mice were irradiated and administered 10^7 DBA/2 bone marrow cells and as yet none relapsed with leukemia within 150 days. Thus, it appears that allogeneic chimerism has prophylactic and therapeutic benefit in virus-related AKR leukemic mice. By a similar procedure, germfree SJL/J mice which develop reticulum cell sarcomas at 9 to 11 months of age, remained disease free beyond age 15 months when chimeric with bone marrow cells from C3H donor mice. At present, conventional AKR mice have been decontaminated by antibiotics, irradiated, and they too survive with allogeneic DBA/2 bone marrow cells. Clinical evidence of GVH disease is not apparent in germfree mice. It appears that the lethal effect of the so-called GVH disease in allogeneic chimeric mice is actually infection-related. It may be that a new control mechanism has been inoculated with the allogeneic bone marrow cells or that a leukemia-prone clone of cells in AKR mice were destroyed by the lethal X-irradiation [*Proc. Soc. Exp. Biol. Med.* **144,** 659 (1973); **145,** 488 (1974); **146,** 153 (1974)].

DR. FUDENBERG: I think that Dr. Pollard's comments are very relevant. For example, there are certain kids with Swiss-type agammaglobulemia (also called severe combined immunodeficiency). We just call them SKIDS (for sick kids) in our laboratory. If these SKIDS are treated with bone marrow transplants, then kept in a germfree

* They often have a high C.I. at the time of diagnosis, but this drops to low levels within 50–100 days, whether or not the primary is removed.

environment for 2 years, they do not come down with the effects of the GVH reaction. They are still surviving but, from what I gather, there is a terrible psychiatric problem, severe financial problem, etc., and I'm not sure this is warranted. With regard to your other comments, I should say that your observations of germfree animals were among the ones that led me to the idea that autoimmune disease (AID) is a T-cell defect, because animals in which GVH reactions are induced, with attendant autoimmune phenomena, die of viral infections; if you give a GVH in germfree animals, they don't die of viral infections.

Dr. Pollard: You see no evidence of autoimmune disease?

Dr. Fudenberg: Not in the germfree animals. T cells handle viruses; that led us to believe that AID is due to a T cell defect; that was part of the reasoning. Moreover, with AKR mice not raised in the germfree environment, we've managed to prevent leukemia by taking out their own thymus and giving them a young thymus from another species that had the same H2 marker. So we think it's a thymic defect.

Dr. Ioachim: Were your patients on chemotherapy or radiotherapy?

Dr. Fudenberg: No. Some had been on radiotherapy before for 2 months without success. I refer to the patients in Rochester. That's why they got six doses. We thought that the radiotherapy or chemotherapy might depress their durability to respond. Otherwise, we just use one dose in patients that are not on chemotherapy, had never been on, or are off for 2 months.

Dr. Ioachim: Are you aware of any studies of changes in populations of T cells after chemotherapy?

Dr. Fudenberg: Yes. We have some data showing that we now can finally predict whether radiotherapy or chemotherapy, if we have adequate base lines, are helping the patient or will harm him. As you know though, radiotherapy, chemotherapy, etc., can extend the life-span of some patients but shorten the life-span of others to a significant extent; perhaps the net balance is nil. I think we do have some studies now where we can predict what will happen to patients, that is, whether they are responding well or not and whether the radiotherapy or chemotherapy are helping or harming the patient.

Dr. Ioachim: When you speak about a decrease in T cells, don't you concomitantly test for B cells? What is the balance made of?

Dr. Fudenberg: The balance is made of B cells and monocytes. We don't count polymorphonuclear cells; we just count mononuclear cells. We classify them as T cells roughly 65%, B cells roughly 25%, and monocytes, roughly 5%.

Dr. Ioachim: Could you just tell us in a few words, what is the active test that you compare to the classic test that we know.

Dr. Fudenberg: Both the active and the passive were developed by Dr. Wybran about 3 to 4 years ago. The active RFC will bind sheep cells in 5 minutes. For total RFC, we use 16 hours overnight because it is simple. The temperature conditions and the medium also vary in the two procedures, but the time element necessary for binding is the important factor. The "active" T cells bind sheep red cells more rapidly than the "nonactive" T cells.

CHAPTER 10

Genetic Transmission of RNA Tumor Viruses

Robin A. Weiss*

> Cancer stands at what might be the meeting-place, but what is in fact the no-man's land, between the three disciplines of heredity, development, and infection.
>
> C. D. Darlington, 1948 (*14*).

I. INTRODUCTION

Whether heredity or infection play a major role in human cancer remains obscure, yet both are deeply implicated in the development of malignant disease in animals. This discrepancy between man and experimental animals may largely be spurious, for much of our knowledge of cancer in animals has arisen from studies using inbred strains, a method which cannot be used with man.

The selection of inbred strains of mice with high incidences of leukemia by MacDowell and by Furth, and of another strain with a high incidence of mammary cancer by Strong, indicated that different genetic factors could strongly affect the development of specific malignant diseases. Heredity was therefore considered to be the overriding factor in murine neoplasia. In chickens, on the other hand, Ellerman and Bang had demonstrated as early as 1908 that leukemia was infectious, and,

* Recipient of the Gustav Stern Award in Virology, 1974.

from 1910 onward, Rous provided mounting evidence that certain chicken sarcomas were also transmissible by cell-free filtrates. At first this work was ignored or disbelieved. When the infectious nature of the chicken sarcoma agents could no longer be refuted, it was rejected as anomalous on the grounds either that the tumors were not truly "malignant" or that the filterable agents were not "real viruses." Then, in 1936, Bittner showed that the "inheritance" of the susceptibility of C3H mice to mammary cancer was largely maternal. The high tumor incidence could be abrogated by foster-nursing newborn mice on mothers of low-incidence strains, and a high rate of mammary cancer could be introduced into certain low-incidence strains by reciprocal nursing. Thus the major etiologic factor proved to be a virus transmitted from mother to offspring via the milk. The inheritance of murine leukemia, however, did not show a maternal influence, and leukemia is expressed in offspring of the high-incidence AK strain following ovum transplantation to low leukemia foster mothers. Yet Gross showed in 1951 that the leukemia of Ak mice was transmissible to low leukemia C3H mice in cell-free filtrates. Gross found that it was necessary to inoculate neonatal animals, and earlier attempts to detect a transmissible agent may have failed because older mice had been used. Not long after the discovery of murine leukemia virus, Cottral, Burmester, and colleagues showed that the agent causing avian leukemia was frequently transmitted through the eggs of normal-appearing hens. The "genetic" leukemia of mice and the "infectious" leukemia of chickens did not appear so different after all.

A detailed history of the discovery of viruses that induce cancer is given in Gross' valuable text "Oncogenic Viruses" (*17*). Here I shall attempt to present a modern perspective of vertical transmission. Gross coined the term *vertical* transmission for the infection of offspring with tumor viruses derived from the parent, in contrast to the *horizontal* transmission from one host to another typical of common contagious diseases. Gross (*16*) has also reminded us that "there exist in nature numerous examples of pathogenic agents transmitted from one generation to another. Pasteur was the first to observe that pébrine, a virus disease of the silkworm, is transmitted in these insects through the eggs." Some *Rickettsiae* are transmitted through the eggs of their vector without causing harm but, as Darlington (*14*) remarked, "the terror of its *infection* for man has closed our eyes to the fact of its *inheritance* for the tick." Several plant viruses are also transmitted vertically and, of course, the temperate bacteriophages are transmitted from one generation to another in lysogenic bacteria. This last example has proved to be a particularly rewarding model for the vertical transmission of tumor viruses, for the most intriguing part of the story is not concerned with congenital infection via milk, uterus, or eggs, but tells of the true inheritance of viral genes as an integral part of the host genome.

II. REPLICATION OF RNA TUMOR VIRUSES

Mammary cancer and leukemia agents belong to the group called RNA tumor viruses. We use the term "RNA tumor virus" in a taxonomic sense for a group of viruses with a similar structure and mode of replication, and it does not mean that all members of the group induce tumors. Ribonucleic acid tumor viruses have now been isolated from numerous mammalian species, including primates, as well as birds and reptiles. Possibly all species of higher vertebrate are natural and frequent hosts to RNA tumor viruses. I shall discuss our understanding of the transmission and control of RNA tumor viruses in chickens, with allusion to mice and other mammals.

First the general characteristics and replication of RNA tumor viruses must be briefly discussed. A detailed account of the biology and molecular biology of these viruses is provided in a recent book on tumor viruses edited by Tooze (*71*). The RNA tumor viruses are classified into three groups—the mammary tumor, leukemia, and sarcoma viruses, according to the type of cancer with which they are most commonly associated. Mammary tumor viruses are morphologically distinct, having particles with eccentric nucleoids called B-type particles. Viruses of the leukemia–sarcoma complex have concentric nucleoids, and are known as C-type viruses. Sarcoma and leukemia viruses of a given host species share common group-specific (gs) antigens, whereas the antigens of the murine mammary tumor viruses do not cross-react with those of murine leukemia viruses. Leukemia or leukosis viruses can be further classified into lymphoid, myeloid, and erythroid leukemia viruses.

Lymphoid leukemia viruses (LLV) and murine mammary tumor viruses (MMTV) occur commonly in nature. The neoplasms associated with LLV and MMTV do not typically appear for many months after infection, if they become manifest at all. In contrast, sarcoma viruses and erythroid and myeloid leukemia viruses can induce malignant tumors within a few days of infection. These strongly transforming viruses occur more frequently in the laboratory than in the field; there must be effective natural selection against such strong carcinogens, and no evidence of vertical transmission has been observed. Strongly transforming viruses carry genetic information that is required for their oncogenic properties but is not required for viral replication (*39*); we may call this information "oncogenes" (*40*). On the other hand, many strongly transforming viruses lack part of the information required for viral replication; such defective, transforming viruses are propagated through complementation by "helper" viruses, usually LLV, and infectious stocks are, therefore, a mixture of both viruses. Sarcoma viruses have been extensively used for experimental studies because they transform fibroblastic cells *in vitro* with high efficiency. A single sarcoma virus particle is sufficient to induce malignant conversion of the host cell within one cell division cycle; indeed, cell transformation is the criterion used in the standard bioassay of these viruses. Lymphoid leukemia viruses replicate in fibroblastic cells without inducing neoplastic transformation.

Figure 1 shows a simplified scheme of the replication of a C-type RNA tumor virus. The virus particle contains a single-stranded RNA genome comprised of 2 to 3 subunits. Associated with this genome are several internal proteins, designated gs antigens. The particle is enclosed in a lipoprotein envelope bearing of the antigens that elicit neutralizing antibodies; these antigens are probably virus-coded glycoproteins and they are required for the virus to penetrate the host cell via receptor sites at the cell surface.

Temin and Bader observed that the replication of RNA tumor viruses shows an early sensitivity to inhibitors of DNA synthesis and a continuous sensitivity to inhibitors of DNA-dependent RNA synthesis. These observations led Temin to postulate in 1964 that RNA tumor viruses replicate via a DNA intermediate, the *provirus*. The provirus hypothesis remained controversial for many years until, in 1970, Temin and Mizutani and Baltimore discovered that the virus particle contains an enzyme, RNA-dependent DNA polymerase or "reverse transcriptase," which is capable of using viral RNA as a template for DNA synthesis (*69*). Subsequent studies with reverse transcriptase mutants indicate that the

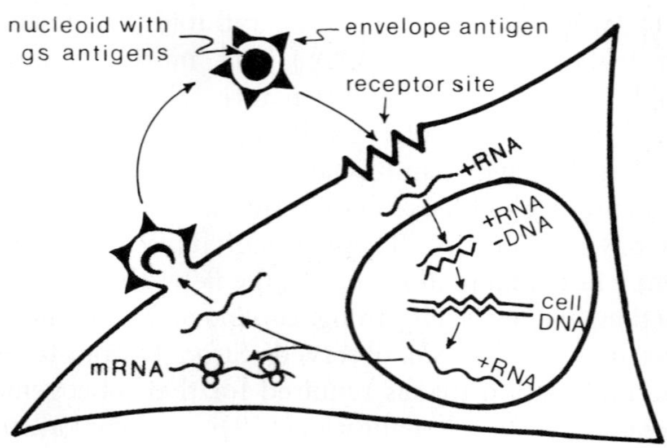

Fig. 1. Replication of a C-type virus.

enzyme is required only for an early step in viral replication. Upon infection it appears that first a DNA–RNA hybrid forms which is then converted into a double-stranded DNA provirus. The provirus integrates with the host chromosomes. To form genomes of progeny virus, RNA is transcribed from the DNA provirus. Since RNA complementary to viral RNA is not found in infected cells, it is assumed that the viral messenger RNA represents the same strand as viral genome RNA, and they may be identical. To complete viral replication, viral proteins and RNA assemble near the cell surface and virus particles are formed by budding from the cell membrane, modified by the insertion of the viral envelope antigens. The process of budding and release of virus particles does not kill the host cell, so that it is possible for the cell to become converted into a malignant cell and at the same time produce viral progeny.

III. CONGENITAL INFECTION AND GENETIC TRANSMISSION

We now recognize two modes of vertical transmission of RNA tumor viruses which are quite distinct at the molecular level and which are consequently subject to different biological controls.

Congenital infection occurs when infectious virus particles released from the mother infect the offspring. The virus may infect the ovum or zygote directly or it may be transmitted via the placenta or milk. However, in each case the virus is transmitted as an infectious particle which must enter a new host cell in which a new DNA provirus is presumably transcribed and integrated into the cell genome. Any interference with the processes of adsorption, penetration, reverse transcription, or integration will restrict congenital infection just as it will restrict horizontal infection.

Genetic transmission, on the other hand, occurs when the viral genome is passed from one generation to the next as a cellular genetic element in DNA form. Bentvelzen has called this viral genome a *germinal provirus* to distinguish it from *somatic proviruses* acquired by infection. Whereas congenital infection usually occurs through the mother, genetic transmission may be either maternal or paternal. Although the provirus could be inherited as a plasmid, the evidence suggests that it is integrated into a chromosome and does not undergo independent replication. The germinal proviruses may remain "silent" for many generations, they may be partially expressed by the synthesis of intracellular viral gene products, or they may give rise to virus particles that can subsequently propagate by infection. Since the viral genome is not transmitted by infection, genetic transmission bypasses any host-range restriction or other interference with infection imposed on the preintegration phases of the replication cycle. It is not yet clear whether all genetically transmitted viruses are actually oncogenic. However, the genetic transmission of LLV and MMTV in mice is now well established, although the lymphoid and mammary neoplasms typically appear only after the viral genomes become activated to replicate as viruses.

Figure 2 depicts these two modes of transmission of RNA tumor viruses in addition to horizontal transmission. In both feral and domestic flocks of fowl, all three modes of transmission commonly occur. Infectiously transmitted LLV belong to serological subgroups A and B, whereas the genetically transmitted virus codes for a serologically distinct envelope protein, E. The host response to infection is characteristically different for each mode of transmission. Infection of the host after hatching typically leads to a strong antibody response to the viral envelope antigens which effectively neutralizes the virus. Congenital infection,

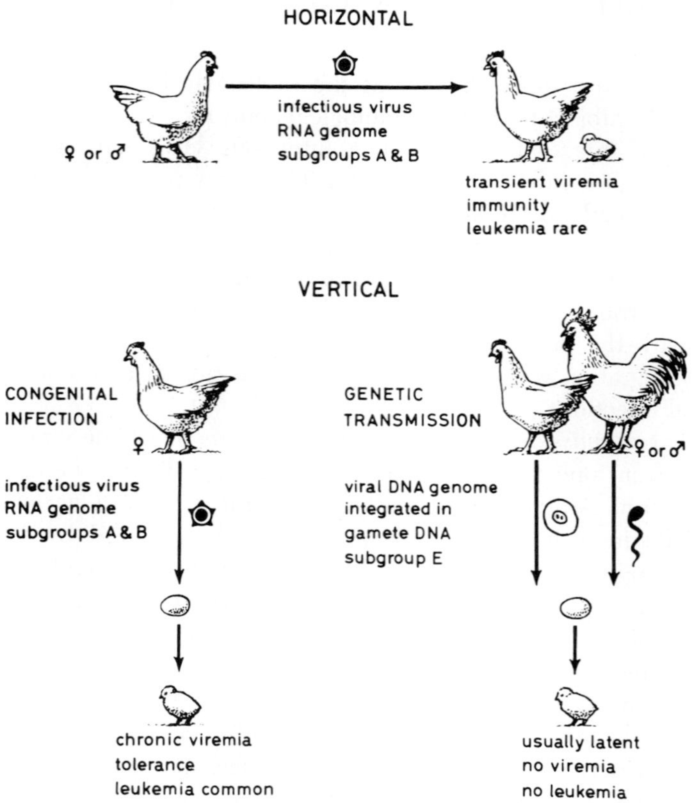

Fig. 2. Transmission of avian leukemia viruses.

on the other hand, usually leads to apparent immunological tolerance (*55*). The host embryo grows normally despite the production of high titers of leukemia virus throughout development, but frequently the birds develop lymphoid leukosis in adult life. In most chicken strains the germinal provirus is partially expressed by the synthesis of gs and envelope antigens without producing virus particles. The level of gs antigen appears to be too low to maintain tolerance (*56*), but there may be genuine tolerance to envelope antigen.

In laboratory mice (feral mice have received little attention so far) the major route of transmission of LLV is genetic. Mice do not usually succumb to the pathological effects of experimental infection if exposed more than a week after birth, probably as a result of immunological responses. With MMTV, the classic milk-borne infection of C3H mice now appears to be an exception to the genetic transmission more commonly observed in other strains.

Recent studies of feline leukemia viruses reveal two interesting differences from the avian and murine viruses. First, horizontal transmission frequently gives rise to malignant and nonmalignant disorders of the hemopoietic system even when adults become infected (*28*), although immunity is the usual response. Second, the genetically transmitted virus of domestic cats is not closely related to the infectiously transmitted feline leukemia viruses (*34*). This will be discussed further in the section on xenotropic viruses.

It must be added here that the different modes of transmission are not as distinct as depicted in Fig. 2. Epidemiologically, there is no separation between horizontal and congenital transmission. Indeed, congenitally infected chickens, being tolerant and shedding high titers of virus, represent the chief source of horizontal infection; they also remain as a reservoir of congenital infection to future generations. Neither is genetic transmission entirely distinct from infectious transmission: once activated to infectious form, the endogenous virus will be maintained by infection provided the host is susceptible to reinfection, and the virus will spread horizontally from cell to cell within the host, and possibly to other hosts. Furthermore, our studies (*90*) on genetic recombination between endogenous and exogenous viruses indicate that genetic markers of the endogenous genome can be acquired by and transmitted with infectiously propagated viruses.

IV. EVIDENCE FOR GENETIC TRANSMISSION

How was the presence of inherited proviruses revealed? In 1959, Lieberman and Kaplan (*32*) reported that thymic lymphomas induced by

X-rays in C57BL mice (a strain with a low spontaneous incidence of leukemia) produced a leukemia virus which induced similar tumors when inoculated into unirradiated hosts. This observation was subsequently confirmed in several laboratories using ionizing radiations or chemical carcinogens. Apparently mice of low cancer strains as well as high-incidence strains commonly harbored leukemia viruses, but in a latent state.

There have been two main views on the activation of latent viruses by carcinogenic agents. First, it was well known that the immunological mechanisms of the host are particularly sensitive to carcinogens, and, under immunosuppressed conditions, any latent virus infections would have an increased probability of becoming virulent. Second, there was the idea dating from the 1930s that tumor viruses rather than being latent infections might be of endogenous origin. This problem was critically reviewed in 1939 by Sir Christopher Andrewes, whom we honor at this symposium, in his lucid article on "Latent virus infections and their possible relevance to the cancer problem" (*3*). It also prompted Boycott to comment, as Rowe (*51*) recently reminded us, that "when it becomes necessary to postulate a normal virus occurring in normal cells one had better call it something other than a virus." It was left to Darlington, who disliked the "high professional status but dubious credentials of the word *virus*," to call it something else, and he called it a "provirus" (*14*). Temin tells me that he was not aware of Darlington's use of the term provirus when he adopted it. In fact Darlington's proviruses are more like Temin's protoviruses (*67*) for he states that "they are not specialized for natural infection by the selection and adaptation of their ancestors," having never been "guilty of infection."

Drawing on his experience of cytoplasmic inheritance in plants, Darlington viewed his cancer proviruses as plasmagenes. True infectious viruses, he stated, could evolve from plasmagenes which are already independent of the chromosomes, by mutation and selection. However, the discovery of lysogeny in bacteria led Lwoff (*37*) and Latarjet (*29*) to suggest that leukemia proviruses, such as those activated by X-rays in Lieberman and Kaplan's experiments, may be carried in an integrated state. With the demonstration that the genome of leukemia viruses was composed of RNA, the lysogenic model was quietly abandoned, leaving Temin and Bader as lone apostles of the provirus hypothesis. Their mission was to convince virologists that a DNA copy was made from viral RNA during replication of exogenous viruses, thus reversing the accepted direction of biological information flow. The idea that a provirus might find itself in the germ line and remain there was not discussed.

It is interesting to note in retrospect how blind we were to indications of genetic transmission until Payne and Bentvelzen drew attention once more to the lysogeny model in 1968. The first three lines of evidence listed in Table 1 were published 2 or 3 years before it was realized that the existence of endogenous viral genetic information was implicated. Nucleic acid hybridization studies were conducted by Harel and Harel (*22*) and Wilson and Bauer (*93*) in order to test the provirus hypothesis which predicted that DNA homologous to viral RNA should appear in cells infected with RNA tumor viruses. Homologous DNA was duly found, but it was also found in the uninfected control cells. The hybridization probe did not seem to have the required specificity. Similarly, Dougherty and Di Stefano (*15*) observed gs antigen in normal chick embryos and concluded that the complement fixation test used was not entirely specific to viral antigens. When Vogt (*75*) and I (*83*) independently observed that cells transformed by the Bryan high-titer strain of Rous sarcoma virus (BH-RSV), hitherto believed to be defective for replication in the absence of a helper virus, actually released infectious virus, we assumed that BH-RSV was not defective after all. Later, I showed that release of infectious progeny of BH-RSV only occurred in gs-positive cells; the defective RSV particles were made competent by phenotypic mixing with viral gene products present in the host cell, analogous to rescue with a helper virus. The complementing factor was not the gs antigen itself, but the envelope glycoprotein (Fig. 3). Hanafusa *et al.* (*21*) also observed the helper effect of normal cells and named it chick-cell-associated helper factory (chf).

The concept of endogenous tumor virus genes did not gain general acceptance, however, until the discovery of reverse transcriptase suddenly transformed the heresy of the provirus hypothesis into the new orthodoxy. Compelling evidence for inherited proviruses emerged the following year from experiments in which normal, uninfected embryonic cells in culture were induced to release virus by treatment with carcinogens and mutagens (*53, 89*). The most effective inducing agents are iodo- and bromodeoxyuridine (*36*). It was the variety of other lines of evidence for genetic transmission, however (summarized for chickens

TABLE 1. EVIDENCE OF VIRAL GENOMES IN NORMAL CHICKEN CELLS

1. DNA homologous to viral RNA
2. Viral antigens inherited as dominant Mendelian trait
3. Complementation of defective Rous sarcoma virus by chick helper factor
4. Rescue of endogenous viral genes by recombination with exogenous virus
5. Spontaneous release of virus
6. Activation of virus by inducing agents

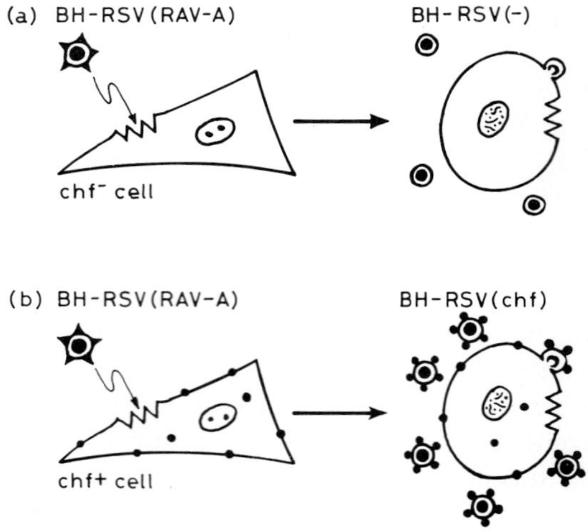

Fig. 3. Complementation of defective Bryan strain Rous sarcoma virus (RSV) by chick helper factor (chf). (a) When a chf⁻ cell is infected with a single particle of BH-RSV pseudotype, the cell transforms and releases noninfectious particles. (b) When a chf⁺ cell is similarly infected, it releases infectious BH-RSV(chf) pseudotypes.

in Table 1) that had led us to attempt to induce viruses. The key observations came from classic Mendelian experiments. In 1968, Payne and Chubb (*45*) published their observations that viral gs antigen in normal chicken embryos was determined as a single dominant trait. These authors suggested that gs-positive embryos might be "lysogenic" for a tumor virus genome or partial genome. In the same year, Bentvelzen (*6*) reported that the development of mammary cancer in GR strain mice, unlike Bittner's C3H mice, was determined by a host Mendelian locus, despite the fact that virus particles were always associated with the tumors. He suggested that all mice carried germinal proviruses for mammary tumor viruses and proposed a scheme for the genetic control of the provirus that was virtually indistinguishable from that for the lambda prophage. The GR mouse strain, according to Bentvelzen, represented a constitutive producer of virus and eventually mammary tumors, whereas lower cancer strains were subject to repression. He and his colleagues proposed (*6*) that a similar model was appropriate for murine leukemia, and this has been amply confirmed by many investigators. It explained the biparental inheritance of virus in high leukemia mouse strains such as AKR, and why certain murine cell lines, after many weeks or months in culture, may spontaneously begin to produce leukemia virus particles (*2*).

In 1969, Heubner and Todaro (*26*), drawing many of these data together, developed a theory similar to Bentvelzen's, and were bold enough to apply it to all cancer. They argued around the fact that most tumors did not release virus particles because the viral gene that acts in malignant transformation (the *oncogene*) may be active when some of the genes for viral replication (the *virogenes*) are repressed. Two experimental systems served as models for transformation by latent viruses: (*a*) transformation of cells that are not permissive for viral replication (e.g., avian sarcoma virus in rat cells, SV40 in mouse cells) in which the presence of the virus is often betrayed by viral antigens (gs antigen for RNA viruses, T antigen for DNA viruses) and from which that virus can be rescued by fusion with permissive cells; (*b*) the nonproducer state of cells transformed by defective sarcoma viruses, where the virus can be rescued by superinfection with helper viruses. Could not spontaneous cancer, Huebner and Todaro reasoned, be explained by changes in the expression of similar genes present in inherited proviruses, the gs antigen being a frequent marker for such gene activity? The "viral oncogene theory," although not particularly original in conception, has proven to be effective propaganda. It has been most useful, perhaps, because the ensuing controversy encouraged lively debate, experiment, and further theories (*40, 67*). Nonvirologists disliked it because it resurrected Lwoff's postulate that carcinogens acted through the derepression of endogenous viral genes. Virologists outside the immediate field initially viewed it with some disparagement too: first, because Boycott's sentiments about normal viruses in normal cells still prevailed; second, because the evidence was and remains wanting that genetically transmitted viruses carry oncogenes distinct from virogenes, analogous to those of exogenous sarcoma viruses (*39*).

V. VERTICAL TRANSMISSION AND HOST ANTIVIRAL MECHANISMS

Let us now consider the different modes of transmission of RNA tumor virus genomes in relation to the main topic of this symposium—antiviral mechanisms. It is immediately apparent that there are more opportunities for antiviral mechanisms to act on genomes transmitted by infection, be it horizontal or congenital, than on genomes transmitted as integrated proviruses.

A. Interferon

Little is known about interferon in relation to RNA tumor viruses. It is not readily induced by these viruses, but many inhibit their replication to some degree (*4*). Whether it acts before or after integration remains to be elucidated. It does not appear to be an important factor in the natural history of RNA tumor viruses.

B. Immunological Responses

The immunological response to infection with avian viruses has already been alluded to (Fig. 2). Congenitally infected chickens appear to be tolerant to viral antigens (*54*), presumably because the virus is present in substantial titers from early in development onward. Expression of the genetic provirus is less effective in inducing tolerance to gs antigens, but antibodies specific to the endogenous viral glycoprotein have, with one exception, been found neither to occur naturally nor to be inducible in chickens (*85*). In high leukemia strain mice, circulating antibody is also difficult to detect. At first this was thought to be due to the induction of tolerance, but investigations of AKR and NZB strain mice indicate that immune complexes of antibody and excess viral antigen remove the free antibody from the circulation (*41*). These positive immunological reactions may be manifest because the endogenous virus does not appear in measurable titers until the perinatal period of development (*53a*) which may be too late to induce tolerance.

The interaction between the immune system and endogenous leukemia viruses is more complex than a failure to protect the host adequately from a virus that is part of "self." Hirsch *et al.* (*25*) have shown in mice that graft-versus-host reactions or immune suppression activate the release of endogenous viruses, and mixed splenocyte cultures similarly activate endogenous viruses. The virus activated by immune reactions is selectively xenotropic (*59*) (see as follows) and the same kind of virus is present in NZB mice exhibiting a spontaneous autoimmune disease. A causal relationship among endogenous viruses, autoimmune disease, and lymphoreticular neoplasms is by no means clear (*31*), and in my opinion there is little to choose between the various permutations.

Immune responses directed against virion antigens should not be confused with responses of nonvirion antigens present on the surfaces of virally transformed cells. However, the presence of virion antigens in tumor cells may enhance their immunogenicity. If malignant transformation relaxes the repression of "virogenes" (*53, 70*) so that viral antigens are newly expressed, one can argue a selective advantage to hosts carrying inherited proviruses that become activated in cells progressing to malignant or other aberrant states.

C. Genetic Control of Host Susceptibility

The susceptibility of the hosts to RNA tumor viruses is controlled by a variety of genetic factors. Regarding infection, the major susceptibility genes in fowl affect the ability of the virus to penetrate into the host cell, whereas in mice the major susceptibility locus, *Fv-1*, affects some subsequent intracellular event in the replication cycle of the virus. There are also genetic factors that affect the transformation of a cell (e.g., the *Fv-2* locus in mice) or the ability of transformed cells to form tumors (e.g., the *H-2* region in mice), but I shall restrict discussion to viral replication.

The RNA tumor viruses of chickens are classified into five subgroups, A, B, C, D, E, according to the properties of the envelope. Each subgroup has distinct envelope antigens and each reflects a different host range. In chickens, four genetic loci, the "tumor virus" or *tv* loci (*tv-a, tv-b, tv-c, tv-e*) have been described (*46*), each affecting susceptibility to a specific virus subgroup. In each case the allele for susceptibility is dominant, and probably codes for a "receptor." Since susceptibility or resistance is expressed at the cellular level, it can be analyzed in fibroblast cultures, using sarcoma viruses (Fig. 4). A subgroup A virus can enter a cell through an A-type receptor only and a subgroup B virus through a B-type receptor only; if the A-type receptor is lacking, a subgroup A virus cannot initiate infection. However, virus adsorbs with equal avidity to genetically resistant and susceptible cells and, therefore, it is most likely that the receptors are required for penetration; resistance can be overcome either by phenotypic mixing with a virus of another subgroup or by using inactivated Sendai virus to promote entry by membrane fusion.

The leukemia viruses of mice fall into four major host range subgroups: B-, N-, NB-, and X-tropic. As with chickens, susceptibility and resistance can be assayed in fibroblast cultures. There are two major types of host cell, N-type and B-type, which are susceptible to infection with and replication of N-tropic and B-tropic viruses, respectively. The NB-tropic viruses replicate in either cell type. Xenotropic or X-tropic viruses replicate in cells of other species but not N- or B-type mouse cells. Crosses between N-type and B-type strains of mice (named after the NIH-Swiss and BABL/c strains) show that a single Mendelian locus controls susceptibility to both subgroups (*33*). This locus has been identified with the *Fv-1* gene, previously described as governing susceptibility to Friend disease *in vivo*. The alleles for susceptibility to N- and B-tropic viruses have been named *Fv-1*^{n} and *Fv-1*^{b}, respectively. Heterozygotes are resistant to infection with either N- or B-tropic viruses and resistance is, therefore, dominant. However, the penetrance of resistance is not as strong as in avian cells, where it is virtually absolute. The kinetics of infection of resistant cells suggests that two "hits" of independent particles on one target cell are required to obtain successful infection,

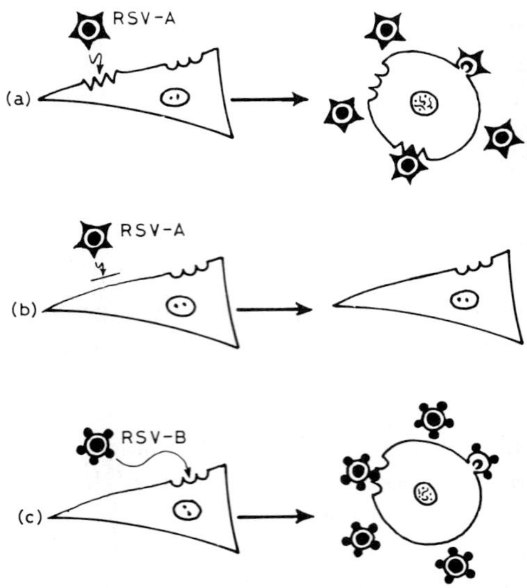

Fig. 4. Susceptibility and resistance to avian RNA tumor viruses. (a) Rous sarcoma virus of subgroup A (RSV-A) enters a susceptible cell through a specific cell surface receptor. As a consequence of infection the virus replicates and the cell or daughter cells become transformed. Progeny virus occupies the specific receptors sites. (b) Virus RSV-A cannot enter a resistant cell because there is no compatible receptor. (c) Virus RSV-B can enter a cell resistant to RSV-A provided there is a receptor specific to RSV-B.

compared to one-hit kinetics in susceptible cells. The mechanism of host cell resistance is not yet understood, but it takes place after virus penetration, possibly after integration.

How do these host range controls affect vertical transmission? The transmission of inherited proviruses will be unaffected by resistant hosts. For instance, B-type mouse strains carry N-type endogenous viral genomes, and chickens resistant to infection with subgroup E viruses carry their genomes genetically. However, as I shall illustrate later, the horizontal spread of virus from cell to cell in the event of endogenous virus activation will be restricted in resistant hosts. Even in strains that carry a relatively high spontaneous activation rate the host range restriction is sufficient to prevent viremia and, if the virus is leukemogenic, reduce the incidence of leukemia.

With congenital transmission the host range restrictions will prevent infection in the same manner as they prevent horizontal transmission. The incidence of leukemia induced by field strains of subgroup A viruses is reduced in flocks of domestic fowl with prevalent *tv-a*^{r} resistance alleles (*44*). It is interesting here to compare the expectation of susceptibility to RNA tumor viruses in outbred populations of mice and chickens. Susceptibility to the common field strains of avian leukemia virus, subgroups A and B, are controlled by independently segregating loci with dominant alleles for susceptibility; heterozygosis at the *tv-a* and *tv-b* loci will predispose the birds to infection with both types of virus. However, resistance alleles are not found to be prevalent among outbred chicken flocks or feral jungle fowl (*88*). By contrast, mice that are heterozygous at the *Fv-1* locus will be resistant to both N- and B-tropic viruses. Heterozygotes should, therefore, have a selective advantage over either homozygote, assuming that susceptibility to infection is disadvantageous. It is not yet known whether balanced polymorphism at the *Fv-1* locus is common in feral mouse populations.

Since NB-tropic viruses will replicate in heterozygous cells and in cells that are homozygous for either allele, these viruses would have a natural selective advantage in polymorphic host populations, yet naturally occurring NB-tropic viruses have not been described. The NB-tropic viruses can arise when N- or B-tropic viruses are deliberately passaged at high titer through resistant cells or animals. Whatever mutational or recombinational event takes place, it does not commonly occur outside the laboratory.

D. Resistance to Superinfection

When avian tumor viruses replicate the host cell becomes resistant to superinfection by viruses of the same subgroup (*81*). This was first observed by Rubin (*54*) who found that certain chick embryos cultures were resistant to infection by Rous sarcoma virus. He showed that the resistance could be transmitted to susceptible cultures by infection. The resistance-inducing factor proved to be a congenitally transmitted leukemia virus. This phenomenon is known as *viral interference*. Its mechanism appears to be the blocking of the cell surface receptor sites by progeny virus (*61*), and there is no interference to viruses that can enter through alternative receptors (Fig. 4). Since the discovery of interference, avian leukemia viruses have been assayed in culture by their capacity to prevent infection and, hence, transformation of cells by sarcoma viruses. Viral interference is also found among murine viruses and feline viruses.

Interference of sarcoma virus infection by leukemia viruses may be viewed as a model antiviral mechanism for excluding a pathogenic virus strain with a relatively nonpathogenic strain. Viral interference may act *in vivo:* Rowe (*49*) showed that preinfection of mice with lymphoid leukemia virus caused resistance to erythroid leukemia upon subsequent challenge with Friend virus. It is not clear whether this type of protection is a result of immunity or viral intereference. It is possible that natural congenital infections with LLV or activation of endogenous LLV will protect the host from more pathogenic viruses such as Friend virus and sarcoma viruses. Such strongly transforming viruses, however, rarely occur in nature. Their rarity might be due in part to viral interference. When nondefective sarcoma viruses are propagated at high multiplicities, they generate nontransforming deletion mutants (*76*) that eventually form the major proportion of the virus population. The nontransforming viruses act as interfering agents and may limit the spread of transforming virus. This phenomenon resembles, at the population level, that of defective interfering particles in other viral systems described by Huang at this symposium. However, the interference acts quite differently at the molecular level and the interfering particles are themselves capable of replication. Helper-dependent transforming viruses, such as murine sarcoma viruses and Friend virus, are similarly subject to interference by their own helper viruses. Whereas the transforming virus depends on the helper virus to produce infectious progeny, further infection by the progeny will also be inhibited by excess helper virus. Such complex populations of helper viruses and transforming viruses do not usually maintain a stable polymorphism under natural conditions.

I have described how partial expression of the inherited provirus in chicken cells provides a helper function (chf) for defective Rous sarcoma virus (RSV). In Section VI,D, it is argued that chf, like exogenous helper viruses, can also function as an interfering agent.

VI. GENETIC CONTROL OF ENDOGENOUS VIRUS EXPRESSION

During the last 4 years we have witnessed a great advance in our knowledge of the interactions of the RNA tumor viruses and host genome, particularly with respect to endogenous viruses. This has been achieved in part as a result of the conceptual clarification of genetic transmission that I described in Section IV, and also by two technical advances: First, the use of reverse transcriptase to synthesize DNA *in vitro* complementary to viral RNA for use in molecular hybridization studies (*69*), whereby complementary DNA provided a useful probe with which to search in host cells for DNA and RNA sequences homologous to the viral genome. Second, the ability to induce virus production in normal cell cultures (*53, 89*), coupled with more sensitive techniques for virus assay, has helped us to gain a better understanding of endogenous viruses and has led to the discovery of more viruses. Most important, however, to the study of endogenous viral gene expression has been the application of *classic Mendelian genetics,* which had already proved to be so fruitful in elucidating the control of host susceptibility and resistance.

A. Genes Affecting Partial Expression

Using inbred strains of chicken, Payne and Chubb (*45*) showed that the presence of gs antigen was determined by a dominant autosomal gene. Later, we found that chf complementing the defective envelope of Bryan strain RSV was also determined in these strains at the same locus (*91*). At first we were inclined to believe that the locus represented the structural genes of the provirus and that those chickens which were negative for gs antigen and chf had not inherited the provirus. However, nucleic acid hybridization studies (*48, 74*) indicated that gs^- chf^- embryos contained the same amount of viral sequences as gs^+ chf^+ embryos. Furthermore, we were able to induce infectious virus production in gs^- chf^- cells (*89*). Thus it appeared that both kinds of chicken inherited the provirus and that the *gs* locus controlled its expression. At what level is the control imposed? Studies on the hybridization of RNA extracted from gs^+ and gs^- cells with DNA complementary to viral RNA (*8, 24*) indicated that viral RNA is not synthesized in gs^- cells. Thus the *gs* locus appears to act as a regulatory gene affecting the transcription of at least two structural genes for viral proteins.

It is interesting to note that the dominant allele, gs^+, allows viral gene expression; therefore the *gs* locus does not act like a repressor gene. Exogenous viruses, moreover, replicate with equal efficiency in gs^+ and gs^- cells. These observations suggest that the *gs* locus may exert a cis effect only and may be an integral part of the provirus. Since we have no independent marker for the provirus, this cannot be tested.

What is the nature of chf? Functionally, it complements RSV that is defective for envelope glycoprotein (*57*) by providing envelope components of the same antigenic specificity and host range as the fully activated virus, RAV-0. Therefore the simplest interpretation is that chf is envelope antigen coded by the endogenous viral genome; RSV is complemented through its ability to assemble chf as a part of its own envelope in the same manner as RSV assembles envelope antigens of exogenous helper viruses. The phenotypically mixed particles are called pseudotypes, and are designated RSV(chf). Antiserum specific to RAV-0 neutralizes the infectivity of RSV(chf). Recently, more direct evidence has been obtained that envelope antigen is present in chf^+ cells (*18, 87*). Antibodies that neutralize RAV-0 are specifically adsorbed by extracts of chf^+ cells but not chf^- cells; adsorption to live cells and immunoelectron microscopy show that envelope antigen is exposed on the surface of chf+ cells. These results indicate that endogenous envelope antigen, like gs antigen, is present in chf+ cells before infection with RSV; its synthesis does not depend on activation by exogenous infection. Independent proof that envelope antigens are present and available for assembly in chf+ cells comes from experiments with vesicular stomatitis virus (VSV), a rhabdovirus unrelated to RNA tumor viruses. Zavada (*94*) has shown that VSV forms pseudotypes with avian and murine leukemia viruses by phenotypic mixing of envelope antigens. Using a mutant of VSV with thermolabile glycoprotein, Love and I (*35*) showed that the thermolabile component was significantly complemented in chf+ cells; the host range of the complemented VSV(chf) particles was identical to that of RSV(chf) pseudotypes, and they were neutralized by antiserum specific to subgroup E virus. The ability of VSV to form pseudotypes with endogenous viral glycoproteins should prove useful in the search for helper factors in mammalian cells.

Table 2 summarizes the degree of expression of endogenous viral genes in gs−, gs+, and virus-producing (V+) cells.

We do not know why gs+ chf+ cells do not relase complete virus. One viral gene product that is lacking in gs+ chf+ cells is the RNA-dependent DNA polymerase. The enzyme activity cannot be detected directly (*92*), and gs+ chf+ cells do not complement mutants of RSV that are defective (*19*) or conditionally defective (*86*) for the enzyme, whereas we know that exogenous helper viruses do. The absence of the enzyme per se, however, is insufficient to explain the lack of virus particles, because Hanafusa and Hanafusa (*19*) have observed that a mutant of RSV that cannot synthesize reverse transcriptase, nevertheless, releases virus particles, although they are noninfectious. Two explanations may be put forward for the lack of complete endogenous virus synthesis in gs+ chf+ cells. The first is that the endogenous provirus is defective, that is, partially deleted, and cannot code for all the necessary functions.

TABLE 2. EXPRESSION OF VIRAL GENES IN NORMAL CHICKEN EMBRYO CELLS

Designation	Viral components					
	DNA provirus	RNA	gs Antigen	Envelope antigen	Poly-merase	Infectious virus particles
gs−, chf−	+	−	−	−	−	−
gs+, chf+	+	+	+	+	−	−
V+	+	+	+	+	+	+

The second is that the genome is complete but is incompletely expressed, just as in gs⁻ chf⁻ cells none of the markers examined are normally expressed. I favor the second interpretation because one can induce complete virus from certain chicken strains with mutagens. However, it does not prove that all strains carry a complete provirus or that the induced provirus is the one controlled by the *gs* locus.

Preliminary evidence for a further locus affecting endogenous virus expression has recently been presented by Hanafusa *et al.* (*20*). Some chickens express exceptionally high levels of chf following infection with RSV, whereas before infection they express only low levels of gs antigen and moderate amounts of envelope antigen. Limited breeding data indicate that this effect may be inherited as a dominant trait by a locus or loci unlinked to the *gs* locus. The recent application of radioimmune techniques to the assay of avian endogenous viral proteins (*11, 72*) promises to yield more quantitative analyses of viral gene expression than have hitherto been possible.

In mice, there is also evidence of genetic loci that affect viral gene expression, although the genetic control and molecular biology are not as well understood as that of the *gs* locus in chickens. Taylor, Meier, and Huebner (*65*) have described a locus that differs from the avian *gs* locus in that it is recessive for expression of gs antigen; the level of its control is unknown. A function analogous to chf has not been observed in murine cells because the defective virus particles released by certain strains of murine sarcoma virus lack RNA-dependent DNA polymerase and, therefore, resemble those mutants of RSV that are not complemented by chf. As chf represents viral envelope antigen, immunological methods should reveal whether envelope antigens are present in murine cells that do not release virus particles. Stockert, Old, and Boyse (*63*) found that a cell surface antigen (G_{IX}) of virus-induced leukemia cells in mice and rats was expressed in normal cells of strain 129 mice in the absence of complete virus. Crosses of 129 mice with other strains showed that the G_{IX} antigen is controlled by two unlinked loci, both of which are required for expression. It is likely that G_{IX} represents a virus-coded antigen. Stockert *et al.* suggest that one of the G_{IX} loci of 129 mice might represent a structural gene of a provirus.

B. Genes Promoting Virus Activation

Vogt and Friis (*80*) observed that some chicken embryos release complete virus particles that bear the same envelope specificity (subgroup E) as chf. This virus is called RAV-0; its spontaneous release represents a third class of virus expression (Table 2). Although RAV-0 contains envelope antigens and RNA-dependent DNA polymerase as abundantly as any other avian tumor virus, RAV-0 is relatively inefficient for replication, at least when assayed on the standard Japanese quail cells. Nevertheless, it may be regarded as a complete infectious virus, since it can be propagated in gs⁻ chf⁻ cells that are susceptible to subgroup E viruses. Using a strain of chicken, line 100, which is related to that of Vogt and Friis, Crittenden *et al.* (*12*) found that all line 100 chickens that are susceptible to infection with subgroup E viruses become viremic with RAV-0. Crosses between line 100 and other strains susceptible to subgroup E virus indicate that the line 100 genotype contains a component that behaves like a dominant trait for complete virus expression; we may designate this genotype, V^+. We do not yet know whether V^+ represents one or more loci and whether it is linked or allelic with the *gs* locus. It appears to acts as a regulatory locus. The same questions arise for the *V* locus as for the *gs* locus; whether it is a "host" or "proviral" gene, and whether V^- birds possess the complete provirus that becomes activated in V^+ birds.

In mice, several genes that affect spontaneous or induced leukemia virus production (*V* loci) have also been identified. In an elegant genetic study, Rowe (*50*) showed that AKR mice have two unlinked V^+ genes; either locus exerts a dominant tendency to virus activation in crosses with low leukemia strains. Stephenson and Aaronson (*62*) have described similarly acting *V* loci in C58 and BALB/c mouse strains, and Rowe finds that three *V* loci, two of them nonallelic with the *V* loci of AKR mice, are present in the C3H/Figge strain. The B-tropic and X-tropic viruses are also known to be endogenous to some mouse strains. In crosses between AKR mice and strains known to carry B-tropic proviruses, Rowe and Hartley (*52*) observed that the *V* locus activated only N-tropic virus, suggesting that the locus has a cis effect. Even when the host inherits V^+ alleles the spontaneous activation of infectious virus in mouse or chicken cells appears to be a rare stochastic event. Teich *et al.* (*66*) showed that, whereas the spontaneous activation rate in AKR embryo fibroblasts was about 1 cell in 10^8 per culture passage, after treatment with iododeoxyuridine (IUdR) about 5% of the cells produce virus and over 50% synthesize viral antigens. To produce this effect, IUdR must be incorporated into DNA. It is not known whether the provirus is excised during induction; transcription and translation of viral RNA should be sufficient for the production of infectious virus.

Because of the apparent cis effect, the effect of incorporated IUdR, and the different sites of V^+ genes in different mouse strains, Rowe (*51*) argues persuasively that *V* loci probably represent the integrated proviruses themselves. The *V* alleles resemble the dominant allele of the locus governing activation of mammary tumor virus in GR strain mice. Bentvelzen and Daams (5) propose that tumor virus activator genes are analogous to constitutive operator mutants of prophage.

C. Genes Affecting Spread of Activated Virus

Since the spontaneous activation of endogenous virus is a rare event even when the host inherits V^{+} genes, viremia does not occur unless the activated virus can be propagated in the tissues of the host. In other words, horizontal transmission of the activated virus from cell to cell within the host is subject to host range restrictions.

The effect of the *Fv-1* locus on spreading infection is revealed in studies of the activation of endogenous N-tropic viruses by IUdR or BUdR in N- and B-type mouse cells in culture (*52*, *62*). Once activated in a few of the cells, the virus can proliferate to high titers in N-type cells; in B-type cells, however, infectious virus reaches a peak titer about 3 days after induction and then falls back to undetectable levels. If N-type cells are added to the induced B-type cells, the virus titer rises as a result of infection of and proliferation in the permissive cells. The disappearance of N-tropic virus a few days after induction in B-type cells remains unexplained but suggests to me that one or more of three events occurs: (*a*) induced cells die within a few days of induction, (*b*) virus activation mediated through the *V* gene is reversed, (*c*) the *Fv-1* gene reasserts its repression of virus production, despite the activation of the *V* gene. If the restriction on viral replication exerted by the *Fv-1* gene acts on a "late" event in the replication cycle, say, on transcription of viral RNA or later, the resistance allele may inhibit virus production in the induced cell itself. Rowe and Hartley (*52*) found that in progeny of AKR mice crossed with $Fv\text{-}1^{b}$ (resistant) strains, the appearance of virus *in vivo* was considerably delayed and did not reach high titer.

The influence of susceptibility genes on endogenous avian viremia is also clearly illustrated in Crittenden's studies (*12*) on the appearance of RAV-0 in line 100 chickens. However, the complex genetics of susceptibility to RAV-0 infection must be explained first. Cellular susceptibility to RAV-0 (subgroup E) depends not only on the presence of suitable receptor sites but also on the absence of a dominant, epistatic inhibitor gene (*47*). Let us first consider susceptibility to RAV-0 in the absence of the inhibitor. Receptor genes have been defined in two separate studies. Payne *et al.* (*47*) identified a *tv-e* locus from crosses between Reaseheath C-line and I-line chickens. Crittenden *et al.* (*13*) and I (*87*) have observed that all chickens that are resistant to subgroup B viruses (i.e., $tv\text{-}b^{r/r}$ homozygotes) are also resistant to subgroup E viruses. This finding is in accord with the ability of subgroup B leukemia viruses to interfere with infection of subgroup E sarcoma virus pseudotypes. Pani (*42*) has recently suggested that the *tv-b* and *tv-e* genes represent distinct loci, but his data also appear to be consistent with identity if the *tv-e*^{r} allele were to permit susceptibility to viruses of subgroup B while excluding those of subgroup E.

Line 100 was initiated by crossing line 7 and line 6 chickens. Line 7 is homozygous for resistance to subgroup A and B viruses (*tv-a*$^{r/r}$, *tv-b*$^{r/r}$), whereas line 6 is susceptible (*tv-a*$^{s/s}$, *tv-b*$^{s/s}$). Line 100 was developed by backcrossing the 7×6 F_1 progeny to line 7. In each generation, doubly susceptible heterozygous males (*tv-a*$^{s/r}$, *tv-b*$^{s/r}$) are backcrossed to line 7 females. Thus line 100 is becoming cogenic with line 7 but segregates resistance and susceptibility alleles at the unlinked *tv-a* and *tv-b* loci. The development of line 100 was initiated before the presence of RAV-0 was known and the observation that by chance line 100 carries the V^{+} determinants has greatly aided our analysis of genetic factors influencing RAV-0 production (*12*). Table 3 shows that only those birds that carry a *tv-b*^{s} allele become viremic for RAV-0. Virus can be readily detected in embryos, and high titers are found in the plasma of birds of all ages. The *tv-b*$^{r/r}$ homozygotes are not viremic and virus cannot readily be detected in embryonic fibroblast cultures of this genotype. However, if these cells are cocultivated with cells that are susceptible to infection with RAV-0, the virus appears in the mixed cultures within two to three passages. We conclude that the *tv-b* locus does not influence the spontaneous activation of RAV-0 but does affect its horizontal spread from cell to cell.

Because line 100 *tv-b*$^{s/r}$ heterozygotes are viremic it may be argued that the presence of RAV-0 in this line is maintained by congenital infection rather than genetic transmission. However, line 100 has been bred at each generation by mating *tv-a*$^{s/r}$, *tv-b*$^{s/r}$ roosters with nonviremic line 7 *tv-a*$^{r/r}$, *tv-b*$^{r/r}$ hens. Since congenital transmission of subgroup A and B

TABLE 3. SUSCEPTIBILITY AND VIREMIA IN LINE 100 CHICKENS[a]

Genotype at *tv* loci		Phenotype: Susceptibility to subgroup: A	B	E	RAV-0 in plasma
a^s/a^r	b^s/b^r	+	+	+	+
a^r/a^r	b^s/b^r	−	+	+	+
a^s/a^r	b^r/b^r	+	−	−	−
a^r/a^r	b^r/b^r	−	−	−	−

[a]Data from Crittenden *et al.* (*12*).

viruses is restricted to hens, it is likely that transmission of RAV-0 is similarly restricted.

D. The Epistatic Inhibitor Gene

When it became apparent that the infectivity of Bryan strain RSV was determined by chf, we began a study of the genetic control of susceptibility of chickens to infection with RSV(chf). Previous studies (*75*, *83*) had shown that most chickens were resistant to RSV(chf) but that Japanese quail (*Coturnix coturnix*) and ringed-neck pheasants (*Phasianus colchicus*) were susceptible. Working with the inbred Reaseheath chicken lines C and I, Payne and I found that about 19% of C $\times$ I F_2 embryos were susceptible to RSV(chf), whereas both parental lines were resistant. It was apparent that host susceptibility to subgroup E virus was not controlled in the same way as susceptibility to subgroups A and B. For 2 years we puzzled over these data until Pani pointed out that the solution was obvious because the segregation in the F_2 generation gave a 13:3 ratio of resistance to susceptibility. This ratio implied that two unlinked loci were involved, one (*tv-e*) dominant for susceptibility, the other (I^e) exerting a dominant epistatic inhibition over the first. Further crosses were set up to test Pani's hypothesis, the crucial ones being the first-generation backcrosses. A summary of the data is shown in Table 4. All matings tested were consistent with the two-loci hypothesis, assuming that line C carried recessive alleles at both loci and that line I carried dominant alleles. Further studies by Pani (*42*, *43*) and by Crittenden *et al.* (*13*) on the *tv-b* locus in other chicken lines confirm the two-loci hypothesis. We assume that the *tv-e* or *tv-b* locus determines cell surface receptors, as do the other *tv* genes, and that both loci control penetration or uncoating of this virus. The basis for this assumption is that infection with subgroup E envelope *pseudotypes*, whether RSV(RAV-0) and RSV(chf) (*47*) or VSV(RAV-0) (*87*), follows the expected pattern of susceptibility and because treatment of resistant cells with inactivated Sendai virus following adsorption of RSV(chf) permits infection of the virus (*84*).

The physiological expression of this complex genetic control of susceptibility to RAV-0 is not yet fully elucidated, but we have strong circumstantial evidence that the endogenous virus itself plays a role. Payne *et al.* (*47*) noted that line I, which carries the I^e allele, also carries the gs^+ allele, and that the segregation frequencies of susceptible embryos in second-generation backcrosses to line C was consistent with linkage of I^e and gs^+. Crittenden *et al.* (*13*) and I (*87*) have found that all individually characterized embryos or birds that carry the I^e allele are positive for gs antigen and chf. Not only is there linkage, but the dominant

TABLE 4. EVIDENCE FOR A DOMINANT EPISTATIC GENE THAT INHIBITS SUSCEPTIBILITY TO RSV (RAV-0)[a]

Line or cross	Proposed genotype		Proposed phenotype[b]	No. of embryos tested	% Susceptible embryos: Observed	% Susceptible embryos: Expected
C	e^r/e^r	e^e/e^e	−	204	0.5	
I	e^s/e^s	I^e/I^e	−	30	0	0
C × I(F_1)	e^s/e^r	I^e/i^e	−	14	0	0
CI × CI(F_2) 1	e^r/e^r	i^e/i^e	−	250	18.80	18.75
3	e^r/e^r	$I^e/-$	−			
9	$e^s/-$	$I^e/-$	−			
3	$e^s/-$	i^e/i^e	+			
C × CI	e^r/e^r	i^e/i^e	−	236	23.70	25.00
	e^r/e^r	I^e/i^e	−			
	e^s/e^r	I^e/i^e	−			
	e^s/e^r	i^e/i^e	+			
I × CI	e^s/e^s	I^e/I^e	−	53	0	0
	e^s/e^s	I^e/i^e	−			
	e^s/e^r	I^e/I^e	−			
	e^s/e^r	I^e/i^e	−			

[a] Data from Payne, Pani, and Weiss (*47*), Pani and Payne (*43*), and Weiss (*87*).
[b] + = susceptible; − = resistant.

alleles, I^e and gs^+, in diverse strains, are always found to be coupled. It seems possible, therefore, that the I^e and gs loci are identical, in which case the partial expression of the endogenous viral genome may itself inhibit susceptibility to exogenous infection. The gene product that would act as the inhibitor substance could be the envelope antigen (chf) which must have a high affinity for subgroup E receptor sites. If present in sufficient amounts, chf could block the sites so that an exogenous virus with the same envelope specificity could not enter the cell (Fig. 5). Envelope antigen has been demonstrated by specific adsorption of antibodies at the surface of chf$^+$ cells of both the e^s/e^r and the e^r/e^r genotypes (*87*). Thus the dominant inhibition of susceptibility resembles viral interference.

The epistatic inhibition of the receptor by the I^e gene is not fully penetrant; that is, cells of the $e^s/—$, $I^e/—$ genotype are not usually as resistant as e^r/e^r cells to infection with RSV(RAV-0). There is also variation in the degree of inhibition. When assayed by inoculation of the chorioallantoic membrane, I^e resistance to RSV(RAV-0) is greater than when assayed in fibroblast cultures that have undergone several passages *in vitro.* Furthermore, not all subgroup E receptor sites are equally subject to inhibition. Crittenden *et al.* (*13*) have described two distinct sus-

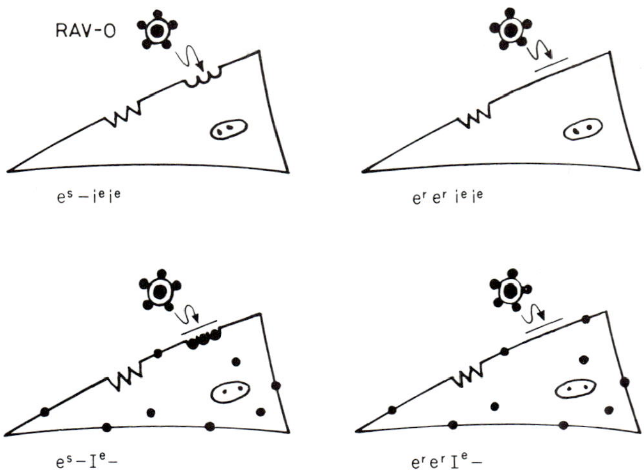

Fig. 5. Interference model for host resistance determined by the I^e gene to horizontal infection with the endogenous virus, RAV-0.

ceptibility alleles at the *tv-b* locus. The *tv-b*^{s1} allele is distinguishable from the *tv-b*^{s2} allele because it is associated with the R_1 erythrocyte antigen; it is the only *tv* receptor known to be antigenic. Crittenden's studies indicating that the receptors coded by *tv-b*^{s1} are relatively unaffected by the I^e gene. Viral interference of RSV(RAV-0) infection by preinfection with RAV-0 is also difficult to achieve in *tv-b*^{s1} cells. These observations support the hypothesis that the dominant epistatic inhibitor represents an endogenous interference phenomenon.

Studies of virus production in cultures derived from $V^+ \times I^e$ crosses indicate that the I^e gene does inhibit virus replication (Table 5). However, when the V^+, I^e cells are mixed with susceptible cells, RAV-0 replicates to high titers. Thus the I^e-*gs* locus does not affect primary

TABLE 5. EFFECT OF RESISTANCE GENES ON PRODUCTION OF RAV-0 IN CHICKENS

Genotype			
Activator	Receptor	Epistatic inhibitor	Viremia
v^-/v^-	$s/-$	I^e/i^e	−
$V^+/-$	$s/-$	i^e/i^e	+
$V^+/-$	r/r	i^e/i^e	−
$V^+/-$	$s/-$	$I^e/-$	−

activation of RAV-0, but, like the *tv-e*^{r} or *tv-b*^{r} gene, it does inhibit the spread of virus from cell to cell.

Virus RAV-0 has not yet been shown to be oncogenic and the incidence of leukemia in viremic line 100 fowl is low. Nevertheless, spreading infection of RAV-0 may be disadvantageous to the host. If so, the immunity to superinfection with homologous virus resulting from the occupation of receptor sites by chf will lead to natural selection of the *gs*$^+$ allele. This is consistent with observation, for it is not only the broiler fowl, selected for fast growth, that express endogenous tumor virus genes; we have found that gas antigen and chf expression is prevalent in feral Red Judge fowl too (*88*).

Table 6 summarizes our knowledge of the genes affecting endogenous virus expression in chickens. The *tv* genes are probably host loci; the *gs* and *V* genes that act as control genes, may well be proviral loci; the I^e gene may be identical with *gs*.

VII. XENOTROPIC VIRUSES

Recently, a new class of endogenous virus has been identified for viruses that are not infectious for the host in which they are genetically transmitted, but are infectious for foreign hosts. Levy (*30*) has proposed the term "xenotropic" or X-tropic for such viruses; he showed that the virus endogenous in NZB mice is xenotropic to rat and human cells.

TABLE 6. HOST GENES AFFECTING ENDOGENOUS VIRUS EXPRESSION IN CHICKENS

Level of control	Locus	Allele	Expression	Phenotype	Possible mechanism
Expression of endogenous genome	*Gs*	Gs^+	Dominant	*Gs* antigen envelope antigen	Regulation of transcription
		gs^-	Recessive	No viral antigens	
	V	V^+	Dominant	Release of infectious virus (RAV-0)	High inducibility
		v^-	Recessive		
Spread of activated virus	*tv-e* *tv-b*	*s*	Dominant	Susceptible to RAV-0 infection	Cell surface receptor sites
		r	Recessive	Resistant	
	I^e	I^e	Dominant	Resistant to RAV-0 infection	Receptor sites blocked by envelope antigen
		i^e	Recessive	Susceptible	

Similar viruses have been detected in cells of the BALB/c and NIH-Swiss mouse strains (*30, 59*); they are infectious in cells of rats, rabbits, and primates, including man. Whether these various isolates represent the same virus species remains to be determined. The xenotropic virus in NZB mice becomes spontaneously activated early in life; if reinfection is completely excluded, the incidence of activation must be much higher than that observed for N-tropic virus in V^+ strains. The X-tropic virus of BALB/c mice is readily induced by IUdR; it differs from the N-tropic virus endogenous in the same strain not only in host range but in being readily inducible by cycloheximide (*1*) or chronic immune reactions (*59*). Thus different leukemia proviruses in the same cells are subject to different control mechanisms.

Domestic cats also carry an endogenous xenotropic virus. From the feline CCC cell line, Livingston and Todaro (*34*) isolated a C-type virus that is infectious for primate cells but not for feline cells. The presence of this type of endogenous virus has been confirmed by a plethora of other isolates and molecular hybridization studies from a variety of feline tissues and cell lines. These studies indicated that RD-114 virus (which had been isolated from human tumor cells that had been inoculated into a fetal cat (*38*) and which had been proposed as a possible human C-type virus) was also an endogenous cat virus. Virus RD-114 had not been considered a feline virus because it was not related antigenically or by nucleotide sequences to the previously described infectious feline viruses. In this respect the endogenous feline virus differs from those of chickens and mice, which share gs antigens with their infectiously transmitted counterparts.

The endogenous viral genome of chickens is not an unsuitable model for mammalian xenotropic viruses. For 4 years, many laboratories studied "nonproducer" Rous cells before it was discovered that such cells released virus [the RSV(chf) pseudotype] infectious for Japanese quail cells (*75*) and, rarely, for chicken cells (*84*). For a further 6 years RSV (chf) and RSV(RAV-0) pseudotypes have been propagated for experimental purposes as xenotropic viruses because of the rarity of susceptible chicken cells. With the observation that line 15 chicken cells are highly susceptible (*13*), experiments involving homotypic cells are now preferred. So a cautionary word that mammalian xenotropic viruses, too, may not prove to be entirely excluded from reinfecting their own species is not out of place; indeed, the SC-1 cell line derived from a wild mouse appears to be permissive for replication of NZB virus (*23*).

The nature of the restriction of homotypic cells to infection with xenotropic viruses is not yet known and may differ for different viruses. In mice, the restriction might be connected with the *Fv-1* locus governing "homotropic" viruses, since the SC-1 cells alluded to above are also exceptional in being susceptible to both N- and B-tropic viruses. However, the genetic control of resistance to RAV-0 in chicken cells, in which I argued that the endogenous virus itself plays a rôle, again provides a useful model. Only 1 out of 12 species of gallinaceous birds tested is resistant to infection with RAV-0. As with RAV-0, the widespread infectivity of mammalian xenotropic viruses for other species lends weight to the idea that specific mechanisms to insure against superinfection have evolved in the host species.

Xenotropism, if viewed as homotypic restriction, may be regarded as an antiviral mechanism. The highly activable nature of the murine and feline xenotropic viruses, however, raises the question whether the hosts in which they are genetically transmitted might provide a reservoir for horizontal transmission to susceptible species. Even if this does not happen in nature, it may well occur in the laboratory, particularly where mixed cell cultures and somatic cell hybridization are employed.

VIII. INHERITANCE OF ACQUIRED GENES

I have described how vertebrate species carry DNA proviruses of RNA tumor viruses as part of their normal genetic constitution. In addition to viruses of chickens, mice, and cats discussed here, endogenous viral genomes have been identified in pheasants, guinea pigs, hamsters, rats, pigs, and baboons. It is pertinent, therefore, to ask where those viruses come from and whether they act as vehicles of genetic exchange.

The primordial origin of animal viruses is, of course, obscure. Some viruses may be derived from parasitic forms of once independent organisms; others may be derived from host genetic elements that have evolved to become independently replicating entities. Nearest-neighbour analysis of their nucleic acids may indicate whether they are of prokaryotic or eukaryotic ancestry (*64*). Ribonucleic acid tumor virus genes are probably derived from escaped cellular genetic elements. Temin proposes (*67, 68*) that there is a continual evolution of viral forms from cellular elements called *protoviruses*. He postulates that "the RNA-directed DNA polymerase activity exists in normal animals; that it plays a role in normal cellular processes like differentiation; that RNA tumor viruses evolved from this activity; and that cancer arises from variational events in the functioning of this activity." There are several attractive features to Temin's protovirus theory, for instance, for the translocation of genetic information in immunological differentiation and for the origin of strongly transforming viruses. The protovirus concept of transferring information from DNA to RNA to DNA provides a scheme for gene duplication and relocation, for genetic reassortment within, and, perhaps, between somatic cells. I wish to propose that the fully fledged virus, like the transducing bacteriophage, provides a mechanism for exchanging genetic information between individuals and perhaps between species too.

A. Viral Genes Acquired by the Host

Temin argues that endogenous viral genomes mark an evolutionary stage between protoviruses and infectious RNA tumor viruses; that the origin of multiple endogenous viruses involves duplication and evolution of protoviruses; and that induction of latent endogenous proviruses represents "the last step in the evolutionary origin" of an RNA tumor virus (*68*). On this issue I disagree with Temin and prefer the lysogeny model. My verdict is that most of the endogenous viruses that have been identified as such have, in Darlington's telling phrase, been "guilty of infection" in the past; they show evidence of specialization for natural infection, presumably by selection and adaptation of their ancestors. The efficiency of virus activation in V^{+} hosts seems too high to be a last stage in the evolution (i.e., mutation and selection) of an infectious virus from a protovirus. The infectious RNA tumor virus particle, like other virus particles, is an efficient machine for genetic transfer, with a complex structure and function. The assortment and packaging of the right genes (in RNA form) to code for its own structural components and for further replication, if not directly incriminating, surely arouses suspicion of previous infectious experience.

There is evidence that the proviruses derived from exogenous infections integrate into host chromosomal DNA (*73*). This could happen in the germ line as well as in somatic cells, giving rise to inherited germinal proviruses. Reinfection of the germ line may well have happened several times during the selection of high leukemia strain mice. It would account for the duplicate or triplicate copies of otherwise indistinguishable viral genomes at different host loci, for instance, in AKR and C3H/Figge mice (*51*). It will be interesting to investigate the presence and location of germinal proviruses in wild mice. In fowl, the germinal provirus has been conserved in evolution, since the expression and antigenic specificity of chf in Red Jungle fowl is indistinguishable from that of most domestic strains (*88*).

We do not know why viral genomes should remain latent in the germ line of susceptible hosts, and little is known about the permissiveness of primordial germ cells and gametes for viral replication. But I wish to cite two examples of the inheritance of acquired proviruses that remain latent.

The first concerns the stable inheritance of avian proviruses in somatic mammalian cells in the laboratory. It has long been known, largely from the pioneering work of Soviet and Czechoslovak scientists, that mammalian cells can be transformed by avian RNA tumor viruses, although the efficiency of transformation is very low compared to chicken cells. The transformed cells are nonpermissive for virus production, that is, there is a block in a late phase of the virus replication cycle. Fusion of transformed cells to chick cells allows the rescue of complete virus from heterokaryocytes. Recent studies indicate that infection and integration of provirus in mammalian cells may be much more efficient than subsequent transformation (*10, 73*). Boettiger (*9*) has recently shown that a majority of rat cells in culture can be stably infected by an avian sarcoma virus. Most of the cells show no sign of virus expression or cell transformation. However, complete virus can be rescued from the rat cells by fusion with chick cells, and this process is enhanced by treatment of the rat cells with BUdR. It will be interesting to investgate the consequences of introducing an avian sarcoma virus into a mammalian germ line.

The second example is speculation on the origin of xenotropic endogenous viruses, based on results emanating mainly from Todaro's laboratory. The virus endogenous to cells of the domestic cat is readily activated by IUdR and replicates in a variety of primate cells (*34*). Molecular hybridization studies (*7*) reveal a substantial degree of homology between the endogenous feline virus and a C-type virus of baboons, indicating some relationship between primate and feline endogenous viruses. More significant, perhaps, is the observation that DNA homologous to the feline virus RNA cannot be detected in the tissues of two feline species, lion and snow leopard, although it is present in the domestic cat and related sequences are found in the DNA of a variety of Old World monkeys. To my mind, these findings indicate that the viral genome endogenous in the domestic cat could be of exogenous origin; it was probably transmitted from a primate, perhaps prior to domestication of cats in ancient Egypt.

Thus, there may be a closer phylogenetic relationship between the viral genomes of different species than between the hosts. Xenotropic endogenous viruses may have come from a foreign source; where they share common gs antigens with homotropic viruses as in mice, they may be recombinants, analogous to recombinant forms of influenza virus arising from human and animal strains (*82*). Perhaps all the "highly inducible" endogenous viruses are of recent exogenous origin, whether from infectious viruses natural to their own host species, to other species, or to recombinants of both types. They do not behave like genetic elements about to become infectious agents for the first time; rather they resemble inherited genomes acquired by infection.

B. Host Genes Acquired by Virus

Since RNA tumor viruses become integrated into and transcribed from the host genome, it is possible that they could act as transducing agents through a DNA to RNA to DNA transfer of information from one cell to another. We may say that this happens if we regard inherited viral genes as host components, for we have firm evidence for the transfer of genes from endogenous to exogenous viruses. This type of genetic exchange is a special case of recombination between genomes that are already largely homologous. However, the recombination process could provide a mechanism for generalized transduction.

Following Vogt's (*77*) demonstration of high-frequency genetic reassortment between avian RNA tumor viruses, we investigated whether similar genetic changes would take place between the endogenous genes coding for chf and nondefective avian sarcoma viruses (*90*). Potential recombinants were selected for their host range markers. Cloned stocks of Schmidt-Ruppin strain Rous sarcoma virus of subgroup B (SR-RSV-B) were propagated in chf^+ cells and the progeny were plated onto Japanese quail cells. Quail cells are resistant to infection by viruses with subgroup B envelope specificity, but are susceptible to subgroup E, including chf. As a result of phenotypic mixing, a proportion of the RSV progeny would be able to infect the quail cells, but could only propagate to high titers if recombinants carrying the gene coding for subgroup E envelope antigen were present among the progeny. After several passages in quail cells, high-titer stocks were obtained which, as expected, possessed the phenotypic character of subgroup E viruses. This character remained stable after cloning the RSV. Thus, passage of virus through xenotypic cells selected for recombinants between exogenous RSV and endogenous chf present in the original stock.

Figure 6 shows a clonal pedigree undertaken to analyze the process of recombination. Six successive generations of infection were analyzed. A single cycle of selection through infective centers was sufficient to exclude any virus particles that did not carry the subgroup E determinant. However, most of the progeny of the first-generation, infective center colonies still expressed the host range of subgroup B in addition to E, suggesting that genes for both subgroups were transmitted. With further cycles of selection for subgroup E host range, the subgroup B allele tended to be lost and, once lost, the progeny bred true for subgroup E alone. Further experiments (*90*) showed that segregation into pure B or E genotype occurred among the progeny of virus bearing

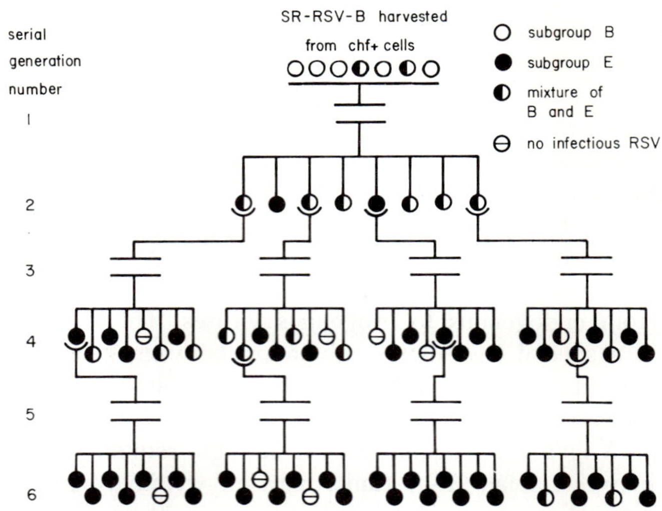

Fig. 6. Selection of recombinant SR-RSV-E among progeny of SR-RSV-B grown in chf⁺ cells. Rous sarcoma virus (RSV) harvested from chf⁺ cells was plated onto quail "initiator" cells. At the end of the eclipse period, when progeny virus was beginning to be released, these "initiator" cells were treated with mitomycin C, which prevented mitosis so that no colonies or foci of transformed cells could develop. However, the initiator cells continued to release virus. When "indicator" cells were added that were selectively resistant or susceptible to subgroup B and E viruses, the foci which formed in the indicator cell layer revealed clonal infective centers producing parental or recombinant RSV. The parallel horizontal bars denote infective center assays using quail initiator and indicator cells. Each dot represents a single colony of RSV cells formed in the indicator layer. Each colony was picked, sonicated to disrupt the cells, and the host range of the virus produced by cells was determined. A cup around a dot denotes those colonies whose virus was subjected to a further cycle of selection. [After Weiss *et al.* (*90*)].

both host range markers according to the selection imposed, but segregation into one or other genotype also occurred under nonselective conditions. The virus expressing dual host range did not appear to be caused by double infections, but this complication cannot be absolutely excluded. These observations suggested that the majority of recombinants arose through the formation of *heterozygous* particles which later segregate into homozygous or haploid parental and recombinant forms. The implication is that RSV can carry redundant information, possibly allowing transduction of nonviral genes too.

Table 7 summarizes the genetic and phenotypic interactions between three strains of RSV with the endogenous viral genome. Genotypic mixing occurred at relatively high frequency with the two nondefective strains (SR-RSV, PR-RSV) grown in chf⁺ cells, but none was observed

TABLE 7. FREQUENCY OF GENETIC REASSORTMENT OF HOST RANGE MARKERS BETWEEN RSV AND THE ENDOGENOUS VIRAL GENOME[a]

Mixed virus harvest				
Source of subgroup E genes	Source of transformation genes[b]	% Phenotypic mixing	% Genotypic mixing	% Heterozygotes among genetically mixed progeny
gs^+ chf^+ cells	SR-RSV-B	8.5	5.5	49
	PR-RSV-B	0.4	17.5	74
	BH-RSV(RAV-7)	100	0	—
gs^- chf^- cells	SR-RSV-B	0	0	—
	PR-RSV-B	0	0	—
	BH-RSV(RAV-7)	0	0	—

[a] Data from Weiss *et al.* (90).

[b] RSV, Rous sarcoma virus; SR, Schmidt Ruppin strain; PR, Prague strain; BH, Bryan high titer strain.

with the same viruses grown in chf^- cells. This indicates that proviral transcription is necessary, and we suggested that the initial genetic interaction was one of reassortment of RSV and chf genome subunits in the assembly of progeny particles (*90*). Reassortment of genetic markers located on different viral subunits would account for the high frequency of recombination observed. However, Vogt (*78*) has suggested that the genome of RNA tumor viruses may be polyploid, composed of two to three identical RNA subunits. There is mounting evidence in favor of this model, and Vogt and Duesberg (*79*) have shown that the genetic markers of cloned recombinants appear to be covalently linked in a single class of subunit. We have proposed (*78, 86*) that true recombinants arise from heterozygote (genotypically mixed) virus particles when the DNA provirus is formed during the next cycle of replication. However, recombination at the forward transcription stage in the initial doubly infected cells is also a possible model, although in the case of RSV and chf it requires integration of the RSV genome in tandem with the endogenous genome.

The first model, genotypic mixing followed by proviral recombination, would provide a general mechanism for transduction of messenger (m)RNA in the virus particle from one cell to the other. The message would be transcribed by viral reverse transcriptase into a DNA element and might integrate into the host genome with the provirus. Endogenous viral mRNA, such as chf mRNA, may be transduced at a relatively high frequency because of its homology to the exogenous viral RNA. Whether cellular species of mRNA that have never been guilty of infection could be similarly transduced is still a matter of speculation. However, Ikawa *et al.* (*27*) have recently found that about 1 in a 1000 leukemia virus particles released from hemoglobin-producing cells carry a globin mRNA molecule. The transmission of cellular genetic elements by RNA tumor virus particles may, therefore, occur, and experiments should be designed to see whether such elements can become adopted by and function in recipient cells.

Genotypic mixing was not observed between chf and the defective Bryan strain RSV (Table 7), although the selective technique would have detected a recombination frequency of 10^{-5}. It is evident that defective viruses would not remain helper-dependent if they acquired the missing information by recombination with their helper viruses. Stable helper-dependent viruses are, therefore, *defective for recombination* in the sense that recombinants containing both the transformation gene(s) and the missing replication or structural genes do not form (*86*). Some kind of restriction is imposed on the formation of stable recombinants of this type, perhaps a limitation of genome size. It is worth noting that most of the strongly transforming C-type viruses are replication-defective; the helper-independent strains of RSV are exceptions. The defectiveness of strongly transforming viruses at first sight appears to weaken the case for transduction, because viruses that cannot transmit genes of their close relatives (the helper viruses) are unlikely to transmit unrelated host genes. However, the strongly transforming viruses may themselves represent transductants, the defective ones being homologous to defective transducing bacteriophages.

The gene or genes of strongly transforming viruses responsible for maintaining the transformed state are not required for viral replication (*39*) and are frequently deleted when stocks are propagated (*76*). Since these genes are not essential to the virus but are essential for rapid transformation of the host cells, we may call them "oncogenes" (*40*). The C-type viruses may have acquired oncogenes from cellular genetic elements. Leukemia viruses that are propagated or activated in sarcoma cells do not normally emerge as sarcoma viruses, but on rare occasions this may happen. Two examples will be cited:

1. Defective murine sarcoma viruses (MSV) have been independently isolated on several occasions following inoculation of high titers of murine leukemia viruses into animals. The genomes of two strains, Moloney-MSV and Kirsten-MSV, have recently been analyzed by Scolnick *et al.* (*58*) using molecular hybridization techniques. Kirsten-MSV, which appeared after passage of murine leukemia virus in a rat, was found to have RNA homologous to sequences present in normal rat cells, whereas Moloney-MSV, which first appeared in mice, showed no homology to rat RNA. Part of the new information in Kirsten-MSV

is homologous to the RNA of endogenous rat C-type virus. In addition, Kirsten-MSV carries sequences that are not homologous to rat leukemia virus or the Kirsten murine leukemia virus but which may be specific to the transforming genes in the MSV. Homologous DNA sequences are also found in rat cells. Therefore, Kirsten-MSV may represent a defective virus transducing rat cell information which acts oncogenically when reintroduced into cells by infection.

2. Purchase, Vogt, and I have recently attempted to generate sarcoma viruses from avian lymphoid leukemia viruses. Sarcomas were induced in chickens by intramuscular inoculation of a high dose of a field isolate of a leukemia virus. Cell-free extracts of the primary sarcomas yielded leukemia virus but no strongly transforming virus. However, after serial transplantation of the sarcoma tissue through several birds, we found that cell-free extracts rapidly induced sarcomas in 3 independent cases. Apparently, a rapidly transforming virus has arisen in the leukemia stock produced by the sarcoma cells, but we must await a thorough virological analysis before drawing further conclusions.

Thus, it is possible that strongly transforming viruses evolve by recombination of weakly transforming or nontransforming C-type viruses and cellular oncogenes. This represents a special case of Temin's protovirus theory, when the viral genes preexist as a functional genomic unit, whether exogenous or endogenous in immediate origin, but when the oncogenes arise from cellular genetic elements. The oncogenes of different strongly transforming viruses may be derived from quite different cellular genes and they may express neoplastic transformation only in specific cell types (*40*). If the oncogenes could be accommodated in addition to all the viral replication genes, helper-independent transforming viruses would arise; if the oncogenes are substituted for viral genes, defective, helper-dependent transforming viruses would arise. It is not clear whether weakly transforming viruses—the lymphoid leukemia and mammary tumor viruses—also carry oncogenes that are distinct from viral replication genes. But we may propose that the RNA tumor viruses have acquired their malignant name by transduction; in short, cancer causes tumor viruses.

IX. EPILOGUE

The widespread occurrence of RNA tumor viruses in animal species including primates (7) shows that these viruses are not restricted to laboratory animals selected for a high incidence of cancer. It would be surprising, therefore, if humans were free from such viruses, although it has been difficult to obtain convincing evidence for the existence of human viruses (*60*), whether transmitted by infection or inheritance. In searching for human RNA tumor viruses several points arising from the animal models might be borne in mind: (1) the inheritance of proviruses in animals has only recently been discovered; (2) host species can contain more than one kind of provirus; (3) some are highly inducible, others are not; (4) separate genetic loci control their expression; (5) some can spread by infection of their own host cells, some are xenotropic, and some might not be infectious at all; and (6) both genotypic mixing and phenotypic mixing occur between endogenous and exogenous viruses so that a virus population may change dramatically upon passage through different hosts. The RNA tumor viruses appear to be a homogeneous group with regard to structure and replication, but they have several modes of transmission and show a wide variety of interactions with each other and with their hosts. Apparently the hosts have been just as resourceful in controlling the viruses, and we might expect host control mechanisms to be strongest in long-lived species such as man.

The no-man's-land between heredity and infection in cancer has indeed become a meeting place, through the discovery of tumor viruses that integrate into the host genome. That meeting is most intimate in the genetic transmission of proviruses. The discovery of host genes that affect virus expression and also susceptibility to virus infection, binds the interaction of viral and host genomes still further, for the distinction between cellular and viral genes is not always clear or useful. The oncogenic transformation of cells is one of the least understood viral functions, and it is dependent on the differentiated state of the cell. The no-man's-land remains, therefore, between the disciplines of infection and development, although the protovirus has staked its claim there. We may hope that studies on the interaction of tumor virus genes with differentiating cells will make it an exciting meeting place in the future.

ACKNOWLEDGMENTS

I thank Warren Levinson for introducing me to tumor virus research, Jim Payne for suggesting the idea of genetic transmission to me, and Peter Vogt for inviting me to work in his laboratory in which many of the studies I have described here were carried out. I am also most grateful to Natalie Teich for helpful criticism of this paper.

REFERENCES

1. Aaronson, S. A., and Dunn, C. Y., *Science* **183**, 422–424 (1974).
2. Aaronson, S. A., Hartley, J. W., and Todaro, G. J., *Proc. Nat. Acad. Sci. U. S.* **64**, 87–94 (1969).
3. Andrewes, C. H., *Proc. Roy. Soc. Med.* **33**, 75–86 (1939).
4. Bader, J. P., *Virology* **16**, 436–443 (1962).

5. Bentvelzen, P., and Daams, J. H., *J. Nat. Cancer Inst.* **43**, 1025–1035 (1969).
6. Bentvelzen, P., Timmermans, A., Daams, J. H., and Van der Gugten, A., *Bibl. Haematol.* **31**, 101–103 (1968).
7. Benveniste, R. E., Heinemann, R., Wilson, G. L., Callahan, R., and Tordaro, G. J., *J. Virol.* **14**, 56–67 (1974).
8. Bishop, J. M., Jackson, N., Quintrell, N., and Varmus, H. E., *in* "Possible Episomes in Eukaryotes" (L. G. Silverstri, ed.), 4th Lepetit Colloq., pp. 61–73. North-Holland Publ., Amsterdam, 1973.
9. Boettiger, D., *Cell* 3, 71–76 (1974).
10. Boettiger, D., Love, D. N., and Weiss, R. A., *J. Virol.* **15**, 108–114 (1975).
11. Chen, J. H., and Hanafusa, H., *J. Virol.* **13**, 340–346 (1974).
12. Crittenden, L. B., Smith, E. J., Weiss, R. A., and Sarma, P. S., *Virology* **57**, 128–138 (1974).
13. Crittenden, L. B., Wendel, E. J., and Motta, J. V., *Virology* **52**, 373–384 (1973).
14. Darlington, C. D., *Brit. J. Cancer* **2**, 118–126 (1948).
15. Dougherty, R. M., and Di Stefano, H. S., *Virology* **29**, 586–595 (1966).
16. Gross, L., *Blood* **9**, 557–573 (1954).
17. Gross, L., "Oncogenic Viruses," 2nd ed., Pergamon Press pp. 1–991. Oxford, 1970.
18. Hanafusa, H., Aoki, T., Kawai, S., Miyamoto, T., and Wilsnack, R. E., *Virology* **56**, 22–32 (1973).
19. Hanafusa, H., and Hanafusa, T., *Virology* **43**, 313–316 (1971).
20. Hanafusa, H., Hanafusa, T., Kawai, S., and Luginbuhl, R. E., *Virology* **58**, 439–448 (1974).
21. Hanafusa, H., Miyamoto, T., and Hanafusa, T., *Proc. Nat. Acad. Sci. U. S.* **66**, 314–321 (1970).
22. Harel, J., Harel, L., and Goldé, A., *C. R. Acad. Sci. Ser. D* **263**, 745–748 (1966).
23. Hartley, J. W., personal communication (1974).
24. Hayward, W. S., and Hanafusa, H., *J. Virol.* **11**, 157–167 (1973).
25. Hirsch, M. S., Phillips, S. M., and Solnik, C., *Proc. Nat. Acad. Sci. U. S.* **69**, 1069–1072 (1972).
26. Huebner, R. J., and Torado, G. J., *Proc. Nat. Acad. Sci. U. S.* **64**, 1087–1094 (1969).
27. Ikawa, Y., Ross, J., and Leder, P., *Proc. Nat. Acad. Sci. U. S.* **71**, 1154–1158 (1974).
28. Jarrett, W. F. H., Jarrett, O., Mackay, L. J., Laird, H., Hardy, W. D., and Essex, M., *J. Nat. Cancer Inst.* **51**, 833–841 (1973).
29. Latarjet, R., and Duplan, J. F., *Int. J. Radiat. Biol.* **5**, 339–344 (1962).
30. Levy, J. A., *Science* **182**, 1151–1153 (1973).
31. Levy, J. A., *Amer. J. Clin. Pathol.* (in press).
32. Lieberman, M., and Kaplan, H. S., *Science* **130**, 387–388 (1959).
33. Lilly, F., and Pincus, T., *Advan. Cancer Res.* **17**, 231–277 (1972).
34. Livingston, D. M., and Todaro, G. J., *Virology* **53**, 142–151 (1973).
35. Love, D. N., and Weiss, R. A., *Virology* **57**, 271–278 (1974).
36. Lowy, D. R., Rowe, W. P., Teich, N., and Hartley, J. W., *Science* **174**, 155–156 (1971).
37. Lwoff, A., *Cancer Res.* **20**, 820–829 (1960).
38. McAllister, R. M., Nicolson, M., and Gardner, M. B., *Nature (London), New Biol.* **235**, 3–6 (1972).

39. Martin, G. S., *Nature (London)* **227**, 1021–1023 (1970).
40. Martin, G. S., and Weiss, R. A., *Proc. Can. Cancer Res. Conf.* **10**, 10–30 (1973).
41. Oldstone, M. R., Aoki, T., and Dixon, F. J., *Proc. Nat. Acad. Sci. U. S.* **69**, 134–138 (1972).
42. Pani, P. K., *J. Gen. Virol.* **23**, 33–40 (1974).
43. Pani, P. K., and Payne, L. N., *J. Gen. Virol.* **19**, 235–244 (1973).
44. Payne, L. N. *In* "RNA Viruses and Host Genome in Oncogenesis" (P. Emmelot and P. Bentvelzen, eds.), pp. 93–115. North-Holland Publ., Amsterdam, 1972.
45. Payne, L. N., and Chubb, R., *J. Gen. Virol.* **3**, 379–391 (1968).
46. Payne, L. N., Crittenden, L. B., and Weiss, R. A., *in* "Possible Episomes in Eukaryotes" (L. G. Silvestri, ed.), 4th Lepetit Colloq., pp. 94–97. North-Holland Publ., Amsterdam, 1973.
47. Payne, L. N., Pani, P. K., and Weiss, R. A., *J. Gen. Virol.* **13**, 455–462 (1971).
48. Rosenthal, P. N., Robinson, H. L., Robinson, W. S., Hanafusa, T., and Hanafusa, H., *Proc. Nat. Acad. Sci. U. S.* **68**, 2336–2340 (1971).
49. Rowe, W. P., *Science* **141**, 40–41 (1963).
50. Rowe, W. P., *J. Exp. Med.* **136**, 1271–1285 (1972).
51. Rowe, W. P., *Cancer Res.* **33**, 3061–3068 (1973).
52. Rowe, W. P., and Hartley, J. W., *J. Exp. Med.* **136**, 1286–1301 (1972).
53. Rowe, W. P., Hartley, J. W., Lander, M. R., Pugh, W. E., and Teich, N., *Virology* **46**, 866–876 (1971).
53a. Rowe, W. P., and Pincus, T., *J. Exp. Med.* **135**, 429–436 (1972).
54. Rubin, H., *Proc. Nat. Acad. Sci. U. S.* **46**, 1105–1119 (1960).
55. Rubin, H., Fanshier, L., Cornelius, A., and Hughes, W. F., *Virology* **17**, 143–156 (1962).
56. Ruoslahti, E., Vaheri, A., Estola, T., and Sandelin, K., *Int. J. Cancer* **11**, 595–603 (1973).
57. Scheele, C. M., and Hanafusa, H., *Virology* **45**, 401–410 (1971).
58. Scolnick, E. M., Rands, E., Williams, D., and Parks, W. P., *J. Virol.* **12**, 458–463 (1973).
59. Sherr, C. J., Lieber, M. M., and Todaro, G. J., *Cell* **1**, 55–58 (1974).
60. Spiegelman, S., Axel, R., Baxt, W., Gulati, S. C., Hehlman, R., Kufe, D., and Schlom J., *in* "Cellular Modification of Genetic Transformation by Exogenous Nucleic Acids" (R. F. Beers and R. C. Tibbman, eds.), pp. 282–302. Johns Hopkins Univ. Press, Baltimore, Maryland, 1973.
61. Steck, F. T., and Rubin, H., *Virology* **29**, 628–641 (1966).
62. Stephenson, J. R., and Aaronson, S. A., *Proc. Nat. Acad. Sci. U. S.* **69**, 2798–2801 (1972).
63. Stockert, E., Old, L. J., and Boyse, E. A., *J. Exp. Med.* **133**, 134–135 (1971).
64. Subak-Sharpe, H., *Brit. Med. Bull.* **23**, 161 (1967).
65. Taylor, B. A., Meier, H., and Huebner, R. J., *Nature (London)*, **241**, 184–186 (1973).
66. Teich, N., Lowy, D. R., Hartley, J. W., and Rowe, W. P., *Virology* **51**, 163–173 (1973).
67. Temin, H. M., *J. Nat. Cancer Inst.* **46**, III–VIII (1971).
68. Temin, H. M., *Annu. Rev. Genet.* **8** (in press).
69. Temin, H. M., and Baltimore, D., *Advan. Virus Res.* **17**, 129–186 (1972).
70. Todaro, G. J., *Nature (London), New Biol.* **240**, 157–160 (1972).
71. Tooze, J., ed., "Animal Cells and their Tumour Viruses." Cold Spring Harbor Lab., Cold Spring Harbor, New York, 1973.

72. Vaheri, A., and Ruoslahti, E., *Int. J. Cancer* **12**, 361–367 (1973).
73. Varmus, H. E., Vogt, P. K., and Bishop, J. M., *Proc. Nat. Acad. Sci. U. S.* **70**, 3067–3071 (1973).
74. Varmus, H. E., Weiss, R. A., Friis, R. R., Levinson, W., and Bishop, J. M., *Proc. Nat. Acad. Sci. U. S.* **69**, 20–24 (1972).
75. Vogt, P. K., *Proc. Nat. Acad. Sci. U. S.* **58**, 801–808 (1967).
76. Vogt, P. K., *Virology* **46**, 939–946 (1971).
77. Vogt, P. K., *Virology* **46**, 947–952 (1971).
78. Vogt, P. K., *in* "Possible Episomes in Eukaryotes" (L. G. Silvestri, ed.), 4th Lepetit Colloq., pp. 35–41. North-Holland, Publ., Amsterdam, 1973.
79. Vogt, P. K., and Duesberg, P. H., *in* "Virus Research" (C. F. Fox and W. S. Robinson, eds.), pp. 505–511. Academic Press, New York, 1973.
80. Vogt, P. K., and Friis, R. R., *Virology* **43**, 223–234 (1971).
81. Vogt, P. K., and Ishizaki, R., *Virology* **30**, 368–374 (1966).
82. Webster, R. G., Laver, W. G., and Granoff, A., *in* "Virus Research" (C. F. Fox and W. S. Robinson, eds.), pp. 513–524. Academic Press, New York, 1973.
83. Weiss, R. A., *Virology* **32**, 719–722 (1967).
84. Weiss, R. A., *J. Gen. Virol.* **5**, 511–528 (1969).
85. Weiss, R. A., *J. Gen. Virol.* **5**, 529–539 (1969).
86. Weiss, R. A., *in* "Possible Episomes in Eukaryotes" (L. G. Silvestri, ed.), 4th Lepetit Colloq., pp. 130–141. North-Holland, Amsterdam, 1973.
87. Weiss, R. A., unpublished observations (1974).
88. Weiss, R. A., and Biggs, P. M., *J. Nat. Cancer Inst.* **49**, 1713–1725 (1972).
89. Weiss, R. A., Friis, R. R., Katz, E., and Vogt, P. K., *Virology* **46**, 920–938 (1971).
90. Weiss, R. A., Mason, W. S., and Vogt, P. K., *Virology* **52**, 535–552 (1973).
91. Weiss, R. A., and Payne, L. N., *Virology* **45**, 508–515 (1971).
92. Weissbach, A., Bolden, A., Muller, R., Hanafusa, H., and Hanafusa, T., *J. Virol.* **10**, 321–327 (1972).
93. Wilson, D. E., and Bauer, H., *Virology* **33**, 754–756 (1967).
94. Zavada, J., *J. Gen. Virol.* **15**, 183–191 (1972).

DISCUSSION

DR. KOPROWSKI: With reference to a virus that may be genetically transmitted in one species but propagated in another, I would like to describe briefly the experiments with SV40 and polyoma in mouse embryos. In contrast to mouse cells obtained from older embryos or adult tissue, mouse embryo cells at the two-celled stage, blastocyst stage, and egg cylinder stage support replication of simian virus (SV40) and some of the embryos through the ninth day of pregnancy (the latest stage still observed in culture). The same was observed in case of polyoma infection of mouse embryos for which the mouse is a permissive host. It is clear from these results that "permissiveness" of a host for infectious oncogenic viruses is not related to the species of the animal but to the stage of differentiation at which infection takes place. It is obviously important to find out whether infection of embryos at the earliest stage of differentiation results in the integration of the viral genome into the host DNA which may persist in animals even after birth.

DR. WEISS: Dr. Jaenisch has recently done some elegant experiments in this field [*Proc. Nat. Acad. Sci. U. S.* **71**, 1250–1254 (1974)]. He has infected early mouse embryos with SV40 DNA and normal mice developed in several cases. Molecular hybridization studies showed that many tissues contained the SV40 genome. He is currently studying whether the viral genome is transmitted to the next generation. So it is interesting to ask why genetic transmission of papovaviruses has not been seen in nature. Perhaps we have not looked carefully enough, for it would be necessary to examine nonpermissive hosts.

DR. SABIN: Dr. Weiss has stated that genetic mechanisms determine the nature of the receptor substance. Is this based on actually measuring the capacity of receptive substance to bind to virus particles or by implication of failure of the virus to multiply? Many years ago it was demonstrated that susceptibility or resistance of mice to certain arthropod-borne viruses was inherited as a dominant which apparently fitted the data of the single gene. The factor that determined the difference between a 100% survival and 100% death was a multiplication-inhibiting factor. I would like to know whether or not it is really demonstrated by the combining capacity of receptor substance for the virus. Of course, there could be a number of different mechanisms, but I'm curious to know the basis of your statement that it is the receptor substance that is influenced by the gene.

DR. WEISS: Piraino and Crittenden have shown that avian RNA tumor viruses adsorb to resistant cells. We think that resistance is expressed at the penetration stage because pseudotypes of Rous sarcoma virus (RSV), wrapped in envelopes of other virus strains to which the cell is susceptible, can infect and transform the resistant cell. Moreover, as Harriet Robinson was the first to show, RSV can penetrate resistant cells if inactivated Sendai virus is added to the cells after adsorption of the RSV. Therefore we think that the receptors of susceptible cells are located at the cell surface. We do not know their biochemical nature. The adsorption of virus by resistant cells has thwarted attempts to isolate receptors by binding capacity.

DR. SABIN: My intention was not to challenge this explanation which is beautifully logical but rather to indicate another way of actually checking the receptors. A mechanism for studying receptors without actually isolating them, which you said was so difficult, was initiated by Dr. Thomas Francis many years ago with polio, before we had plaque-counting methods. When I was working with polio (with different strains and in different tissues), we had a very simple method to measure receptors for a certain virus, by determining simply how many plaque-forming units of virus and tissue would remove from a known virus concentration, without ever knowing the chemistry of it. I thought that perhaps a direct check of this theory might be possible by measuring the capacity of certain chicken cells to remove plaque-forming units of the particular virus with a particular coat. Has that been done?

DR. WEISS: Yes. It's a nice idea. It was done many years ago by Piraino, and has been repeated by others. If you look at the removal of a focus-forming virus or at the absorption of radioactive-labeled virus, you find that it binds equally well to susceptible and to resistant cells. Absorption and penetration are two different phenomena with the avian tumor viruses, and it seems to be the penetration step which is specific. We are trying to determine whether we can get a differential elution of the viruses. So far we have not found any. Simple absorption or binding which would have been so nice has not provided the answer.

DR. FJELDE: Do agents such as phytohemagglutinin and concanavalin A effect the attachment, penetration and ultimate reproduction of viruses in your studies?

DR. WEISS: Concanavalin A agglutinates some strains of RNA tumor virus, so one may see a reduced titer because the virus particles have clumped together. However, polycations are widely used to enhance the adsorption of many strains of virus to the host cells.

DR. FJELDE: Does it affect the provirus also?

DR. WEISS: I don't think so. Once the provirus is stably associated with the cell, then the infection is stable. This is an intracellular form of the virus not an extracellular form.

CHAPTER 11

Type C Virus Inhibitor Associated with Cells Cultivated from New Zealand Black Mice

Jay A. Levy

I. INTRODUCTION

We have demonstrated biological activity of the C-type virus associated with New Zealand Black (NZB) mice by cocultivating NZB embryo cells with nonvirus-yielding, murine sarcoma virus (MSV)-transformed hamster and rat cells (*6, 7, 9*). The recovery of an NZB pseudotype sarcoma virus indicated the presence of a murine C-type virus in the NZB cells that could not propagate in mouse cells. It could grow and replicate, however, in cells from several other animal species including rat, cat, rabbit, guinea pig, cow, and man. Because of this unusual tropism of this mouse virus, we have called it a *xenotropic* virus, from the Greek inferring it turns toward a foreign species (*7*). During the course of these studies we noted that the titer of our NZB pseudotype viruses increased two- to threefold as we diluted the supernatant fluid containing the virus (*6*). Subsequent experiments showed that supernatants from NZB mouse embryo cultures when added to mouse cells prevented focus formation by other MSV and blocked XC plaque formation by standard murine leukemia virus (MLV). We could use these supernatants directly because they contained very low titers of the NZB xenotropic virus and, moreover, the virus could not be propagated in mouse cells. It was tantalizing to speculate that this inhibitory activity was in some way responsible for the low yields of infectious virus from NZB cells as well as related, perhaps, to the resistance of mouse cells to infection with NZB virus. Let me add that supernatants from NIH Swiss and BALB/c mouse embryo cells grown in the same media did not show the presence of any inhibitory factors.

II. BIOLOGICAL AND PHYSICO-CHEMICAL PROPERTIES OF THE INHIBITOR

In collaboration with Drs. Chermann and Jasmin in France, this "inhibitor" was extracted in a crude form from the supernatant from an NZB cell culture by the following method (*8, 11*): 5% polyethylene glycol (PEG) was added to the supernatant to precipitate the virus. The PEG was then eliminated by adding an equal volume of a potassium phosphate buffer at pH 7.0. At this point, two distinct liquid phases formed. The upper phase contained the PEG; the lower phase contained the inhibitor. This lower phase was dialyzed to remove any possible remaining PEG, and then lyophilized. The quantity of lyophilized material obtained amounted to approximately 3 mg/ml of the original NZB cell culture supernatant.

By using this lyophilized material containing the inhibitor, we were able to repeat results first observed with supernatant fluids of the NZB tissue culture cells and study in a more quantitative way the biological and chemical characteristics of the inhibitor. Some of these results have been reported elsewhere (*8*).

The biological properties of the inhibitor preparation include the following (Table 1):

1. This inhibitor is associated with most NZB mouse cell cultures, but not with uninfected cultures of other mouse cells.
2. It prevents focus formation and replication by MSV in mouse cells

TABLE 1. BIOLOGICAL ACTIVITY OF THE INHIBITOR ASSOCIATED WITH NZB CELLS

1. Associated with NZB cell cultures.
2. No direct effect on C-type viruses.
3. Inhibits focus formation by murine sarcoma viruses in mouse, rat, and human tissue culture cells.
4. Inhibits focus formation by Rous sarcoma virus in chick cells.
5. Inhibits murine leukemia virus XC plaque formation in mouse cells.
6. Inhibits replication of murine and avian C-type viruses.
7. Active after virus penetration.
8. Active primarily during the first hour after virus infection.
9. Not neutralized by antiserum prepared against mouse interferon.
10. Not active against VSV, EMC, or SV40 viruses.

at 25 μg/ml. Assays were conducted using standard techniques (*3, 5*). The reduction in progeny MSV correlates directly with the extent of reduction in focus formation. At a five- to tenfold increase in concentration (125–250 μg/ml), it suppresses focus formation by MSV in rat and human cells (Table 2).

3. It inhibits focus formation by Rous sarcoma virus in chicken cells at 125 μg/ml.

4. It prevents the replication of MLV as measured by XC plaque formation (*12*). The dose–response relations for the inhibitor assayed against 50 plaque-forming units (PFU) of MLV in mouse cells shows 50% inhibition with about 18 μg/ml of the inhibitor preparation, and even at 5 μg/ml some inhibition can be noted (Fig. 1). Using the same input of MLV in biological assays, we can use this dose–response curve to estimate the amount of inhibitor detected in supernatants of NZB cells. In general, positive NZB cultures contain between 5–20 μg/ml. Since 3 mg of crude material is obtained from 1 ml of original supernatant, it appears that antagonists to the inhibitor must also be present in the supernatant and are removed, perhaps, by the PEG treatment and/or dialysis.

5. There is no direct effect of the inhibitor on the virus itself. When you incubate MLV with 100 μg of inhibitor for 30 minutes at room temperature, no inactivation or neutralization of virus occurs (Table 3).

6. The inhibitor is active after virus penetration. In these experiments, virus is allowed to adsorb for 30 minutes at 37°C. The culture is then trypsinized to remove any residual virus which has not penetrated the cell (*10*), and the cells are replated in the presence of inhibitor (Table 3).

TABLE 2. EFFECT OF NZB INHIBITOR ON MSV FOCUS FORMATION[a]

Virus	NIH-ME	NRK	W138
H-MSV	41	500	—
+NZB-I (250 μg/ml)	0	143	—
+NZB-I (25 μg/ml)	7	500	—
M-MSV	—	—	140
+NZB-I (250 μg/ml)	—	—	24

[a] Values expressed as the average number of foci induced by murine sarcoma virus (MSV) in duplicate plates of mouse (NIH-ME), rat (NRK), and human (W138) cells. Assays were conducted as previously described (*3, 5*). The Moloney sarcoma virus (M-MSV) employed is a variant which grows efficiently in human cells (*1*). Two separate experiments with the Harvey strain of MSV (H-MSV) are presented in which H-MSV was inoculated at a different input multiplicity onto the mouse and rat cells.

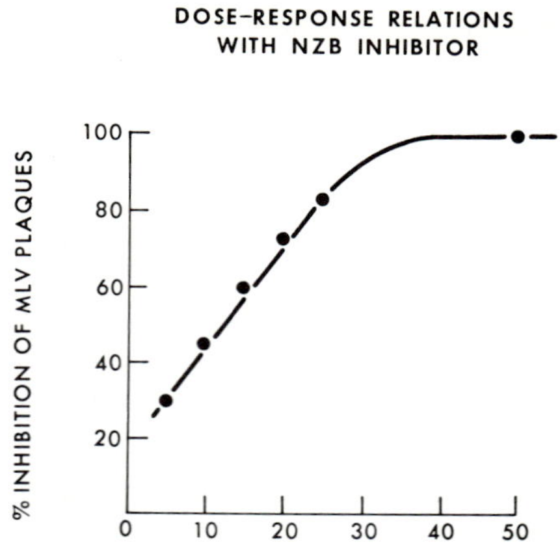

Fig. 1. The Gross strain of murine leukemia virus was added at an input multiplicity of 50 plaque-forming units/culture to NIH Swiss mouse embryo cells in the presence of varying concentrations of the NZB inhibitor preparation. The degree of MLV inhibition is expressed as the percent in reduction of XC plaques formed (*12*).

7. In experiments in which the inhibitor is added to cultures at various time periods after virus adsorption, the inhibitor demonstrates its most significant effect primarily during the first hour after virus infection (Table 4). It is interesting in fact that this inhibitor can have an effect even after only 1 hour exposure to cells.

8. The activity of this inhibitor does not depend on pretreatment of cells with diethylaminoethyl (DEAE)-dextran.

9. It is not neutralized by antimouse interferon.*

TABLE 3. BIOLOGICAL PROPERTIES OF THE NZB INHIBITOR[a]

Assay	NIH-ME
A. AKR-MLV alone	125
AKR-MLV incubated with NZB-I (100 µg) for 30 min at 20°C	150
B. AKR-MLV absorbed 30 min at 37°C, cells trypsinized and replated:	
alone	150
+NZB-I (50 µg/ml)	45

[a] Values expressed as the average number of XC plaques induced by mouse leukemia virus (MLV) in duplicate plates of NIH mouse embryo (NIH-ME) cells. The assays were conducted according to the method of Rowe *et al.* (*12*).

* Supplied by Dr. George Galasso, National Institutes of Health.

TABLE 4. TIME OF ACTIVITY OF NZB INHIBITOR AFTER VIRUS INOCULATION

G-MLV[a]	Time inhibitor was added (hr)[b]	Time culture was washed and refed (hr)	NIH-ME[c]
+	None	1	160
+	0	1	75
+	0	2	50
+	0	3	50
+	0	6	45
+	1	6	125

[a] The Gross strain of murine leukemia virus (G-MLV) was absorbed to cell cultures for 30 min, 37°C, before the timed experiments were begun.

[b] 50 μg/ml of the NZB inhibitor preparation was used.

[c] Values expressed as the average number of XC plaques (*12*) induced by G-MLV in duplicate plates of NIH mouse embryo (NIH-ME) cells.

10. Finally, the inhibitor is not active against vesicular stomatitis virus (VSV), encephalomyocarditis virus (EMC), nor simian virus 40 (SV40) as determined by Dr. Lois Epstein, Mr. Jack Weaver, and Mr. Ed Teisher, respectively.

We have evaluated some of the chemical characteristics of this inhibitory substance. Although these results may not all be confirmed with purified material, they do give some indication of its chemical nature (Table 5).

1. It is not dialyzable and, as a rough estimate as determined by sucrose density gradient separation, has a molecular weight of about 80,000.
2. It is resistant to heat at 56°C for 1 hour (Table 6).
3. It is resistant to heat at 80°C for 5 minutes. At this temperature, a precipitate is formed, and inhibitor activity can be found in the supernatant, although somewhat diminished.

TABLE 5. PHYSICAL AND CHEMICAL PROPERTIES OF THE NZB INHIBITOR

1. Not dialyzable.
2. Molecular weight ~80,000.
3. Resistant to heat, 56°C for 60 min.
4. Resistant to heat, 80°C for 5 min.
5. Resistant to chloroform.
6. Resistant to pH 2.0.
7. Sensitive to pH 12.0.
8. Sensitive to trypsin.

4. It is resistant to chloroform (Table 6).
5. It is stable in acid at pH 2.0 (Table 6) but inactivated at pH 12.0.
6. It is sensitive to trypsin.

The inhibitory effect is not due to the presence of RNAse or DNAse in our crude preparation since neither of these substances when used in our assays inhibit C-type viruses (*8*). The NZB cells that are producing this substance in low concentration can be infected by standard MLV, and the titers of the virus are equal to those obtained in other mouse cells. Additional amounts of inhibitor, however, will suppress MLV infection of these cells.

The inhibitor added to cells chronically infected with MLV does not significantly alter the amount of progeny MLV produced. These experiments suggest that the inhibitor does not affect RNA synthesis. To determine whether it affects DNA synthesis, cell counts and tritiated thymidine uptake were measured in cultures after the addition of the inhibitor. No difference in cell counts was detected, nor any toxicity of the inhibitor preparation noted even at doses as high as 1 mg/ml. However, after 4 and 24 hours, a 10–30% decrease in thymidine uptake into acid-precipitable material was noted in those cultures receiving 100–200 μg of the inhibitor. In these experiments, 1 μC of tritiated thymidine was used per culture, and DNA was extracted and measured by the technique of Scott *et al.* (*13*). It is doubtful that this degree of inhibition of cellular DNA synthesis is responsible for the C-type virus inhibition observed since the amount of virus inhibition with this concentration of inhibitor was significantly greater. Moreover, based on Dr. Humphries'

TABLE 6. CHEMICAL CHARACTERISTICS OF THE INHIBITOR ASSOCIATED WITH NZB CELLS

Assay	NIH-ME[a]
A. AKR-MLV alone	185
AKR-MLV + NZB-I (250 μg/ml)	
heated 56°30′	10
heated 56°60′	16
B. AKR-MLV alone	135
+NZB-I (250 μg/ml) (extracted with chloroform)	1
+ medium (extracted with chloroform)	145
C. G-MLV alone	43
+NZB-I (50 μg/ml), held at pH 2.0 for 20 hr	0

[a] Values are expressed as the average number of XC plaques (*12*) induced by mouse leukemia virus (MLV) in duplicate plates of NIH mouse embryo (NIH-ME) cells.

comments this morning (*4*), it appears C-type viruses can infect a cell whose DNA synthesis is blocked and can remain in a latent state within the cells. Once cellular DNA synthesis is resumed, the viral genome can be expressed, and progeny virus is produced (*4*). We have found that when the NZB inhibitor is removed after 1 hour, complete recovery of the cellular DNA synthesis occurs. Nevertheless, as I have shown by the timed experiments, this short exposure to the inhibitor preparation is enough to give more than 50% inhibition of mouse leukemia virus replication in tissue culture. These observations suggest that the inhibitor directly affects the expression of the virus. It probably acts in the early stages of viral infection, perhaps at the transcription of viral RNA to DNA. The NZB inhibitor preparation inhibits reverse transcriptase activity *in vitro* (*4a*) but no conclusions can be drawn from these experiments since the crude preparation was used.

Our inhibitor resembles in some ways the biological inhibitor described by Raynaud and his associates (*11*). It differs from the inhibitor described by Cassingena and Tournier which specifically inhibits SV40 and is thermolabile (*2*). Perhaps, however, many viruses induce the production of viral inhibitors that are specific for the particular infecting virus. The detection of this inhibitor so readily in NZB cells, then, may be directly related to their endogenous production of the xenotropic virus. We would imagine, in fact, that this NZB-type inhibitor might be found in many if not all cells producing C-type viruses and, perhaps, in many nonvirus-producing MSV-transformed cells. Human cells propagating the xenotropic NZB virus also appear to product this inhibitor. Moreover, we have noted some inhibitory activity in supernatants taken from nonvirus-producing rat cells transformed by Harvey sarcoma virus.

I recognize that certain of the properties described may change after purification of this inhibitor. Nevertheless, it is clear that a substance, most likely a protein, which inhibits C-type viruses is associated with NZB cell cultures. It is not a result of the extraction procedure since supernatants from normal uninfected mouse cell cultures and even some NZB cell cultures prepared in the identical way do not contain any inhibitory activity. In fact, in an analysis of lines established from 33 one-cell clones of NZB embryo cells, 27/33 or 77% produce enough inhibitor in the supernatant that we can detect it by our tissue culture assays, but 23% do not. Our results illustrate that the existence of a virus-inhibitory substance or substances in cell culture supernatants should be considered since they may prevent detection of C-type viruses produced by other cells suspected of containing these viruses. Careful dilution of the samples or separation of the viruses from the cell culture medium may be a necessary prerequisite for their eventual identification.

ACKNOWLEDGMENTS

I would like to thank Drs. Mel Goldberg and Ron Rasmussen for their advice and assistance in this work.

The work reported was supported by U. S. Public Health Service grant CA 13086 and a grant from the Hobson Foundation.

REFERENCES

1. Bernard, C., Lasneret, J., Boucher, M., and Boiron, M., *C.R. Acad. Sci.* **268**, 624–627 (1969).
2. Cassingena, R., and Tournier, P., *C.R. Acad. Sci.* **267**, 2251–2255 (1968).
3. Hartley, J. W., and Rowe, W. P., *Proc. Nat. Acad. Sci. U. S.* **55**, 780–786 (1966).
4. Humphries, E., *Perspect. Virol.* **9**, 215–231.

4a. Levinson, W., unpublished data.

5. Levy, J. A., *J. Nat. Cancer Inst.* **46**, 1001–1007 (1971).
6. Levy, J. A., *Amer. Soc. Microbiol. Abstr.* p. 222 (1973).
7. Levy, J. A., *Science* **182**, 1151 (1973).
8. Levy, J. A., Chermann, J.-C., Jasmin, C., and Raynaud, M., *C.R. Acad. Sci.* **277**, 1421–1423 (1973).
9. Levy, J. A., and Pincus, T., *Science* **170**, 326–327 (1970).
10. Levy, J. A., and Rowe, W. P., *Virology* **45**, 844–847 (1971).
11. Raynaud, M., Chermann, J.-C., Plata, F., Jasmin, C., and Mathé, G., *C.R. Acad. Sci.* **272**, 2038–2040 (1971).
12. Rowe, W. P., Pugh, W. E., and Hartley, J. W., *Virology* **42**, 1136–1139 (1970).
13. Scott, S. F., Fraccastoro, A. P., and Taft, E. P., *J. Histochem. Cytochem.* **4**, 1–10, (1956).

DISCUSSION

DR. DEINHARDT: A clear distinction should be drawn between the inhibitor found in NZB cell culture supernatants and the previously reported inhibitors in JLSV-5 cell supernatants. The latter have been shown clearly to be nonspecific, i.e., they were cytotoxic and did not selectively inhibit viral multiplication. Similar inhibitors were found in preparations of fresh unused cell culture media, and the activity of these inhibitors probably was due to a carryover of phosphotungstic acid used at that time for the preparation of the inhibitors (Chirigos, M. A., Deinhardt, F., and Marszynska, B., unpublished data).

DR. TENNANT: I would like to ask Dr. Levy, Does inhibitor alter titration pattern (in 1 or 2 hit) in permissive cells?

DR. LEVY: It hasn't been tried.

CHAPTER 12

Inhibition of Rous Sarcoma Virus Replication by Inhibitors of Cell Replication

Eric H. Humphries*

I. INTRODUCTION

Rous sarcoma virus (RSV) is the prototype RNA tumor virus. The virion of RSV is enveloped and contains an RNA genome. The usual host for RSV is the chicken.

A model of the replication of RSV is presented in Fig. 1. After entrance into a cell, the DNA polymerase associated with the viral RNA genome transcribes the viral RNA into DNA. The DNA copy of the virus genome, the provirus, becomes stably associated with the infected cell. The DNA provirus is replicated with host cell DNA and, upon division of the infected cell, the provirus is transmitted to both daughter cells in the same fashion as normal cellular genes. The production of progeny RSV requires transcription of the DNA provirus into RNA.

Early studies (*1, 10–12, 14, 16*) with RSV demonstrated that its replication was sensitive to a number of inhibitors. These inhibitors have been placed into three groups (Table 1). Group I includes inhibitors of DNA synthesis, for example, excess thymidine and cytosine arabinoside. The sensitivity of RSV replication to these Group I inhibitors involves an early stage in the replication of the virus and is transient.

* Present address: Imperial Cancer Research Fund, London, England.

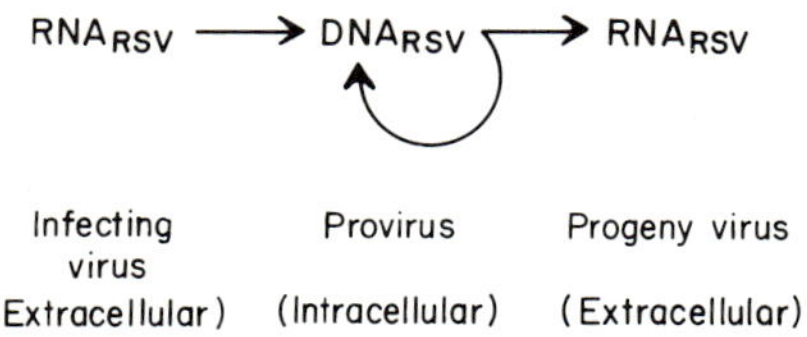

Fig. 1. A model for the replication of Rous sarcoma virus in permissive hosts.

Group II contains actinomycin D, which inhibits transcription of DNA into RNA. The sensitivity of RSV replication to this drug involves a late stage in the replication of the virus and is not transient. Group III includes a number of inhibitors of cell replication such as colchicine and serum-free medium. The inhibitors found in Groups I and II also inhibit cell replication. Experiments using these inhibitors demonstrated that when cell division was prevented after infection with RSV, progeny virus did not appear or was delayed (*5, 13*). This requirement for cell division after infection for production of progeny virus has been called "activation" (*13*).

The molecular nature of the action of the inhibitors in Groups I and II is understandable in terms of the model of the replication of RSV presented in Fig. 1. Early in the replication of RSV, inhibitors of DNA synthesis prevent formation of the DNA provirus, whereas later treatment of RSV-infected cells with actinomycin D prevents transcription of the DNA provirus into viral RNA. However, it was not clear how inhibition of cell division prevented the replication of RSV. The work described

TABLE 1. INHIBITORS OF ROUS SARCOMA VIRUS REPLICATION

Group I (Inhibitors of DNA synthesis)	Excess thymidine Cytosine arabinoside 5-Bromodeoxyuridine Amethopterin
Group II (Inhibitor of RNA transcription from a DNA template)	Actinomycin D
Group III (Inhibitors of the replicative cell cycle)	Ultraviolet radiation X-Irradiation Mitomycin C Colchicine Serum-free medium Group I Group II

here was performed to determine the nature of the sensitivity of RSV replication to inhibitors of cell replication.

II. ABSENCE OF VIRAL ANTIGENS IN ROUS SARCOMA VIRUS-INFECTED STATIONARY CHICKEN EMBRYO FIBROBLASTS

To study the relationship between the replication of RSV and cell division, a synchronized cell culture system was developed using serum-free medium. This system provided both a reversible inhibition of cell division and cell synchrony. In the absence of serum, chicken embryo fibroblasts become stationary, coming to rest in interphase or I-G_1 (Fig. 2). In the presence of serum-containing medium, these stationary cells are stimulated to move into S-G_1, where they are committed to DNA synthesis, then into S, then into G_2, and then into mitosis (*15*).

Such stationary chicken cells, in serum-free medium, were exposed to RSV. After infection, the medium on some cultures was changed to serum-containing medium. For both serum-stimulated and stationary cells, the number of cells per culture and the virus titer in the culture fluid were determined at different times. Cells were also examined for the presence of viral antigens using indirect immunofluorescence. The results (Fig. 3) show that the dividing cells were 80% positive for viral antigens at 48 hours after infection, while the control stationary cells were only 5% positive. Those stationary cells initially negative for viral antigens were 80% positive for viral antigens at 140 hours after infection after stimulation with serum at 84 hours after infection. The titer of RSV in the culture fluid of stationary cells at 48 hours after infection was 1% of that found in the culture fluid from dividing cells.

These results demonstrate that, although the DNA provirus of RSV is stably associated with infected stationary cells, viral antigens and progeny virus do not appear until after cell multiplication as a result of stimulation with serum.

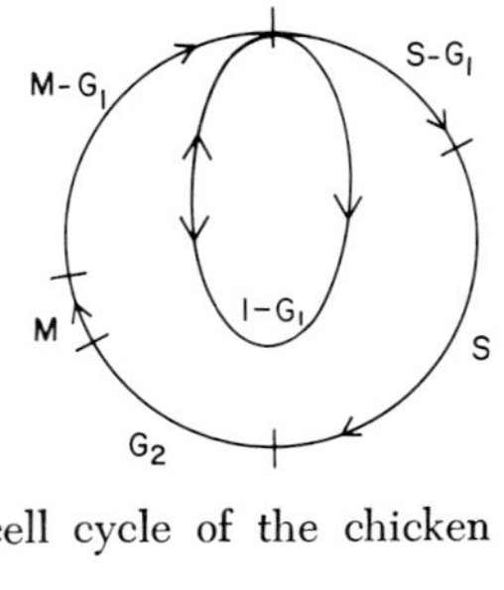

Fig. 2. A model of the cell cycle of the chicken embryo fibroblast. Taken from Temin (*15*).

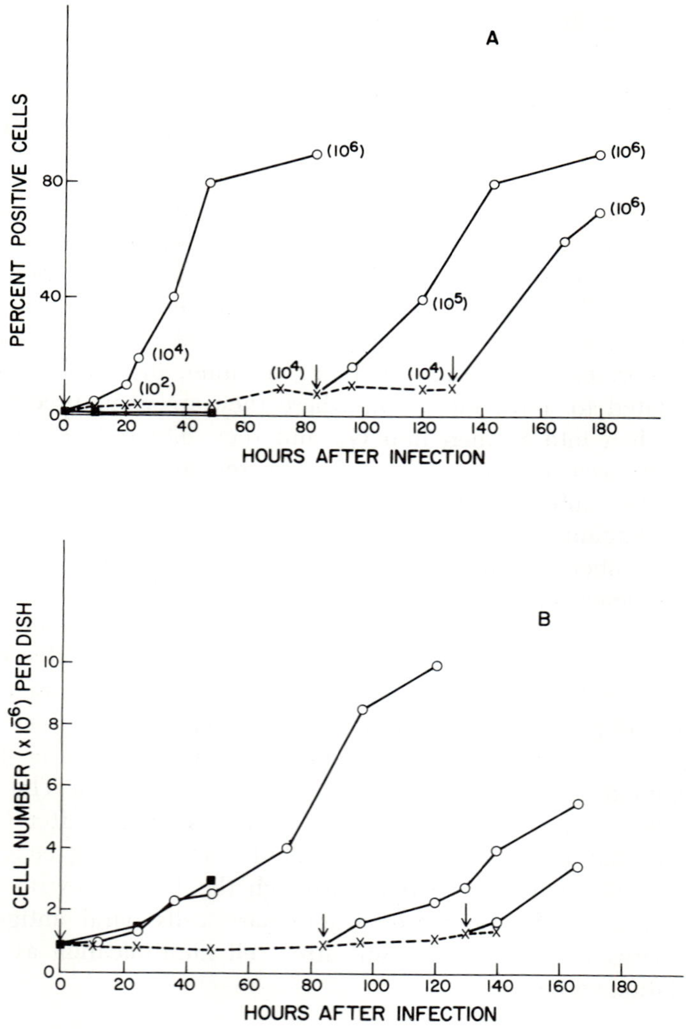

Fig. 3. Absence of viral antigens in Rous sarcoma virus-infected stationary chicken embryo fibroblasts. Stationary chicken embryo fibroblasts were exposed to Rous sarcoma virus. Serum-containing medium was added to some cultures (↓) at 0, 84, and 130 hours after infection. (A) At different times, stationary (×) and dividing (○) RSV-infected cells and (■) uninfected cells were examined by indirect immunofluorescence for viral antigens, and culture fluids were assayed for RSV. Virus titers, FFU* per milliliter are given in parentheses. (B) Cell number per culture was determined for stationary (×) and dividing (○) RSV-infected cells and (■) uninfected cells at different times. Taken from Humphries and Temin (*6*).

* Focus-forming units.

III. EXPERIMENTAL SYSTEMS

To examine the nature of the requirement for cell division in the replication of Rous sarcoma virus more closely, infected cells were examined for the presence of virus-specific RNA at different times before and after stimulation with serum. To detect virus-specific RNA, RNA–DNA hybridizations were performed between RNA prepared from RSV-infected cells and the DNA-^{3}H made in the *in vitro* endogenous DNA polymerase reaction of RSV virions. In the presence of actinomycin D and labeled substrates in the *in vitro,* endogenous, DNA polymerase reaction, the virion DNA polymerase synthesizes DNA-^{3}H complementary to the majority of the sequences in the virus RNA genome (*4*). This DNA was purified and hybridized to cellular RNA. The hybridization reaction was analyzed with a single strand-specific nuclease, S_1, to determine the percentage of the input DNA hybridized (*8*).

The hybridization data are expressed as the percent of input RSV DNA-^{3}H resistant to S_1 nuclease at a particular C_rt value [where the C_rt value is determined by the RNA concentration in moles of nucleotides per liter times the duration of the hybridization in seconds (*2*)]. The results of a standard hybridization are shown in Fig. 4 where RSV RNA

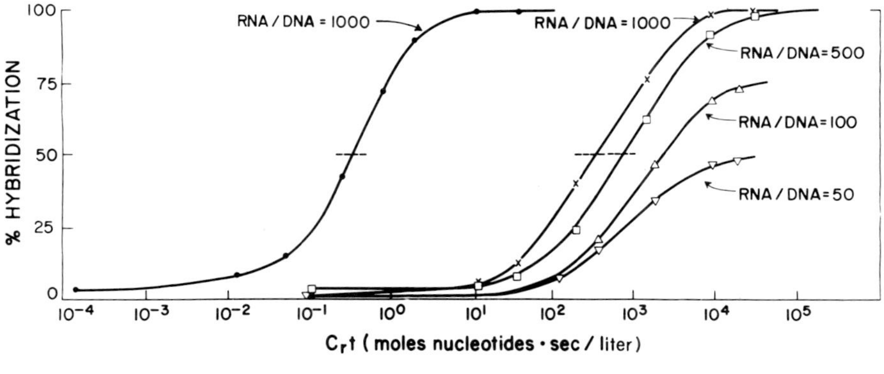

Fig. 4. Hybridization of Rous sarcoma virus (RSV) RNA to RSV DNA-^{3}H. The RSV/RNA was hybridized at 6.6 μg/ml (×), 3.3 μg/ml (□), 0.66 μg/ml (△) and 0.33 μg/ml (▽) in the presence of 7.5 mg/ml uninfected chicken cell RNA and at 6.6 μg/ml (●) in sealed 10-μl micropipettes to 7500 dpm of RSV DNA-^{3}H product, 120 dpm/pg, at 63°C in 0.6 *M* NaCl, 0.05 *M* Tris-HCl (pH 7.5), 0.001 *M* ethylenediaminetetraacetate, 0.1% sodium dodecyl sulfate, and 1% phenol. Samples were withdrawn at different times and frozen at −60°C. The extent of hybridization was determined by S_1 nuclease digestion at the end of the incubations. The results are expressed as the proportion of total DNA-^{3}H hybridized at a certain C_rt value. All C_rt values were determined from the total of RNA in the hybridization mixture and corrected for ionic strength (*3*).

was hybridized to RSV DNA-^{3}H. Fifty percent hybridization was achieved at a C_rt value (the C_rt_{50} value) of 1–2 $\times$ 10^{-1} moles sec/liter. The C_rt_{50} value depends on the genetic complexity of the RNA being examined (*2*). In the results to be presented later, the genetic complexity of the RNA was constant (only RSV RNA was examined). Therefore, the differences in the C_rt_{50} values obtained when different RNA preparations were hybridized to DNA-^{3}H were reflections of differences in the concentration of the virus RNA in the samples. For example, compared to the C_rt_{50} value of RSV RNA 1–2 $\times$ 10^{-1} mole sec/liter, a sample of cell RNA containing 0.1% virus RNA hybridized with a C_rt_{50} value of 1–2 $\times$ 10^2 moles sec/liter as shown in Fig. 4.

One technical difficulty with this method of analysis is that the relative abundances of the DNA copies of the different RNA sequences vary. Since the hybridization reaction is being driven to completion by the RNA excess, and the RNA/DNA ratio for different sequences varies, the DNA copies of different sequences will hybridize to different degrees of completion. Consequently, the amount of RNA excess will affect the percent hybridization observed. The importance of the RNA/DNA ratio is demonstrated in Fig. 4 where a constant amount of viral DNA-^{3}H has been hybridized to different concentrations of virus RNA. Assuming that the different virus sequences are equally represented in virus RNA, these data demonstrate that, by changing the concentration of the RNA without changing the genetic complexity, the percentage hybridization is affected. Therefore, in the experiments to be presented later, it must be realized that the observed differences in percent hybridization may reflect differences in the amount of viral RNA in the cell as well as differences in the genetic complexity of this RNA.

This hybridization technique made it possible to examine stationary RSV-infected chicken embryo fibroblasts for the appearance of virus RNA after stimulation of the cells with serum-containing medium. The hybridization technique is estimated to be sensitive enough to detect less than 1 virus genome equivalent of RNA per cell (*7*). This sensitivity required a well synchronized population of chicken embryo fibroblasts in order to be able to determine the time of the appearance of virus RNA. This synchrony was established by making chicken embryo fibroblasts stationary by plating cells in medium with low serum and allowing the cells to deplete the medium of multiplication-stimulating activity (*9, 15*). Addition of fresh serum-containing medium to these stationary cells stimulated the cells, resulting in the cell cycle shown in Fig. 5. After an initial delay, the rate of DNA synthesis increased reaching a maximum at 12 hours after addition of serum. The cell number was

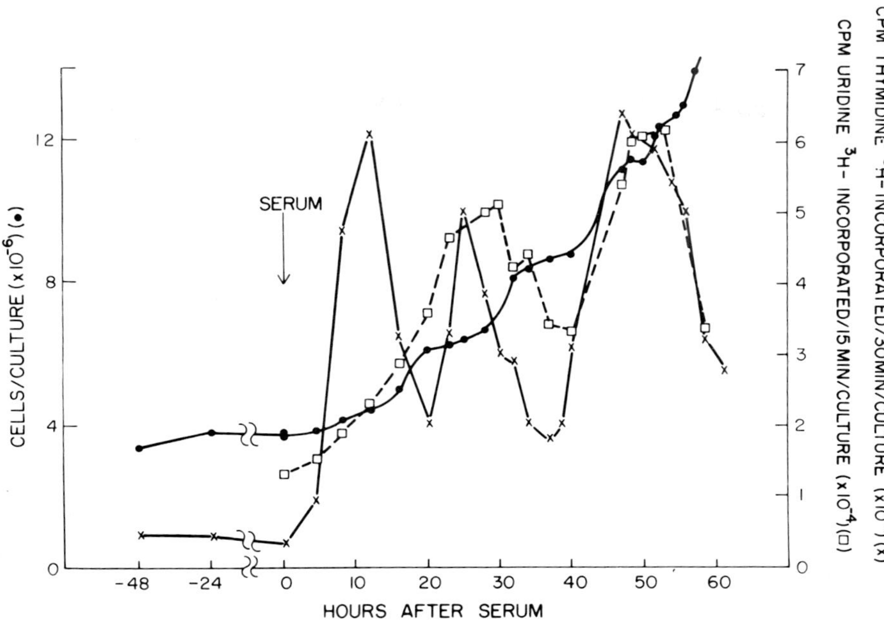

Fig. 5. Cell cycle of chicken embryo fibroblasts. Chicken embryo fibroblasts were made stationary in multiplication-stimulating activity-depleted, serum-containing medium. Eight days after plating, the media were changed to serum-containing media. At different times, the number of cells per culture, the amount of incorporation of thymidine-^{3}H in 30 minutes, and the amount of incorporation of uridine-^{3}H in 15 minutes were determined.

first observed to increase at approximately 12 hours after addition of serum, with an increase of approximately 5 to 10 percent. The major increase in cell number was observed at approximately 16 to 18 hours after addition of serum. The rate of cell RNA synthesis was observed to increase shortly after the major period of DNA synthesis. Thereafter, the rate of RNA synthesis increased and decreased in parallel with the rate of DNA synthesis. Synchronous increases in cell number were observed at approximately 30 and 45 hours after addition of serum.

IV. APPEARANCE OF VIRUS RNA IN ROUS SARCOMA VIRUS-INFECTED STATIONARY CHICKEN EMBRYO FIBROBLASTS AFTER STIMULATION WITH SERUM

Chicken embryo fibroblasts made stationary in multiplication-stimulating activity depleted, serum-containing medium were exposed to RSV, and, 12 hours after infection, fresh serum-containing medium was added.

At different times after addition of serum, cells were harvested, and cytoplasmic and nuclear RNA's were prepared. The RNA's were then hybridized to RSV DNA-^{3}H. Results from a typical experiment are presented in Fig. 6. The results show that nuclear RNA from uninfected cells hybridized 2–3% of the RSV DNA-^{3}H at a C_rt value of 10^4 moles sec/liter. Nuclear RNA prepared from RSV-infected cells harvested at 0, 6, and 1 hours after addition of serum showed no hybridization above that detected in uninfected cells. Increased hybridization was first detected with the RNA from cells harvested at 18 hours after addi-

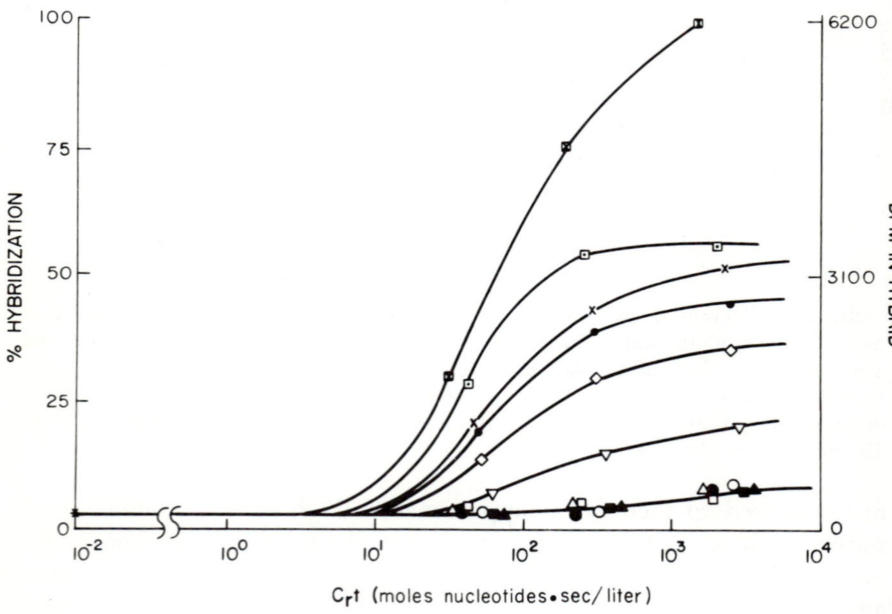

Fig. 6. Appearance of virus-specific RNA in Rous sarcoma virus (RSV)-infected chicken embryo fibroblasts. Chicken embryo fibroblasts were made stationary in multiplication-stimulating activity-depleted, serum-containing medium. Seven days after plating, the cells were either mock-infected or exposed to RSV at a multiplicity of 0.3 to 0.4 FFU/cell for 40 minutes at 37°C. The inoculum was removed, and 7 ml of multiplication-stimulating activity-depleted, serum-containing medium were added to each culture. Twelve hours after infection, the medium was changed to serum-containing medium. Cells were harvested at different times, washed, and frozen. Nuclear RNA was prepared and hybridized to 6200 dpm of RSV endogenous DNA product-^{3}H in 10-μl micropipettes, as described in the caption of Fig. 5. Uninfected cell RNA (●) 0 hours after serum (HAS), (▲) 12 HAS, and (■) 72 HAS. Rous sarcoma virus-infected cell RNA (○) 0 HAS, (△) 6 HAS, (□) 12 HAS, (▽) 18 HAS, (◇) 24 HAS, (•) 29 HAS, (×) 36 HAS, (⊡) 53 HAS, and (⊠) 73 HAS. S_1 nuclease background (*).

tion of serum. Ribonucleic acid prepared from cells harvested at later times, 24, 29, 36, 53, and 73 hours after addition of serum, showed a further increase in the amount of hybridization detected.

To determine if the increase in hybridization reflected an increase in the genetic complexity of the virus RNA detected, RNA was prepared from cells harvested at 24 and 72 hours after addition of serum and was hybridized to RSV DNA-^{3}H (Fig. 7). The RNA in these hybridizations was approximately 10 times more concentrated than in the hybridizations of Fig. 6 and was able to hybridize 100% of the virus DNA product. Therefore, the RNA detected in cells harvested at 24 hours after addition of serum contained all the sequences represented in the DNA product. In addition, comparison of the C_rt_{50} values demonstrates

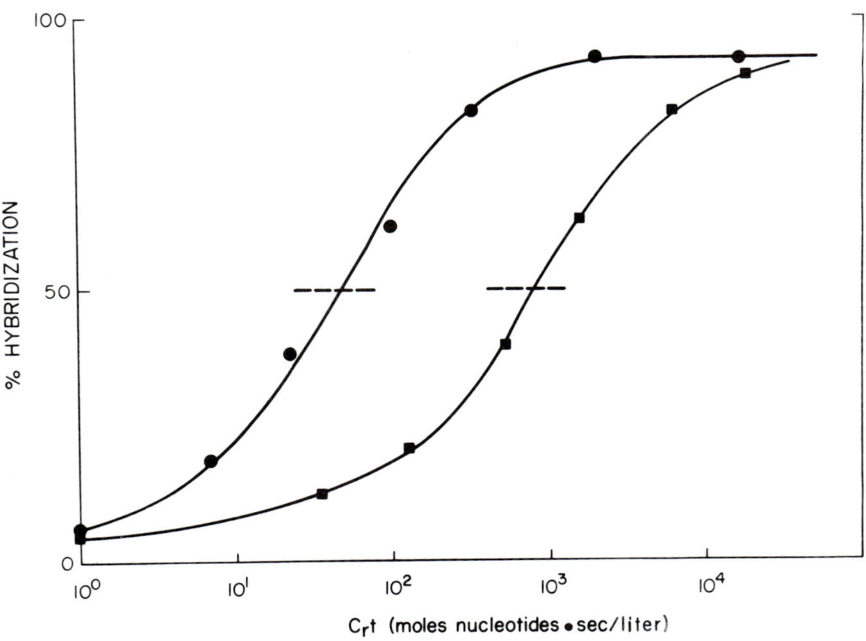

Fig. 7. Quantitation of virus-specific RNA in Rous sarcoma virus (RSV)-infected stationary chicken embryo fibroblasts at 24 and 72 hours after addition of serum. Stationary chicken embryo fibroblasts were prepared and exposed to RSV at a multiplicity of approximately 0.3–0.4 FFU/ml, and the medium was changed to serum-containing medium 12 hours after infection. Cells were harvested at 24 and 72 hours after addition of serum, and cytoplasmic RNA was prepared and hybridized to RSV DNA-^{3}H product in 10-μl capillary pipettes as described in the caption of Fig. 4; 2000 dpm was annealed to 24 hours after serum RNA (■), and 5000 dpm was annealed to 72 hours after serum RNA (●).

that, by 72 hours after the addition of serum, the concentration of these virus RNA sequences in the infected cell increased ten-fold over the concentration of these virus RNA sequences in the infected cell 24 hours after addition of serum.

A summary of the experiment presented in Fig. 6, including the rates of RNA and DNA synthesis, the change in cell number, the virus titers, and the hybridization of the cytoplasmic RNA preparations is presented in Fig. 8. The cell cycle was similar to that described in Fig. 5. The maximum rate of DNA synthesis was observed at 12 hours after addition of serum, and, while a 10% increase in cell number was observed at this time, the major increase in cell number occurred at approximately 20 hours after addition of serum. The first hybridization detected was with RNA prepared from cells harvested 18 hours after addition of serum. The percent hybridization gradually increased until all the DNA was hybridized by RNA from cells harvested at 48 hours after addition

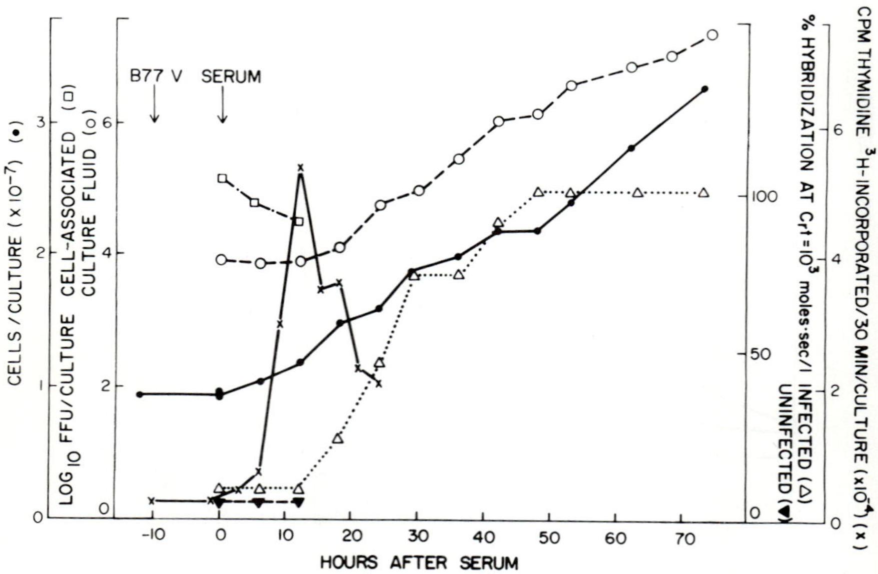

Fig. 8. Appearance of virus-specific RNA and progeny virus in Rous sarcoma virus-infected stationary chicken embryo fibroblasts after addition of serum. The number of cells per culture, the amount of incorporation of thymidine-^{3}H in 30 minutes, the titer of virus in the culture fluid, and the titer of cell-associated virus were determined at different times. The percent hybridization plotted represents the extent of hybridization observed at a C_rt value of 10^3 moles sec/liter for the cytoplasmic RNA samples from uninfected and infected cells from the experiment described in Fig. 6. B77 virus was used in the infection.

of serum. The increase in the percent hybridization observed represents, as shown in Fig. 7, an increase in the concentration of the virus RNA and not an increase in its genetic complexity. Virus production was first detected at approximately 24 hours after addition of serum. The high background at earlier times was apparently due to the large amount of cell-associated virus.

These results indicate that virus RNA does not appear until after the first mitosis. Therefore, although the DNA provirus is present and stable in infected stationary chicken embryo fibroblasts, it is not transcribed into RNA until after the cell undergoes mitosis.

To determine what effect inhibitors of mitosis would have on the appearance of virus RNA, colchicine was added to RSV-infected chicken embryo fibroblasts after addition of serum. Figure 9 shows the effects

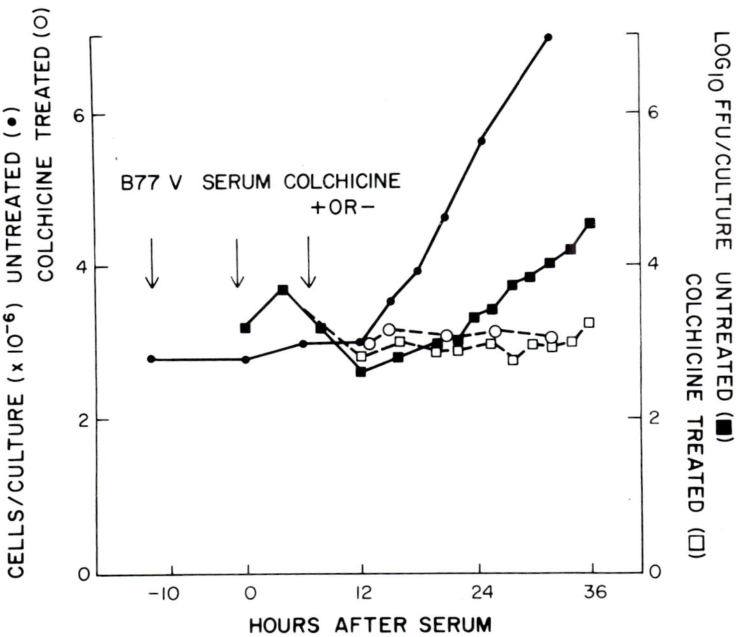

Fig. 9. The effect of colchicine on cell division and production of Rous sarcoma virus (RSV). Stationary chicken embryo fibroblasts were prepared, and infected with RSV at a multiplicity of infection of 0.5 FFU/cell, and serum-containing medium was added at 12 hours after infection. At 6.5 hours after addition of serum-containing medium, some cultures were treated with 5×10^{-7} M colchicine. The cell number per culture for untreated and colchicine-treated cultures and the virus titers for untreated and colchicine-treated cultures were determined at different times. B77 virus was used in the infection.

on cell number and virus production when colchicine, 5×10^{-7} *M*, was added to RSV-infected chicken embryo fibroblasts 7 hours after addition of serum to stationary cells. In the presence of colchicine, an increase in neither cell number nor virus production was detected by 36 hours after addition of serum. In a similar experiment, colchicine was added at the time of addition of serum, and, at different times after addition of serum, cells were harvested for preparation of RNA. These RNA samples were hybridized to RSV DNA-^{3}H. The results (Fig. 10) demonstrate that virus RNA was first detected in untreated infected cells at 24 hours after addition of serum. Comparison of the C_rt_{50} values for the hybridizations of the RNA samples from untreated infected cells

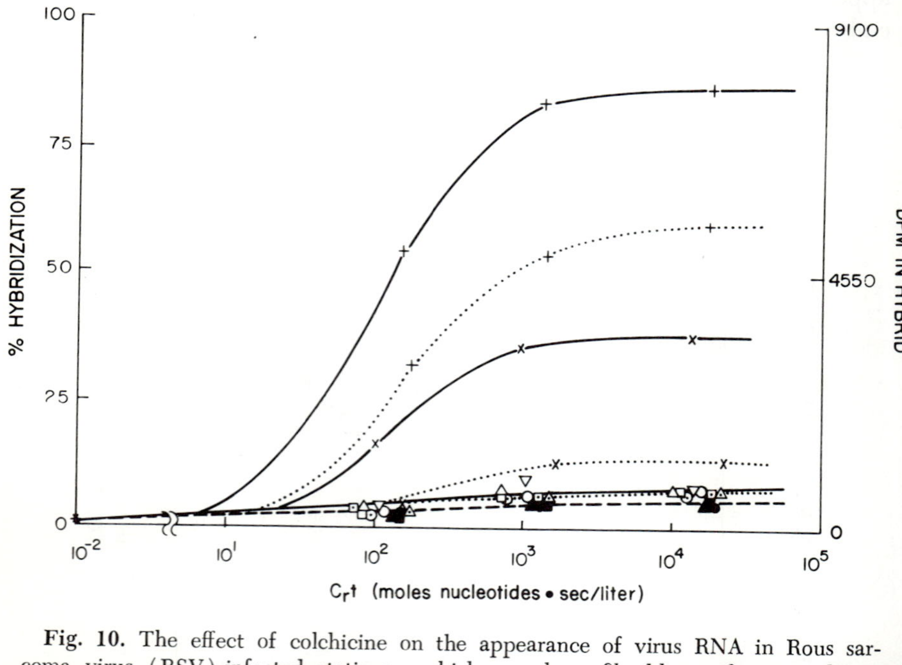

Fig. 10. The effect of colchicine on the appearance of virus RNA in Rous sarcoma virus (RSV)-infected stationary chicken embryo fibroblasts after stimulation with serum. Stationary chicken embryo fibroblasts were prepared, infected with RSV, and serum-containing medium or serum-containing medium containing 5×10^{-7} *M* colchicine was added at 12 hours after infection. Cells were harvested at various hours after addition of serum (HAS), and cytoplasmic RNA was prepared. Ribonucleic acid preparations were hybridized to RSV product DNA-^{3}H in 10-μl micropipettes as described in the caption of Fig. 4. ^{3}H-Labeled DNA (6300 dpm) was hybridized to cytoplasmic RNA prepared from uninfected cells (----) harvested 0 HAS (●), 6 HAS (■), and 48 HAS (▲); infected cells (——) at 0 HAS (○), 6 HAS (□), 9 HAS (▽), 12 HAS (△), 24 HAS (×), and 48 HAS (+); and from infected colchicine-treated cells (···) at 0 HAS (⊙), 6 HAS (⊡), 12 HAS (◬), 24 HAS (×), and 48 HAS (+).

at 24 hours after addition of serum and for colchicine-treated infected cells at 48 hours after addition of serum demonstrates that colchicine-treated cells at 48 hours after addition of serum contained approximately the same amount of virus RNA present in untreated infected cells 24 hours after addition of serum. Therefore, addition of colchicine to RSV-infected stationary chicken embryo fibroblasts at the time of addition of serum-containing medium inhibited cell division and delayed the appearance of virus RNA approximately 24 hours. Virus production was delayed at least 12 hours. Since colchicine is known to bind to microtubules at metaphase and to prevent mitosis, it appears that the event required for transcription of the DNA provirus, which is delayed by treatment with colchicine, occurs after cell division.

V. ROUS SARCOMA VIRUS PRODUCTION IN CHICKEN EMBRYO FIBROBLASTS MADE STATIONARY AFTER ACTIVATION

Initialy transcription of the DNA provirus appears to require that the infected cells proceed through mitosis, that is, that the infected cells become "activated." To determine whether the replication of RSV required continuous cell replication, cells were exposed to RSV, transferred 3 times until they were all infected and producing virus, and then made stationary by depleting serum-containing medium of multiplication-stimulating activity as described previously. Cells infected with and producing RSV were made stationary by using a mutant of RSV temperature-sensitive for transformation but not replication and keeping the infected cells at 41°C, the nonpermissive temperature. The stationary cells were shifted to 37°C, and fresh serum-containing medium was added to some of the cultures. The medium on other cultures was changed to fresh serum-free medium. The rates of DNA and RNA synthesis and the cell number per culture were determined at different times for the serum-stimulated cells. Virus production was determined for both serum-stimulated and stationary cells by changing the medium at different times. The results (Fig. 11) show that, although the serum-stimulated RSV-infected cells proceeded through a cell cycle very similar to that for uninfected stationary cells after stimulation with serum, virus production was no different in stationary cells from that observed in serum-stimulated cells. Other experiments (7) showed that the virus produced by the stationary cells contained newly synthesized RNA. Therefore, although the RSV-infected chicken embryo fibroblast must proceed through mitosis in order to begin to produce virus, RSV production does not require continuous cell division.

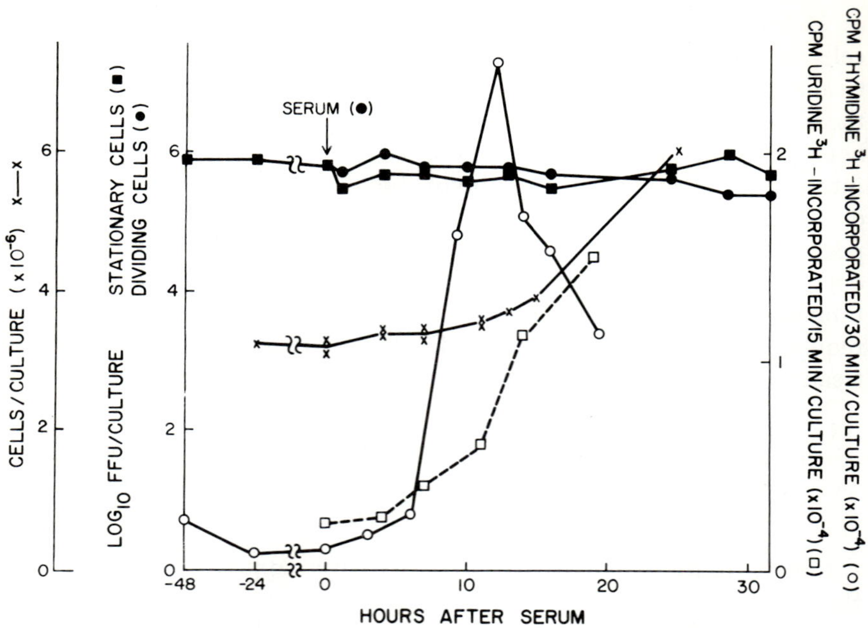

Fig. 11. Virus production in Rous sarcoma virus (RSV)-infected chicken embryo fibroblasts made stationary after infection and subsequent cell division. Chicken embryo fibroblasts were infected with a mutant of RSV temperature-sensitive for transformation, grown for ten to fifteen generations at the nonpermissive temperature, 41°C, and made stationary by depletion of serum-containing medium of multiplication-stimulating activity. Seven days after plating, the medium on some cultures was changed to serum-containing medium and on others to serum-free medium. All cultures were switched to the permissive temperature, 37°C. At different times, the number of cells per culture, the amount of incorporation of thymidine-^{3}H in 30 minutes, the amount of incorporation of uridine-^{3}H in 15 minutes, and the titer of virus in the culture fluids of stationary and dividing cells were determined.

VI. A MODEL FOR THE REPLICATION OF ROUS SARCOMA VIRUS

These results demonstrate that the replication of RSV requires division of the infected cells to initiate transcription of the DNA provirus. It had previously been reported that cell division was required for the synthesis of progeny virus (*5, 13*). These results provide more detailed information on the requirement for cell division. Figure 12 presents a model for the replication of RSV that incorporates these findings. Virus RNA is transcribed into a DNA provirus which is stably associated with the infected cells. This synthesis may be inhibited by inhibitors of DNA synthesis. The DNA provirus must then be activated before

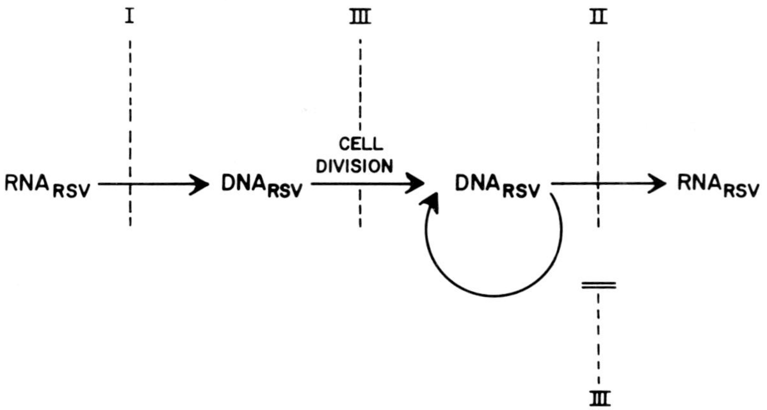

Fig. 12. A model for the replication of Rous sarcoma virus. Group I, Group II, and Group III inhibitors are as described in Table 1. Dashed line—inhibits; dashed line with parallel bars—does not inhibit.

transcription into RNA. The activation is accomplished by division of the infected cells. Thus, activation may be prevented by inhibitors of cell division. Transcription of the DNA provirus into RNA leads to production of progeny virus and may be inhibited by actinomycin D. However, once RNA transcription has begun, activation is no longer required, and inhibitors of cell division are unable to prevent transcription.

VII. SUMMARY

The replication of RSV has been studied using synchronized chicken embryo fibroblasts and a sensitive hybridization technique for detection of RSV RNA. The replication of RSV requires division of the infected cell for production of progeny virus. Cell division is required for the appearance of virus-specific RNA. Inhibitors of cell replication that prevent division of the infected cell delay the appearance of both virus-specific RNA and progeny virus. Once the infected cell has undergone division, the production of progeny virus is no longer sensitive to inhibition of cell division.

ACKNOWLEDGMENTS

I thank Dr. Howard M. Temin for his support and guidance during this investigation. I think R. Grimstad, S. Hellenbrand, V. Kassner, and T. Lewandowski for technical assistance.

This investigation was supported by U. S. Public Health Service research grant CA-07175 from the National Cancer Institute. I was supported by training grant T01-CA-5002 from the National Cancer Institute.

REFERENCES

1. Bader, J. P., *Virology* **26**, 253–261 (1965).
2. Britten, R. J., and Kohne, D. E., *Science* **161**, 529–540 (1968).
3. Britten, R. J., and Smith, J., *Carnegie Inst. Wash., Yearb.* **68**, 378–386 (1970).
4. Garapin, A. C., Varmus, H. E., Faras, A. J., Levinson, W. E., and Bishop, J. M., *Virology* **42**, 264–274 (1973).
5. Hobom-Schnegg, B., Robinson, H. L., and Robinson, W. S., *J. Gen. Virol.* 85–93 (1970).
6. Humphries, E. H., and Temin, H. M., *J. Virol.* **10**, 82–87 (1972).
7. Humphries, E. H., and Temin, H. M., *J. Virol.* **14**, 531–546 (1974).
8. Leong, J. A., Garapin, A. C., Jackson, N., Fanshier, L., Levinson, W. E., and Bishop, J. M., *J. Virol.* **9**, 891–902 (1972).
9. Pierson, R. W., and Temin, H. M., *J. Cell. Physiol.* **79**, 319–330 (1972).
10. Rubin, H., and Temin, H. M., *Virology* **7**, 75–91 (1959).
11. Temin, H. M., *Virology* **20**, 577–582 (1963).
12. Temin, H. M., *Virology* **23**, 486–494 (1964).
13. Temin, H. M., *J. Cell. Physiol.* **69**, 53–63 (1967).
14. Temin, H. M., *Cancer Res.* **28**, 1835–38 (1968).
15. Temin, H. M., *J. Cell. Physiol.* **78**, 161–170 (1971).
16. Vigier, P., and Goldé, A., *Virology* **23**, 511–519 (1964).

DISCUSSION

Dr. Ginsberg: Your data, which were so beautifully presented, imply that the virus-specific DNA must be integrated as an initial event. Are there unambiguous data to substantiate this?

Dr. Humphries: That answer really isn't available yet. Most people interpret it to be integrated. Experiments in Michael Bishop's laboratory suggests this strongly. It is possible to argue that it is not integrated. I believe it is integrated in the fully infected and dividing cell. The question is whether or not it is integrated in the stationary cell. There is presently no direct evidence on that question. We are working on that now. The evidence from additional experiments suggests that it is integrated in a stationary cell. This comes from experiments in which attempts were made to isolate the provirus. What happens is that, by various labeling procedures and by various separation procedures, you cannot isolate the provirus from the rest of the chromosome. With the sophistication of the experiments that we have done in Dr. Temin's laboratory, I believe strongly that it is integrated in the stationary cell. There are at least two techniques available now which will let us answer that question in the very near future. Whether it's integrated before or after division, I think, is still open to debate.

Dr. Levy: Are your stationary cultures completely blocked in DNA synthesis when you infect them with Rous virus? If so, your results indicate that the mechanism of viral DNA synthesis is separate and unaffected by cellular DNA synthesis.

Dr. Humphries: That is correct. Cell DNA synthesis is inhibited. This is the very critical part of the system which allows the experiment to be done. This lets you differentiate between viral DNA synthesis and cell DNA synthesis. The serum proteins that are required for cell division are not required for virus DNA synthesis. Their absence does not prevent viral DNA synthesis. This is a distinction between cell and viral DNA synthesis.

DR. MAHY: Presumably RNA synthesis is mediated by a DNA-dependent RNA polymerase. As the RNA synthesis occurs after cell division and DNA integrations, cellular polymerase may be involved. Is there evidence as to the α-amaniten sensitivity of virus-specific RN production?

DR. HUMPHRIES: Yes. That work has been done, but has not yet been published. This is work that was done in Zurich, in Dr. Charles Weissmann's laboratory. By means of a very sophisticated hybridization technique, the results indicate that the α-amaniten-sensitive RNA polymerase is involved in this transcription. That work has been submitted for publication.

CHAPTER 13

Mechanism of Interferon–Induced Block in mRNA Translation

Michel Revel*
J. Content
A. Zilberstein
B. Lebleu†
E. Falcoff
R. Falcoff

I. INTRODUCTION

Addition of mouse interferon to L cells makes these cells unable to support the multiplication of viruses by blocking viral gene expression (*23*). Interferon is also known to reduce the rate of cell division and the intracellular multiplication of some parasites (*3*). To determine the changes induced by interferon in the cell, studies of gene expression in cell-free systems have been undertaken in several laboratories. Evidence for an effect of interferon treatment on mRNA translation is now well established *in vitro* (*5–7, 10, 19*). Our work (*5*) has shown that cell-free extracts from interferon-treated mouse L cells have lost their ability to translate into protein exogenously added natural mRNA's as Mengo RNA or hemoglobin (Hb) mRNA. This block in translation appears in uninfected L-cell extracts only under conditions in which interferon induces the antiviral state (*6*). Endogenous protein synthesis

* Present address: Department of Biology, Yale University, New Haven, Connecticut.

† Present address: Department of Molecular Biophysics and Biochemistry, Yale University, New Haven, Connecticut.

and polyurydilic acid (poly U) translation are not significantly inhibited in interferon-treated cell extracts (*5*).

We have shown that an inhibitor of translation, which appears to be a protein associated with the ribosomes, accumulates in extracts of interferon-treated L cells (*6, 19*). This inhibition of translation can be eliminated by the addition of purified fractions of mammalian transfer RNA (tRNA) (*4*). The requirement for tRNA in interferon-treated cell extracts is caused by the ribosome-associated inhibitor. The tRNA species that restore the translation of Mengo RNA and Hb mRNA in interferon-treated cell extracts have been purified and are not identical, suggesting a possible basis for the selective effects of interferon in virus-versus-host protein synthesis.

II. MATERIALS AND METHODS

A. Cell Cultures and Preparation of Cell-Free Extracts

Mouse L cells (strain CCL 1) were grown in suspension cultures in Eagle's MEM (Gibco) supplemented with 8% decomplemented calf serum, 0.2% bactotryptose phosphate, 0.1% methyl cellulose, 5.5 gm/liter glucose, 40 mg/liter gentamycin, 10^5 U/liter penicillin, and 0.1 gm/liter streptomycin. Routinely 2 liters of culture to a density of 10^6 cells/ml were treated with 150–200 units/ml of mouse interferon. Interferon was prepared from Newcastle disease virus (NDV)-infected L cells and purified on carboxymethyl (CM)-Sephadex as described previously (*6*). Specific activities ranged from $1\text{–}5 \times 10^6$ reference units per milligram protein.

After 35 hours at 37°C, cells were harvested, washed, and a cell-free extract (S10) was prepared as described before (*6*) except that 20 m*M* Hepes buffer, pH 7.5, was used instead of Tris–HCl buffer. Preincubated S10 was used to translate Mengo virus RNA and rabbit Hb mRNA as before (*6*). Fractionation of S10 into ribosomes, ribosomal wash fraction, and high-speed supernatant has been detailed before (*6*). Cell-free extracts from Krebs ascites tumor cells and reticulocyte initiation factors were obtained as before (*16*). ^{14}C-Radioactive protein hydrolyzate (50 μC/μAt C) or methionine-^{35}S (over 150 mC/μmole) were purchased from Radiochemical.

B. Preparation and Purification of tRNA

Krebs ascites cells and L cells tRNA were prepared through phenol extraction, 1 *M* NaCl and isopropanol precipitation as outlined by Aviv *et al.* (*1*). Chromatography on BD-cellulose* was carried out according

* Benzoylated diethylaminoethyl-cellulose.

to Petrissant *et al.* (*18*). Purified rabbit liver $tRNA_{Met\ F}$ after BD-cellulose and DEAE–Sephadex was a gift from Dr. G. Petrissant. Total rabbit liver tRNA was a gift from Dr. D. Hatfield, and *E. coli* B tRNA was from Schwarz Co. Purified *E. coli* methionyl tRNA synthetase was a generous gift from Dr. J. P. Waller. The RNA-free, *E. coli,* total supernatant enzymes were prepared to charge tRNA, as described by Muench and Berg (*15*) and for formylation (*8*). Methyldinitroanilinomethyl propionamide (MDMP) was a generous personal gift from Dr. Baxter.

III. RESULTS

A. Inhibition of mRNA Translation in Interferon-Treated L-Cell Extracts

Cell-free cytoplasmic extracts were prepared from L cells, clone CCL1, treated for 18 hours with mouse interferon. These extracts (interferon S10) translate very poorly Mengo RNA as compared to control S10. There is a good correlation between this effect and the antiviral effect on the multiplication of Mengo virus (Table 1). If actinomycin D is given together with interferon, both effects are abolished (*5*). Translation of rabbit reticulocyte Hb mRNA is also inhibited, although consistently

Table 1. Effect of Various Doses of Interferon on mRNA Translation Activity

L-Cell treatment[a]	Antiviral effect[a] Mengo virus multiplication *in vivo* (Pfu/ml)	(%)	Translation *in vitro* of[b] Mengo RNA leucine-^{14}C (cpm)	Hemoglobin mRNA leucine-^{14}C (cpm)	Poly U phenylalanine-^{14}C (cpm)
No interferon	1.7×10^8	100	3,725 (100)	2,820 (100)	8,100 (100)
Interferon, 390 ref. units/ml	2.1×10^5	0.1	225 (6)	490 (17)	5,780 (71)
Interferon, 39 ref. units/ml	7.3×10^7	43	1,135 (30)	2,110 (75)	7,770 (94)
Interferon, 3.9 ref. units/ml	2.0×10^8	117	3,980 (106)	2,920 (103)	10,205 (125)

[a] Purified interferon specific activity: 1.3×10^6 reference units/mg protein; 39,000 reference units/ml determined by inhibition of cytopathic effects of VSV. The antiviral effect is the Mengo virus yield measured as described in (*6*).

[b] Cell-free mRNA translation in S-10 extracts of L cells treated for 18 hours as indicated. Background incorporation in the absence of mRNA was subtracted. Percent of control activity is given in parentheses.

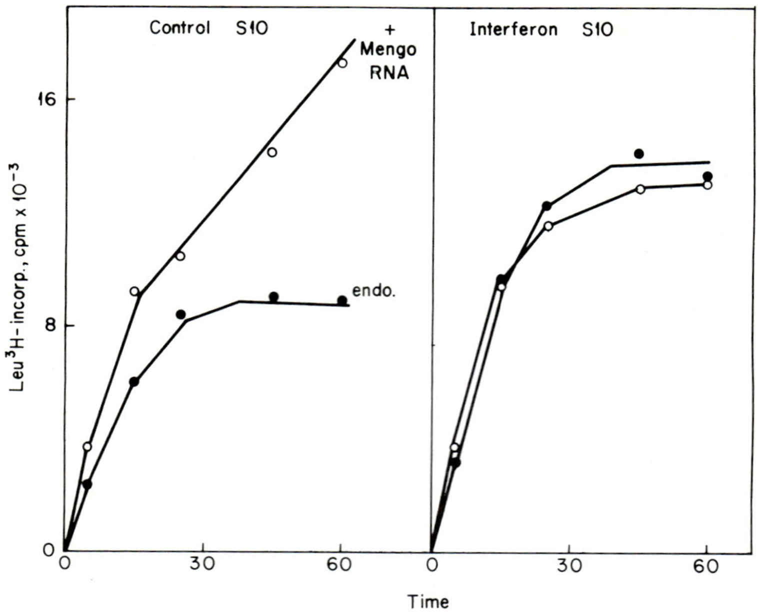

Fig. 1. Endogenous and exogenous protein synthesis in nonpreincubated extracts of control and interferon-treated L cells. Activity was measured by leucine-^{3}H incorporation into proteins without added mRNA (endogenous) or with 1 μg Mengo RNA.

to a lesser extent than that of Mengo RNA. Poly U translation is not blocked.

When S10 extracts are not preincubated, the translation of endogenous L-cell mRNA's can be measured (Fig. 1). Endogenous protein synthesis is active in interferon-treated L-cell extracts; however, whereas Mengo RNA stimulates leucine incorporation in nonpreincubated S10 from control cells, it does not stimulate in the nonpreincubated interferon S10 (Fig. 1).

Routinely we have used preincubated S10 from L cells treated with 200 units of mouse interferon (partially purified; 10^6 units/mg protein) for 24 to 48 hours to investigate the mechanism of the translation block for exogenous natural mRNA's. An important finding is that the block is reversible: the ability of interferon S10 to translate Mengo RNA can be restored by adding transfer RNA. Figure 2 shows that the addition of Krebs ascites cell tRNA restores over 90% of the translation activity. The control S10 is not significantly stimulated by tRNA.

Periodate treatment of tRNA, which destroys the amino acid acceptance, abolishes its activity to restore mRNA translation (*19*). *Escher-*

ichia coli tRNA has very little activity (Fig. 2A) as compared to mammalian tRNA, unless it is first aminoacylated with *E. coli* synthetases. Preparations of tRNA from L cells were found to be significantly less active than tRNA from Krebs cells or from rabbit liver; tRNA extracted from whole L cells treated with interferon was found to have some restoration activity (*4*).

B. Analysis of tRNA That Restores Translation in Interferon S10

The nature of the restoring tRNA was investigated by fractionating Krebs ascites cells tRNA on a column of benzoylated DEAE cellulose (*18*). Figure 3A shows that the activity restoring Mengo RNA translation in interferon S10 is separated from the bulk of the tRNA and eluted late from the BD-cellulose column. The chromatogram in Fig. 3 shows also the position of initiator $tRNA_{Met\ F}$ and of $tRNA_{Met\ M}$. These species do not restore Mengo RNA translation (*19*). The active species for Mengo RNA translation were further purified by rechromatography on BD-cellulose (Fig. 3B). The most active fractions, which eluted again after the main peak of tRNA, restored when added at a dose of 0.03 μg to the interferon S10 which by itself already contains 0.8 μg of endogenous tRNA. The peak of Mengo restoring activity corresponds to the peak of a minor Leu-tRNA species. No other significant amino acid acceptance besides leucine was found in the purified fractions. By filtration on Sephadex G75 and by electrophoresis on SDS-polyacrylamide gels, the purified fractions were shown to contain 4 S RNA.

In several preparations of tRNA and using several cell-free extracts from interferon-treated L cells, the peak of restoring tRNA for Mengo RNA translation was found to elute at the same position from the BD-cellulose columns as in Fig. 3A. Some variations in the width of the peak were, nevertheless, observed, and it cannot be excluded that more than one tRNA species stimulates Mengo RNA translation in interferon S10.

We have carried out a similar study for the translation of Hb mRNA. The block in translation of this mRNA in interferon S10 is also overcome by the addition of tRNA. However, as for control S10 the presence of reticulocyte initiation factors is also required (*16*). We analyzed the fractions from the BD-cellulose chromatography for their activity to restore globin synthesis in interferon S10. The pattern obtained is clearly different from that seen with Mengo RNA (Fig. 3A). Although there is some overlap, the peak for Hb mRNA translation elutes much earlier from BD-cellulose. The peak of activity is broad and may comprise different tRNA species. When only the formation of complete globin

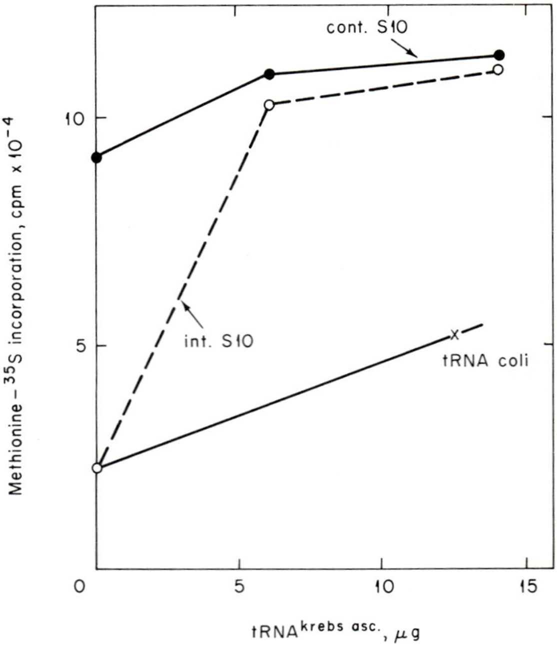

Fig. 2A. Restoration by tRNA of Mengo RNA translation in interferon-treated L-cell S10. Conditions of assay as in Section II with 0.25 A_{260} units of S10 from control or interferon-treated L cells and 1 μg Mengo RNA in 0.05 ml. Incubation for 60 minutes at 30°C. Hot TCA insoluble radioactivity of a 0.04-ml aliquot was determined.

TABLE 2. RELATIVE TRANSLATION OF HEMOGLOBIN mRNA AND MENGO RNA BY INTERFERON-TREATED L-CELL S10[a]

Additions	Amino acid-^{14}C incorporation with: Hemoglobin mRNA (cpm)	Mengo RNA (cpm)	Translation ratio, Hemoglobin / Mengo
None	440	720	0.6
+ $tRNA^{rabbit\ liver}_{total}$	5,106	4,505	1.1
+ $tRNA^{Krebs}$ fractions from BD-cellulose:			
Fraction 110	4,265	779	5.5
Fraction 142	3,525	2,604	1.3

[a] Conditions as in Fig. 2. Only Interferon S10 was used with 0.5 μg rabbit reticulocyte hemoglobin mRNA and ribosomal wash fraction from reticulocytes (*16*) or with 1 μg Mengo RNA. BD-cellulose tRNA fractions refer to the chromatography in Fig. 3A.

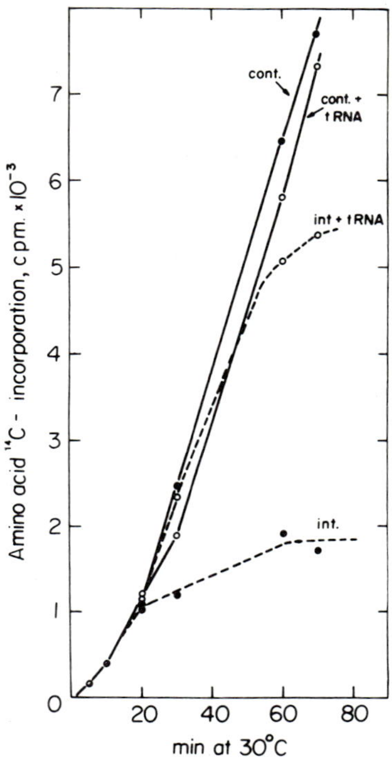

Fig. 2B. Kinetic study of the tRNA effect. Conditions as in Fig. 2A but with amino acids-^{14}C. L-Cell extracts used here were either treated with 150 units mouse interferon per milliliter culture for 18 hours, or untreated. Rabbit liver tRNA (15 μg) was used, and 0.15 A_{280} units ammonium sulfate-precipitated L cell-S100 were added.

chains is measured (by electrophoresis on Cellogel or polyacrylamide gels), the peak is found at the same position but is much narrower. The other tRNA fractions that stimulate Hb mRNA-directed amino acid incorporation were found to promote the synthesis of incomplete globin chains [Fig. 4; (*4a*)]. Here again, it is possible that more than one tRNA species is needed to allow the interferon S10 to synthesize complete globin molecules.

The comparison between Mengo RNA and Hb mRNA translation is very important since it establishes a possible basis for the discriminatory effects of interferon treatment on different mRNA's. With some fractions from the BD-cellulose chromatography, it is possible to restore completely the activity of an interferon S10 for globin synthesis but not for Mengo RNA translation (Table 2). The specificity of the cell-free extracts can, therefore, be determined according to the tRNA added. Some tRNA

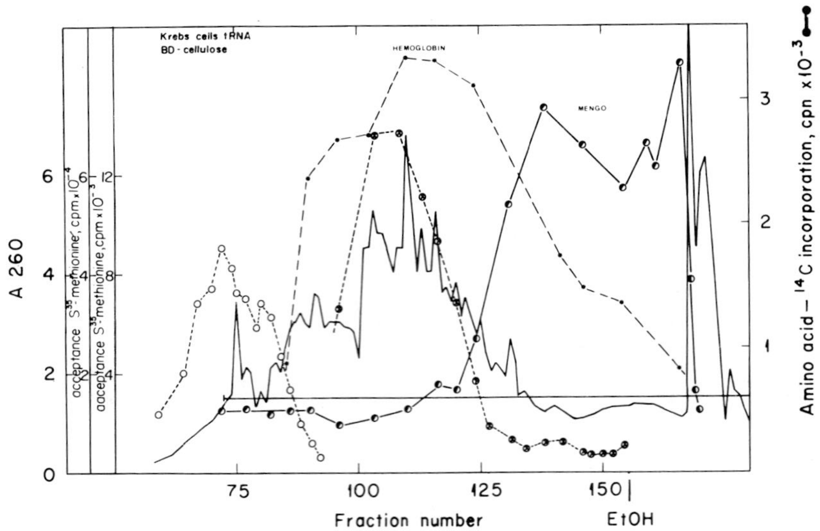

Fig. 3A. Chromatography of restoring tRNA on BD-cellulose. A 0.9 × 12.5 cm column of BD-cellulose (Schwartz) was loaded with 12.5 mg tRNA extracted (*1*) from Krebs ascites cells. The column was washed with 0.35 *M* NaCl, 0.01 *M* $MgCl_2$, 0.01 *M* Na acetate buffer, pH 4.5, and eluted with a 100-ml gradient of 0.35–1.0 *M* NaCl in the same buffer. Fractions, 0.5 ml, were collected, dialyzed against Hepes buffer, pH 7.5, 0.02 *M*, KCl 0.12 *M*, $MgCl_2$ 0.005 *M*, β-mercapto-ethanol 0.007 *M*, and glycerol 10%, and assayed for their ability to restore Mengo RNA translation in S10 extracts from interferon-treated L cells as in Fig. 2 (◐). Restoration of hemoglobin mRNA translation was assayed as in Table 2 (●). Fractions were also assayed for acceptance of methionine-^{35}S with purified *E. coli* methionyl tRNA synthetase (○) ($tRNA_{Met\ F}$) (*20*) or L-cell S100 (⊗) ($tRNA_{Met\ M}$).

species corresponding to minor codons may limit the translation of mRNA's which use these codons frequently without affecting other mRNA's which do not use them. Interferon treatment, therefore, makes the protein-synthesizing machinery tRNA dependent in a rather specific way: not all the tRNA species become limiting, but only certain defined species. These may correspond to minor tRNA's, which exist in low concentration in the cell. If there is adaptation of the tRNA population of the cell to the type of protein it produces (*14*) the translation of mRNA foreign to the cell may be particularly affected.

C. Nature of Block in Interferon-Treated L-Cell Extracts

What is the reaction which is blocked in interferon S10? A kinetic study of Mengo RNA translation (Fig. 2B) shows that in interferon S10 the reaction is initiated as in the control but stops after 20 to 30 minutes at 30°C. With both Mengo RNA and Hb mRNA, the polypeptide

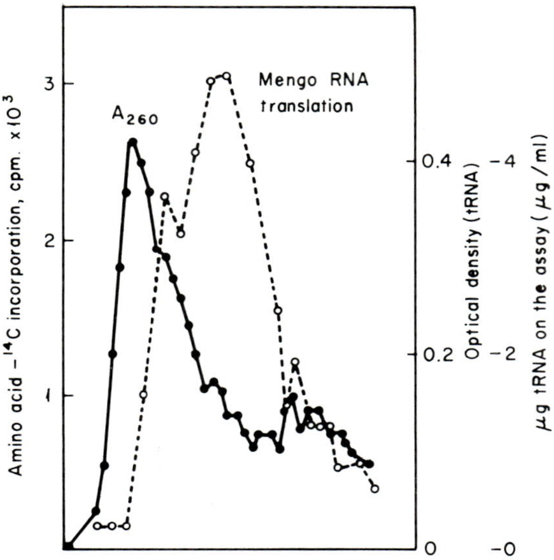

Fig. 3B. Rechromatography of tRNA fractions from Fig. 3A, tubes 128–166, on a 0.75-ml BD-cellulose column. An aliquot of 5 μl from tube 82 fully restored Mengo RNA translation in interferon S10.

products that accumulate during the first 30 minutes correspond to short chains of about 5–10 $\times$ 10^3 mol wt. Addition of tRNA to interferon S10 allows protein synthesis to proceed for a much longer time at the same rate as the control S10 and longer chains, including complete globin, are formed (Fig. 4). These results suggest that the block is in elongation.

If amino acid incorporation is measured between 30 and 60 minutes after the addition of Mengo RNA, a complete arrest in translation is

TABLE 3. EFFECT OF DELAYED ADDITION OF TRANSFER RNA[a]

Extract	Addition at 30 min	Methionine-^{35}S incorporation from 30 to 60 min at 30°C (cpm)	(%)
Control S10	None	60,000	(100)
Interferon S10	None	3,750	(6)
	+ tRNA, 6 μg	62,810	(105)
	+ Mengo RNA, 1 μg	3,650	(6)

[a] Conditions as in Fig. 2 except that reaction mixture contained from 0 to 30 min cold methionine, 1 μM. At 30 min, 2.3 μC of methionine-^{35}S (195 mC/μmole) were added. Mengo RNA, 1 μg, was present from time zero.

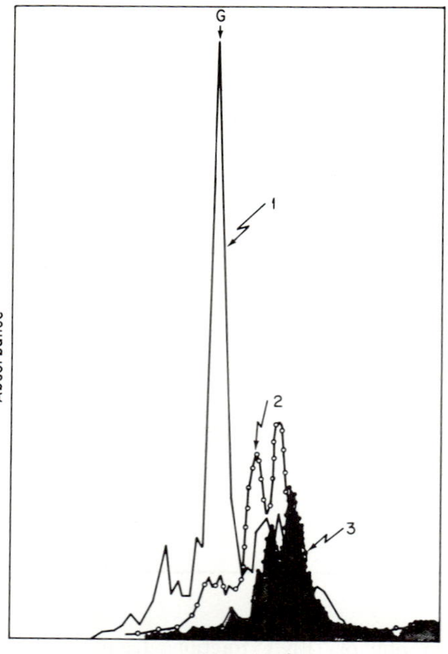

Fig. 4. Hemoglobin mRNA products analyzed by electrophoresis on SDS polyacrylamide gels. Interferon S10 was used as in Fig. 3A to measure methionine-^{35}S incorporation with hemoglobin mRNA and reticulocyte ribosomal wash fraction (*16*). Curve 1, with tRNA fraction 116 from Fig. 3A; curve 2, with tRNA fraction 86 from Fig. 3A; curve 3, in the absence of added tRNA. The arrow on abscissa shows the direction of migration. After radioautography of the gel, the X-ray film was scanned in a Gilford spectrophotometer. G indicates the position of the rabbit globin marker (16,000 MW).

observed (Fig. 2B and Table 3). (If the system is incubated without mRNA, this arrest does not occur: upon addition of mRNA the translation still starts and stops after 20 to 30 minutes.) The delayed addition of tRNA, 30 minutes after Mengo RNA, fully restores translation activity (Table 3), demonstrating that no mRNA degradation takes place in the interferon S10. In line with this, addition of more Mengo RNA at 30 minutes does not restore activity (Table 3). If inhibitors of polypeptide chain initiation such as aurintricarboxylic acid (ATA) or MDMP (*2*) are added at 30 minutes together with tRNA, no restoration is observed (*13, 19*). This implies that new initiations are required to see restoration by tRNA. This result may seem paradoxical since restoration is observed with tRNA fractions devoid of initiator tRNA (Fig. 3A). We, therefore, investigated directly if the block, 30 minutes after adding mRNA, affects initiation of translation.

When the translational block is fully established, 30 minutes after the addition of Mengo RNA to the interferon S10, newly added radioactive Mengo RNA-^{3}H still binds to ribosomes; monosomes and disomes are formed although heavier polysomes do not appear (*13*). Binding of methionyl-^{35}S $tRNA_{Met\ F}$ to ribosomes is, in contrast, much decreased in the interferon S10. We find, however, that if formylmethionyl-^{35}S tRNA is used, its binding to initiation complexes is not inhibited (*13*). The presence of the formyl group overcomes the inhibitory effect on the formation of initiation complexes. Nevertheless, incorporation of formylmethionine-^{35}S into protein [hot trichloroacetic acid (TCA)-precipitable] is markedly inhibited at 30 minutes after adding the mRNA and, at the same time, we find an accumulation of short, soluble formylmethionine peptides in the interferon S10 (*13*). This is in agreement with the fact that formylmethionyl tRNA by itself does not restore translation (Table 4). We can, therefore, conclude that even if new initiation complexes are formed the polypeptide chains do not become elongated.

The decreased binding of methionyl $tRNA_{Met\ F}$ to initiation complexes, which is observed, (*13*) may result from a mechanism other than that blocking elongation. But, when the elongation block is abolished by adding purified restoring tRNA (which does not contain initiator tRNA), the block on initiation disappears also and new chains are initiated, which explains the sensitivity of the restoration reaction to inhibitors of initiation. We propose that the block of initiation is *secondary* to the block on elongation and results from a feedback which may be similar to the initiation block observed in cells starved for an amino acid other than methionine (*22*). The resistance of the formylated initiator tRNA species to this block may be due to a higher affinity of this molecule to ribosomes or to its resistance to enzymic deacylation (*9*).

TABLE 4. ABSENCE OF RESTORATION BY *N*-FORMYLMETHIONINE-^{35}S $tRNA_{Met\ F}$[a]

	Methionine-^{35}S incorporation from 20 to 60 min at 30°C		
Additions	Control-S10 (cpm)	Interferon S10 (cpm)	(%)
None	33,600	7,000	21
Fmet tRNA, 0.3 μg	33,100	8,200	25
Fmet tRNA, 0.6 μg	23,100	8,750	38
Fmet tRNA, 0.15 μg + restoring tRNA, 0.5 μg	35,800	29,600	83

[a] Conditions as in Fig. 2A. Formylmethionine (Fmet)-^{35}S $tRNA_{Met\ F}$ was prepared as in Section II. Free ^{35}S-methionine (1 μC and 55 pmoles for 0.05 ml) was added. Without Mengo RNA, 600 cpm was incorporated from 20–60 min at 30°C.

D. Mechanism of tRNA Dependency in Interferon S10

We have shown that an inhibitor of translation accumulates in interferon-treated cells. This inhibitor can be obtained from the ribosomes by extraction with 0.5 *M* KCl. It is precipitated between 70–90% ammonium sulfate (Fig. 5), elutes from DEAE-cellulose at 0.3 *M* KCl, and is inactivated by heating for 7 minutes at 50°C, which suggests that this factor is at least in part a protein. When added to extracts from untreated L cells or from Krebs ascites cells, this factor inhibits the translation of Mengo RNA and Hb mRNA. It does not affect endogenous protein synthesis in these extracts (Table 5). No effect on Poly U translation is seen.

The inhibition produced by the interferon-induced factor can be overcome by the addition of tRNA (Table 5). Actually, the inhibitor makes the control S10 tRNA-dependent, which is normally a characteristic of the interferon S10. This suggests that this factor causes both the block

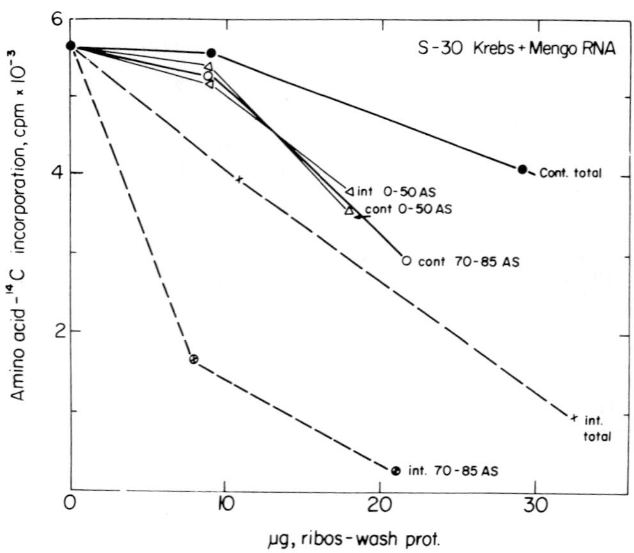

Fig. 5. Protein synthesis-inhibitory activity from ribosomes of interferon-treated L cells. Ribosomes from L cells (*6*), normal or treated by 200 units/ml interferon for 38 hours, were treated for 4 hours at 0°C by 0.5 *M* KCl in Hepes buffer, pH 7.5, 0.02 *M*; $MgCl_2$, 0.005 *M*; β-mercaptoethanol, 0.007 *M*; glycerol 20%. After centrifugation, 3.5 hours at 50,000 rpm in Spinco R65, the ribosomal wash fraction was dialyzed against the same buffer but with 0.12 *M* KCl. Fractionation by ammonium sulfate at 4°C, was followed by dialysis with the same buffer. Fractions were assayed at the indicated protein concentration for their effect on the translation of Mengo RNA in Krebs ascites tumor cell S-30 (*6*).

TABLE 5. EFFECT OF RIBOSOMAL WASH FRACTION ON L-CELL TRANSLATION[a]

Additions	L-Cell endogenous protein synthesis (%)	Mengo RNA translation (%)
None	100	100
+ tRNA$^{retic.}$, 5 μg	—	85
+ Control ribosomal wash, 20 μg	72	74
+ Interferon ribosomal wash, 25 μg	63	27
Interferon ribosomal wash, 25 μg + tRNA$^{retic.}$, 5 μg	—	83

[a] Conditions as in Fig. 2 Amino acid -^{14}C incorporation was measured in control L-cell S10: nonpreincubated for endogenous protein synthesis (100% = 8950 cpm in 60 min) or after preincubation and addition of 1 μg Mengo RNA (100% = 2330 cpm in 60 min). The indicated amount of ribosomal wash protein was added.

in mRNA translation and the tRNA deficiency. The factor could act by preventing the correct function of tRNA at the polysome level. Alternatively, it could cause the degradation of some tRNA species. Phenol extraction of intact interferon-treated L cells yields tRNA that can restore Mengo RNA translation in interferon S10 (*4*); however, preincubated S10 extracts from interferon-treated L cells were found to be completely deficient in restoring tRNA activity (*4a*). It should, therefore, be possible now to study the fate of these specific restoring tRNA species in the interferon-treated cell extracts.

Endogenous L-cell protein synthesis is active for about 30 to 45 minutes in interferon-treated nonpreincubated extracts (see Fig. 1), but in the same extracts translation of exogenous mRNA is inhibited.* The relative resistance of endogenous mRNA to what appears to be a block on elongation due to lack of some tRNA's has to be explained. Two possibilities will now have to be explored: (1) that cellular polysomes are less sensitive to the loss of certain species of tRNA (see above) or (2) that endogenous cellular polysome have a compartmentalized supply of tRNA which is not available to exogenously added mRNA's.

IV. CONCLUSIONS

Although many questions are still unanswered, we can state that in uninfected mouse L cells, interferon treatment induces a block in translation of exogenous mRNA's. The block appears to result from the accumulation of an inhibitory factor whose action can be reversed by the addition of tRNA. Different mRNA's require the addition of different

* We have verified that both naked Hb mRNA and messenger ribonucleoprotein (*17*) are equally inhibited in interferon S10.

tRNA species to be translated in the interferon S10. For Mengo RNA, a minor Leu-tRNA species seems to be required. Lack of this tRNA stops polypeptide chain elongation and subsequently inhibits new initiation. The tRNA dependency of interferon S10 may be due to degradation of some tRNA species or to another mechanism which makes these tRNA unavailable for translation. This has to be a rather specific mechanism since only certain species of tRNA are affected; it is relevant to note that T4 phage infection induces degradation of Leu-tRNA in *E. coli* (*11*).

The significance of these findings for the antiviral effects of interferon will have to be further verified. It has to be shown, if *in vivo*, changes in tRNA occur that affect the translation of viral cistrons. Interferon has an effect also on cellular functions (*3*): the rate of cell division is slowed down and changes in the cell surface membrane are observed. It is possible that changes in cellular protein synthesis also occur as a result of interferon action on the tRNA population. Evidence for variations in the tRNA population from one tissue to another is now well established (*14*), and these variations may be related to synthesis of specific proteins in these tissues. With respect to virus growth, Subak-Sharpe *et al.* (*21*) have suggested that viruses use other tRNA than the host cell, and experimental evidence for virus-specific tRNA's is rapidly accumulating (*14*). It will be interesting to determine if interferon treatment affects in particular the tRNA species that control viral protein synthesis.

Variations in tRNA populations from one type of cell to another may certainly explain why the sensitivity of different viruses to interferon varies from one cell type to another (*3*). Practically, it may be possible to change the tRNA content of cells by starvation procedures or by drug treatments. Such treatments may lead to an increased sensitivity of cells to interferon (*12*) and to a reduction in the amount of interferon needed to achieve a certain level of antiviral protection. For this reason, we believe that the elucidation of the mechanism of action of interferon and of the inhibitor it induces in cells will be of great interest to devise better ways for the clinical use of the antiviral effects of interferon.

REFERENCES

1. Aviv, H., Boime, I., and Leder, P., *Proc. Nat. Acad. Sci. U. S.* **68**, 2303 (1971).
2. Baxter, R., Knell, V. C., Sommerville, H. J., Swain, H. M., and Weeks, D. P., *Nature (London)* **243**, 139 (1973).
3. Carter, W. A., *in* "Selective Inhibitors of Viral Functions," p. 81. Chem. Rubber Publ. Co., Cleveland, Ohio, 1973.
4. Content, J., Lebleu, B., Zilberstein, A., Berissi, H., and Revel, M., *FEBS (Fed. Eur. Biochem. Soc.) Lett.* **41**, 125 (1974).

4a. Content, J., and Revel, M., unpublished results.

5. Falcoff, E., Falcoff, R., Lebleu, B., and Revel, M., *Nature (London), New Biol.* **240,** 45 (1972).
6. Falcoff, E., Falcoff, R., Lebleu, B., and Revel, M., *J. Virol.* **12,** 421 (1973).
7. Friedman, R. M., Metz, D. H., Esteban, R. M., Tovel, D. R., Ball, L. A., and Kerr, I. M., *J. Virol.* **10,** 1184 (1973).
8. Groner, Y., and Revel, M., *Eur. J. Biochem.* **22,** 144 (1971).
9. Gupta, N. K., and Aerni, R. J., *Biochem. Biophys. Res. Commun.* **51,** 907 (1973).
10. Gupta, S. L., Sopori, M. L., and Lengyel, P., *Biochem. Biophys. Res. Commun.* **54,** 777 (1973).
11. Kano-Sueoka, T., and Sueoka, N., *Proc. Nat. Acad. Sci. U. S.* **62,** 1229 (1969).
12. Kerr, I. M., Friedmann, R. M., Esteban, M., Brown, R. E., Ball, L. A., Metz, D. H., Risby, D., Tovell, D. R., and Sonnabend, J. A., *Advan. Biosci.* **11,** 109–123 (1973).
13. Lebleu, B., Content, J., Berissi, H., and Revel, M., *Eur. J. Biochem.* (in preparation).
14. Littauer, U. Z., and Inouye, H., *Annu. Rev. Biochem.* **42,** 439 (1973).
15. Muench, K., and Berg, P., *Procedures Nucl. Acid Res.* **1,** 375 (1966).
16. Nudel, U., Lebleu, B., and Revel, M., *Proc. Nat. Acad. Sci. U. S.* **70,** 2139 (1973).
17. Nudel, U., Lebleu, B., Zehavi-Willner, T., and Revel, M., *Eur. J. Biochem.* **33,** 314 (1973).
18. Petrissant, G., Boisnard, M., and Puissant, C., *Biochimie* **53,** 1105 (1971).
19. Revel, M., Content, J., Zilberstein, A., Berissi, H., Kimhi, Y., Lebleu, B., Falcoff, E., and Falcoff, R., *Proc. Gulbenkian Meet. Interferons, 1973* (in press).
20. Smith, A. E., and Marcker, K. A., *Nature (London)* **226,** 697 (1970).
21. Subak-Sharpe, H., Bürk, R. R., Crawford, L. V., Morrison, J. M., Hay, J., and Kerr, H. M., *Cold Spring Harbor Symp. Quant. Biol.* **31,** 737 (1966).
22. Vaughan, M. H., and Hansen, B. S., *J. Biol. Chem.* **248,** 7087 (1973).
23. Vilcek, J., *Virol. Monogr.* **6,** (1969).

DISCUSSION

DR. HUANG: Dr. Revel, as you know, there are several virus systems where pretreatment of cells with interferon results in reduced transcription by these viruses. In a search for a possible common mechanism for antiviral action by interferon, there is a lead from the elegant studies on Q_{β} polymerase where translational factors affect both the species and amount of RNA synthesized. In light of this, have you added your inhibitory factor or the tRNA's to those viruses that have virion-associated polymerases and tested *in vitro* for any effect on polymerase activity?

DR. REVEL: It's a good suggestion. We have not done this type of experiment yet. The only comment I would like to make is that in certain virus systems (vaccinia, for example) it seems that the RNA synthesis is not reduced to a significant level by interferon treatment. Therefore, assuming a common mechanism of interferon action on all viruses, one might wish first to solve the problem of the block in translation, and then try to ask the type of question that you are suggesting. As for the example of the Q_{β} replicase, although three subunits are certainly common to transcription and translation, it is not clear yet whether, in the cell, there is a control of RNA replication by the protein.

DR. HO: It is gratifying to hear finally of the possible isolation of so-called translation-inhibitory protein, or the intrinsic antiviral substance induced by interferon. Perhaps the transfer RNA described besides reversing interferon action *in vitro* may inhibit interferon active in intact cell systems. Has this experiment been done?

DR. REVEL: It is only *in vitro* that we have evidence for the recovery of translation activity. The *in vivo* experiment is in progress, but has not been completed.

CHAPTER 14

Clinical Utilization of Exogenous Human Interferon*

Thomas C. Merigan
George W. Jordan†
Richard P. Fried

I. INTRODUCTION

In the 17 years since its discovery, much effort has been put into defining interferon's mechanism and range of action against viruses. In fact, it is possible the present clinical trials with it have available a greater background of basic work than any antimicrobial agent studied previously. Understandably, because of its appealing properties, there has always been a desire to employ it in human medicine. Its successful use in a wide variety of animal model viral infections has been counterbalanced by technical difficulties and great expense in the preparation of suitable quantities of human interferon for clinical studies. In recent years, due to the efforts of many workers, a number of clinically acceptable cell sources have been studied (9). In particular, due to the effort of Dr. Kari Cantell and his group at the State Serum Institute in Helsinki, suitable conditions (5) have been defined for production of human leukocyte interferon in adequate quantities for limited clinical trials. Most of the past studies and ongoing work that will be described in this paper are only possible because of that effort.

As a result of the rather modest successes achieved by the "antiviral

* Supported by a U. S. Public Health Service grant AI-05629 and contract NIH-73-2501.

† Present address: Department of Internal Medicine, School of Medicine, University of California, Davis, California 95616.

screen" approach in industry for the identification of clinically useful compounds, serious questions have been raised as to the possibilities for development of such drugs (*12*). Of course, this must be contrasted with the great successes in recent years in vaccine development, particularly the live virus vaccines. However, there are areas where the use of live virus vaccines does not seem feasible, either because of the numerous virus types involved (rhinoviruses, influenza A and B, etc.) or because of the possible dangers of latent infection by attenuated strains (herpesviruses and papovaviruses). It is in these areas, in particular, that antiviral chemotherapy and prophylaxis must be further explored. In addition, it may be possible to interrupt ongoing viral disease with adequate doses of interferon in the more serious acute viral diseases such as hepatitis, encephalitis, or variola. In the spirit of a symposium on perspectives this paper will assess the progress with interferon toward these goals, including certain work in progress, as well as examine future possibilities.

Table 1 presents the important and generally agreed on properties of interferon. Most specifically, the broad antiviral spectrum, the lack of toxicity, and the lack of antigenicity in a homologous species are the key features in its clinical appeal. Its widespread presence phylogenetically and the many stimuli that trigger its production suggest it may have a more fundamental cellular regulatory role. The inhibitory effects of interferon on rapidly dividing cells support this idea. The latter finding has both good and bad aspects. In particular, it suggests that interferon might have a clinically useful antitumor role but also demands careful monitoring for hemopoietic and immunosuppressive side effects during systemic high-dose or long-term interferon application.

TABLE 1. SUMMARY OF THE PROPERTIES OF INTERFERON

1. A glycoprotein of 20 to 30,000 mol wt with additional higher molecular weight species of 40–60,000, 80–90,000, etc. Specific activity $\geq 10^9$ units/mg protein. Nonantigenic in the homologous species.
2. Synthesized *in vivo* and *in vitro* by all vertebrates studied in response to viral infection as well as to other intracellular parasites, polyanions (polycarboxylates, polynucleotides, etc.), endotoxin, and certain low molecular weight bis basic organic compounds. Also produced as a part of cell-mediated immune (mitogenic) response by lymphocytes.
3. Acts intracellularly in a characteristic species-specific fashion against both RNA and DNA viruses (inhibiting viral mRNA function and/or production) as well as inhibiting more complicated intracellular parasites (Protozoa, *Rickettsia*, etc.). Antiviral action depends on the cellular production of RNA and protein.
4. Nontoxic in that it does not effect normal cell functions (energy metabolism or macromolecular synthesis), but it inhibits rapidly dividing cells, including tumor cells.

II. PRIOR STUDIES WITH EXOGENOUS INTERFERON IN MAN

A. Local Application

For certain infections, direct application of interferon to the virus-infected tissues may be useful through such routes as intranasal, sub- or intracutaneous, or even by direct injection into the eye, conjunctival sac, or subarachnoid space. Obviously, the rationale for such an approach depends on an accurate knowledge of the timing of events in the infectious process at the sites to be treated. This is because we are usually attempting to alter the outcome of an already established viral infection. However, antivirals must be applied at a time when continued pathology is dependent on continued viral spread and growth, because their action ideally must be as a prophylactic for areas as yet uninvolved in infection. Therefore, early initiation of treatment is desirable if the course of infection is to be altered.

The first studies with local exogenous interferon (Table 2) were with monkey kidney cell-derived material and showed convincingly that prior application of interferon to skin would block vaccinia lesion formation in the injected area (*23*). However, an attempt to extend the approach to treatment of a single case of vaccinia gangrenosa failed (*6*). Although many explanations could be advanced for this, it seems that systemic, as well as local, application of the much higher titered human interferon preparations now available should be further pursued in such disease. It is not clear that any important aspects of cutaneous viral infection can be dealt with by employing local injections. It is likely that the interferon has to be injected into the skin in regions where lesions are to appear shortly; for example, the advancing edge of a large zoster lesion. This is not really technically feasible in generalized cutaneous viral disease, and, furthermore, the major danger to the patient with such infections comes from the visceral lesions. Also, with herpes zoster, it is not certain that such injections would be expected to modify the incidence of postherpetic neuralgia which depend as likely on events in the sensory ganglia as in the skin.

The early studies of corneal instillation of monkey (*16*) or human (*31*) interferon were also conducted with materials of uncertain but in all probability much less activity than are available today. The effect on vaccinia was judged beneficial but not striking. The number studied was small, and it appears that they represent the full range of field disease and were not selected for the earliest disease. Further work, therefore, is indicated here as well.

It is in the respiratory tract application that the most persistent and

TABLE 2. USE OF EXOGENOUS INTERFERON IN HUMANS—LOCAL APPLICATION

Site	Interferon dose, routes, and timing	Infection	No. of patients	Result	Ref.
Skin	Monkey interferon (0.1 ml) introduced 1 day prior to challenge	Vaccinia	38 Volunteers (1 site treated, other site control)	In 50% of treated sites, vaccinia lesion prevented	Scientific Committee on Interferon (*23*)
Skin	Monkey interferon (4 ml) injected into and beneath the skin at several sites	Vaccinia	1 Lesion in a single patient with vaccinia gangrenosum	No difference between treated and untreated sites	Connally *et al.* (*6*)
Cornea	Monkey interferon (0.05 ml) into eye every ½ hour	Vaccinia	5 Patients	Healing good, but stromal keritits developed in some	Jones *et al.* (*16*)
Cornea	Human amnion interferon given hourly into the eye	Herpes simplex	17 Patients with interferon, 17 controls	Interferon treatment appeared to speed healing but not prevent recurrence	Tommila (*31*)
Respiratory tract	Monkey interferon (approx. 10^4 units) by frequent nasal spray 1 day before and up to 5 days after viral challenge	Rhinovirus (M type), Coxsackie A21, or parainfluenza I	20 Patients with interferon, 17 controls	No effect on viral shedding, seroconversion, or symptoms	Scientific Committee on Interferon (*24*)
	Human leukocyte interferon (10^4 units) by nasal spray 24 and	Influenza A2/M21/65	191 Patients with interferon, 110 controls	Interferon prevented viral shedding, seroconversion, and symptoms	Soloviev (*26*)

2 hours prior to infection				
Human leukocyte interferon (1.8 $\times$ 10^5 units) by frequent nasal spray 1 day before and 2 days after viral challenge	Rhinovirus 2	4 Patients with interferon, 5 controls	No effect on viral shedding, seroconversion, or symptoms	Tyrrell and Reed (*32*)
Human leukocyte interferon by frequent nasal spray 1 day prior to viral challenge (10^6 units with flu) or 1 day prior to and 3 days after viral challenge (1.4 $\times$ 10^7 units with rhinovirus 4)	Influenza B/Han/1/70	11 Patients with interferon 11 controls (with flu)	Disease delayed 24 hours in flu study	Merigan *et al.* (*20*)
	Rhinovirus 4	16 Patients with interferon, 16 controls (with rhinovirus 4)	Symptoms, seroconversion, and virus shedding prevented with rhinovirus	

thorough efforts have been made. The first study at this site (*24*) may have failed for at least two reasons—relatively small quantities of monkey interferon were utilized and the challenge virus doses were large. The later trial (*32*) with rhinovirus 2 improved on both these deficiencies, but failed to demonstrate significant viral inhibition. Early studies by Soloviev (*26*) suggested that low doses (10^4–10^5 units per patient) of human leukocyte interferon will inhibit influenza in volunteers or in normal individuals during epidemics in controlled trials. His observations have led to widespread use of this material in the Soviet Union during the last two winter influenza epidemics. For example, 2 million doses of leukocyte interferon were utilized in 1972 in the U. S. S. R. in clinical application and 100,000 vials used in Moscow alone for anti-influenzal effect in December 1972 (*17*). Recently, the efforts of Soviet investigators have turned to evaluating the effect of larger doses of interferon in treatment of early symptomatic cases of influenza (*17*). These trials were not double-blind or placebo controlled, but they suggest to the investigators that certain systemic manifestations, particularly fever and headache, resolve immediately, whereas local respiratory symptoms are somewhat slower to resolve. However, even the latter were thought to disappear more rapidly in the treated patients than in controls. Undoubtedly, further efforts to confirm the efficacy of low-dose interferon prophylaxis during influenza epidemics and therapeutic effects with higher doses will be undertaken by these and other investigators. Several viral diseases of the eye and respiratory tract have also been reported to respond favorably to human leukocyte interferon in other controlled trials in the U. S. S. R. (*14*). Unfortunately, few details of these studies are presently available, and they are therefore difficult to evaluate.

Recently, in collaboration with Drs. Sylvia E. Reed, Thomas S. Hall, and David A. J. Tyrrell at the Common Cold Unit in Salisbury, England, one of us (TCM) participated in a study successfully employing interferon as a prophylactic in the human respiratory tract (*20*). We first determined the interferon sensitivity by several techniques of a number of respiratory viruses both in tissue and tracheal organ cultures. The rhinoviruses as measured by yield inhibition technique were among the most sensitive. Since their replication appears to be confined to the superficial respiratory mucosa and during such infections the interferon concentrations found in nasal washings were low, a local interferon prophylaxis trial appeared to be reasonable. The dose of human leukocyte interferon we employed (10^7 units per patient) represents a 100-fold increase over the last Salisbury rhinovirus–interferon trial which was unsuccessful (*32*). In addition, because we noted a suggestion of late breakthrough of infection with 1 day of prophylaxis against an influenza challenge in a preliminary controlled trial involving 22 people (10^6 units per patient), we decided to administer our interferon both 1 day before and 3 days after virus challenge to cover any possible critical period or requirement for sustained presence of interferon. The interferon was also given by thirty-nine separate sprays into each nostril in order to allow for any imperfect distribution due to the crudeness of the delivery system (a sprayer that filled the nostril and produced a fine aerosol of 0.25 ml). Although the expense and effort in preparing the interferon were great, it was gratifying that the dosage of interferon we employed both prevented symptomatology and virus shedding in the 16 treated volunteers compared to the 16 controls who received a placebo. It is not surprising that side effects were absent in this trial as the interferon applied directly to the cells was only minimally absorbed systemically. Direct application to the virus target organ also avoided the rapid losses from the circulation that had been observed upon systemic administration. Successful inhibition of rhinovirus infections in man by human interferon seemed to be dependent on building up high local concentrations by frequent applications. The fact that measurable quantities of interferon persisted in the respiratory tract for up to 17 hours after spraying suggests that dosing need not be so frequent in future studies, especially with higher titered interferon preparations. Furthermore, the amounts of interferon employed could be decreased if a method could be evolved for retaining the interferon in the respiratory tract. This might be done by either application of interferon by a vehicle that allowed slow release or otherwise helped it to remain present in a biologically active form in the vicinity of the cells to be treated.

B. Systemic Application

In any clinical application, it is critical by which route one is applying the interferon and at what site one expects its antiviral effects to occur. During natural virus infection, it has yet to be shown whether serum levels or local tissue or body fluid levels are more important in controlling spread of infection. For example, with disseminated zoster, we feel that local vesicle fluid levels are quite important as they correlate with recovery (28) where circulating levels are usually undetectable. But one cannot generalize from these observations as each virus infection has its own individual pathogenic course in the host. Hence, it is likely that different analyses would apply in different infections. Obviously, interferon cannot be expected to act unless it can reach in adequate concentration the appropriate cells it is to protect. Therefore the route of delivery is critical as well as the concentration and timing. Intramuscular dosing appears to offer a real advantage if parenteral administration is required: observations both in rabbits (*4*) and monkeys (*25*) indicate that they produce a greater mean titer of circulating interferon over a longer period than with intravenous injection where a rapid falloff from initially high circulating levels is noted. Evidence from tissue culture and animal models suggests prolonged blood levels would seem to be useful in producing good tissue resistance to infection.

The previously reported studies with human amnion and leukocyte interferon administration to patients with herpesviral infections or malignancies (Table 3) were not designed to be critical tests of antiviral or antitumor efficacy. For example, it is difficult to draw any conclusion regarding infants with cytomegalovirus and herpes simplex infections who were treated with systemic interferon injections as the study was not controlled and their clinical courses were not presented in detail (*8*). The earliest efforts at administration of interferon to patients with malignancy were those of Falcoff and associates (*8*), who gave modest doses of amnion-produced human interferon to 18 patients with acute leukemia. Although the authors were favorably impressed with the results (including lack of toxicity), they are careful to point out their study was not controlled and allows no conclusions regarding efficacy.

More recently, data have been accumulated on the results of repeated intramuscular injection of human leukocyte interferon into humans with

TABLE 3. USE OF EXOGENOUS INTERFERON IN HUMANS—SYSTEMIC APPLICATION

Interferon dose, route, and timing	Disease	No. of patients treated with interferon	Result	Ref.
Human amnion interferon-varying quantities given intravenously for periods up to 1 year	Cytomegalovirus, herpes simplex, acute leukemia, etc.	24 Patients	No limiting side effects	Falcoff *et al.* (*8*)
Human leukocyte interferon 5×10^6 units IM 3 times weekly for up to 2½ years	Various malignancies	28 Patients	No limiting side effects	Strander *et al.* (*29*)
Human leukocyte interferon up to 25×10^6 units IM or IV daily for 3 days	Herpes zoster, disseminated	9 Patients	No side effects other than fever	Jordan *et al.* (*16a*)

a wide variety of late-stage tumors. Strander has treated 28 patients (*29*) with up to 3×10^6 units (3 times weekly) for 2½ years with no limiting side effects. Fever was observed in most patients for the first few injections. In addition, late in the course of treatment, antibodies were noted to some of the contaminating inactive virus used to stimulate the leukocytes and to the proteins of the chick cells in which the interferon-inducing virus was grown, but not to the interferon itself. The wide assortment of patients studied and the uncontrolled nature of this trial precludes any conclusions regarding efficacy; but it does indicate this dosage is virtually free of toxicity. The patients, still under treatment by Strander at the Karolinska Hospital, include several with Hodgkin's disease (*4*), acute leukemia (*3*), osteogenic sarcoma (*10*), as well as others with various solid tumors.

III. PRESENT STUDIES OF VARICELLA ZOSTER INFECTIONS TREATED WITH INTERFERON

At Stanford University, we are currently in the midst of an evaluation of systemic application of human leukocyte interferon in varicella zoster infections in immunosuppressed hosts. We first defined a large dose of interferon that was tolerated without significant side effects in a small series of patients (*19*) before going on to our currently in progress double-blind, randomized, placebo controlled trial. We have recently completed a similar controlled trial with intravenous cytosine arabinoside (*27*). Here, in certain patients with late-stage lymphomas, we found that the purportedly effective antiviral therapy actually prolonged disease with a demonstrable concomitant immunosuppressive effect on the host responses to the virus infection. In addition, the remaining patients in our series showed no evidence of an antiviral effect of that drug. Herpesvirus infections are particularly appealing for study of antiviral agents in that they are serious and the host appears to be still at risk from further difficulties due to continuing virus replication when he is referred for therapy. But a clinical effective antiviral agent most preferably should work in concert with and not block host defenses.

The first 9 patients we treated with human leukocyte interferon gave us experience in pharmacokinetics as well as a chance to evaluate possible side effects. Since those studies were reported (*19*), we have confirmed these observations in another 8 patients, 5 of whom have been treated as part of the above-mentioned ongoing controlled trial to evaluate the systemic antiviral activity of interferon in this infection. The greatest total dose and longest duration of treatment we have used so far is 105×10^6 reference units given over 3 days.* When the intravenous and intramuscular routes were compared directly, a higher peak was seen after a slow intravenous infusion; however, serum levels greater than 25 reference units were obtained more rapidly and were sustained longer following intramuscular administration. There was no evidence of local losses on intramuscular administration. In fact, this route was much more convenient and just as efficient in production of serum levels of interferon as continued intravenous infusion.

In contrast with early work on the pharmacokinetics of interferon in mice (*2*) and rabbits (*13*) indicating very rapid clearance from the circulation after a rapid intravenous injection, our results showed that the half-life of circulating interferon is prolonged following cessation of continuous intravenous infusion. This appeared to be due to better tissue saturation in our human interferon studies, with slow release from tissué sites maintaining levels longer when the infusion is stopped. When an intramuscular injection was given to patients on 3 successive days, peak daily titers rose progressively, suggesting tissue accumulation. The rise of varicella zoster complement-fixing antibodies and the increase in vesicle fluid interferon and cells were unimpaired in the interferon-treated patients. Fever of 38° to 40°C was seen in association with interferon administration; however, no limiting toxicity was observed. Figure 1 shows the clinical course of a patient who very clearly demonstrates an association between interferon administration and fever. The average level of circulating interferon attained by either the intramuscular or intravenous route is higher than reported levels in naturally occurring human viral diseases and is approximately 17 times the minimal inhibitory concentration for varicella zoster virus in human cell culture (*1*). This is in contrast to all metabolic antagonists where the highest clinically tolerated doses produced serum levels that are barely inhibitory to herpesviruses in tissue culture systems. In keeping with its high molecular weight, systemically administered interferon appears to penetrate very poorly into cerebral spinal fluid or skin vesicles, and it appeared in the urine of only 1 of the 9 patients studied. Interestingly enough, that individual had a rare form of renal disease (amyloidosis) which caused him to excrete plasma proteins in the urine as well.

The patient shown in Fig. 2 was particularly encouraging as to the necessity for further controlled evaluation of systemic application of

* Throughout the remainder of this paper, all unitage is expressed in terms of standard reference units as determined by titration of individual laboratory standards in comparison to reference research standard A 69/19 as obtained from the National Institutes of Health, Bethesda, or the National Institutes of Medical Research, London.

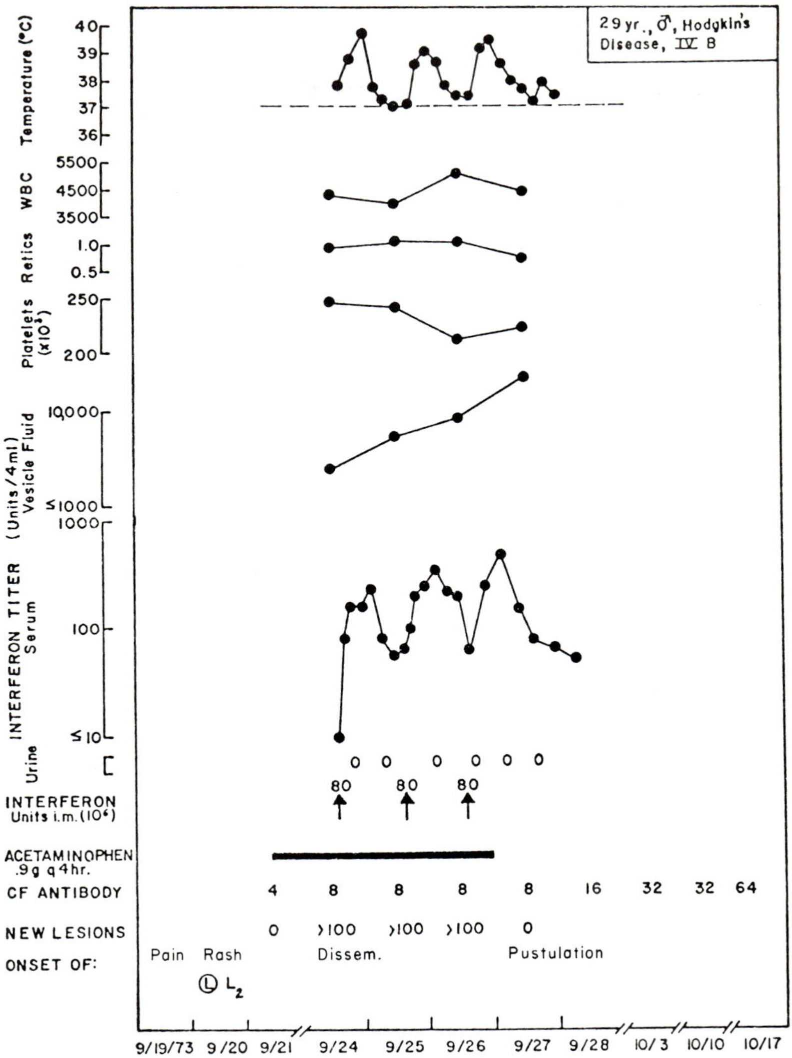

Fig. 1. Clinical course of patient with disseminated zoster given leukocyte interferon. The unitage of interferon referred to here is presented in terms of our own laboratory units which should be divided by 4 to yield standard reference units. CF, complement-fixing.

leukocyte interferon in varicella. She had just been put in remission from acute lymphocytic leukemia with combined chemotherapy and radiotherapy prior to onset of her chickenpox. On initiation of interferon therapy she was undergoing severe cutaneous and visceral dissemination with bilateral varicella pneumonia. This cleared rapidly and new skin

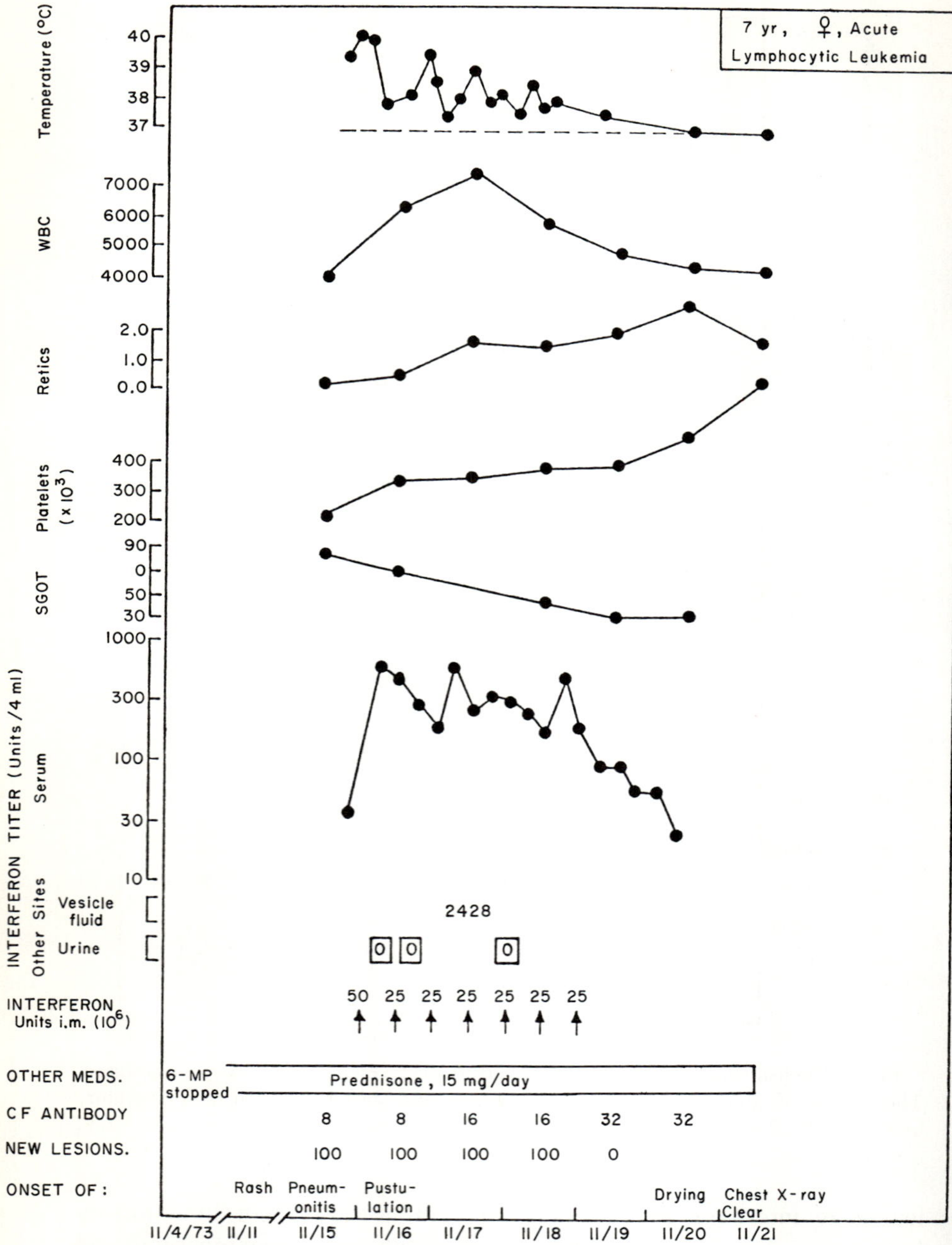

Fig. 2. Clinical course of patient with varicella given leukocyte interferon. The unitage of interferon referred to here is presented in terms of our own laboratory units which should be divided by 4 to yield standard reference units. CF, complement-fixing; 6-MP, 6-mercaptopurine.

lesions stopped forming over the interferon treatment period. We have chosen a dosage regimen similar to hers for our double-blind controlled study as it has rapidly produced and sustained high levels in the serum. This is an intramuscular dose of 0.42×10^6 units/kg followed by twice daily intramuscular doses of 0.21×10^6 units/kg.

We are presently studying four categories of diseases with a separate prospectively randomized, placebo-controlled trial for each: (I) early disseminated zoster in malignancy (first 2 days), (II) localized zoster in malignancy (first 7 days), (III) varicella in malignancy, and (IV) varicella in congenital immune deficiency. In the last two categories we are accepting patients for treatment in the first 3 days of their disease. The first group is treated for between 1 and 3 days, depending on dissemination in the prior 24 hours, whereas the other three groups are arbitrarily treated for 3 days. The localized zoster protocol would permit delineation of a suppressive effect on occurrence and severity of cutaneous and visceral dissemination even if a latent period of several days is required.

Our present experience with disseminated zoster and varicella is summarized in Table 4. Only in the randomized and controlled trial phase was therapy started at the same stage of disease in the treated and untreated portion of the series. Therefore, as patterns of patient referral

TABLE 4. STANFORD UNIVERSITY EXPERIENCE WITH SEVERE VARICELLA ZOSTER INFECTIONS IN PATIENTS WITH MALIGNANCY AND IMMUNOSUPPRESSION

Treatment category	Year	No. of patients	Days of active disease (mean)	Mortality
Disseminated zoster				
No treatment	1968–1971	23	10.8	2/23
Ara C-treated[a]	1971	13	9.5	0/13
Placebo-treated	1971–1973	20	9.1	0/20
Ara C-treated	(controlled trial)	19	11.1	1/19
Interferon-treated[a]	1973	10	9.4	0/10
		85	10.1	3/85
Varicella				
No treatment	1969	1	8	0/1
Poly (rI) · poly (rC) treated[a]	1970	1	14	1/1
Ara C-treated[a]	1970–1972	7	7.5	2/7
Interferon-treated[a]	1973	1	7	0/1
		10	8.1	3/10

[a] Uncontrolled open treatment series.

and the underlying disease and therapy in our patients are constantly changing, these results are presented only to show the increased mortality in varicella (30%) versus disseminated zoster (3.5%) in malignancy. We cannot make any statement about interferon's efficacy as during these studies we had no controls and the mean time of starting therapy was over 7 days after localized zoster began. We will improve upon that as we now have more knowledge of the safety of the material. Our randomly selected placebo-treated controls in each of our four study groups should allow precise estimation of efficacy.

IV. PERSPECTIVES FOR EXOGENOUS INTERFERON

The following discussion of future possibilities must be recognized as only conjecture. However, it does seem useful to review ideas on strategy of use of interferon as the principles demonstrated apply to other antivirals as well. Whether any or all of these possible uses are critically evaluated depends on continued verification of interferon's lack of toxicity in the studies currently in progress. However, the major deterent to the conduct of trials is the limited availability of material. Leukocyte interferon is a by-product of blood collected for normal transfusion purposes. Cantell's interferon production facilities will allow a very limited number of demonstration experiments, but are difficult to envision being extended to other blood transfusion centers because of the quality control, special methodology, and training required for optimum production. In addition, the Finnish source is unique in that it depends on a centralized blood bank for the entire country handling large quantities of blood each day in contrast to the decentralized, locally supervised, smaller volume systems functioning elsewhere in the world. Therefore, two major alternative means for production are being explored. The first and most immediately implementable is the use of diploid or other human cell substrates. The latter might include EBV-infected lymphocytes which multiply at a reasonably rapid rate (*10*) or even transformed human fibroblastic cells analogous to the non-virus-producing C-243-C3 line of 3T3 murine cells infected by Moloney sarcoma virus (*21*). The latter cell produces higher quantities of mouse interferon than the best continuous line, i.e., L_{929} cells, available prior to its discovery. Unfortunately, the lymphoblastoid cells infected by EBV are only moderately good interferon-producing cells (*10*). Thus, the difficult consideration of safety of a partially purified product of virus-transformed cells is avoided. On the other hand, the efficiency of production of interferon in human diploid cells has been improved in recent years (*3, 11*), and its cost of production now approximates that of leukocyte interferon.

The use of such cells is more acceptable than leukocytes to many authorities as they are currently in use in the production of certain widely employed viral vaccines.

The second solution to the problem of availability is less immediate, but perhaps more satisfactory for its ultimate widespread clinical applications. Efforts are currently under way to determine interferon's structure by groups who would like to have such information for its possible organic synthesis. Even though the expense of such synthetic material for routine clinical use might be great, its high specific activity suggests only minute quantities may be required in each individual.

Table 5 lists some clinical settings where use of exogenous interferon warrants consideration. Local applications have already been discussed and have the most clear basis in terms of successfully treated animal model infections (*9*). In at least one of these situations, that is, herpes keritits, a controlled double-blind trial is presently under way with high-titered, human leukocyte interferon. It is still quite possible that local application of synthetic interferon inducers may have use in one or more of these settings as the problem of systemic inducer toxicity would be minimized. However, in regard to systemic application in individuals who are already infected, inducers appear to have less potential because of the hyporeactivity to induction present in such individuals (*30*). Based on the experience with interferon in relevant animal infections (*7, 15, 22*), early postexposure prophylaxis with human interferon

TABLE 5. POTENTIAL CLINICAL SETTINGS FOR APPLICATION OF EXOGENOUS INTERFERON

Local application
  Viral respiratory tract infections,[a] herpes simplex keritits,[b] herpes zoster or vaccinia keritits
Systemic application
Viral:
  Herpes zoster or varicella,[c] herpes simplex,[c] cytomegalovirus infections,[c] vaccinia gangrenosum, variola,[d] viral hepatitis (IH or SH),[e] Rabies,[e] Arbo- or arenovirus infections,[e] kuru, Jacob-Creutzfeldt disease, progressive multifocal leukoencephalopathy, subacute sclerosing panencephalitis
Other:
  Malignancies (leukemia, lymphoma, or solid tumor)[f]

[a] As prophylaxis or early therapy, particularly in individuals with cardiopulmonary disease in whom there may be serious consequences.

[b] To prevent or alter recrudescence or chronic disease.

[c] Particularly in immunocompromised hosts.

[d] In established disease.

[e] In established disease or as supplement to prophylaxis.

[f] Where no other therapy is available or to supplement other therapy.

might be of benefit with rabies or severe otherwise lethal viral disease such as that caused by arenoviruses. Whereas the circulating levels of interferon we have attained in our patients would reasonably be expected to inhibit replication of the cytocidal viruses listed in Table 5, the susceptibility to interferon of most slow viruses is unknown. In fact, a relevant animal virus, scrapie, has been judged resistant to interferon action *in vivo* by one group (*33*). In regard to the application of interferon to human malignancy, the success in animal malignancies demands future clinical study, and we have reviewed that problem elsewhere (*18*). The most frankly speculative of all areas that have been considered for the application of interferon is human neurological disease suspected to be of viral origin, such as multiple sclerosis or amyotrophic lateral sclerosis. Despite the compelling nature of these clinical problems and our present lack of therapy, it seems most reasonable to follow through on evaluation of the other listed applications of interferon where we have a firmer base of animal and tissue culture evidence for possible efficacy.

Although much has been put forth in this final section which is yet far from accomplishable, it is hoped that the effort at bringing together our ideas on rationale and priorities will stimulate others to improve on them in the future.

REFERENCES

1. Armstrong, R. W., and Merigan, T. C., *J. Gen. Virol.* **12**, 53 (1971).
2. Baron, S., Buckler, C. E., McCloskey, R. V., and Kirschstein, R. L., *J. Immunol.* **96**, 12 (1966).
3. Billiau, A., Joniau, M., and De Somer, P., *J. Gen. Virol.* **19**, 1 (1973).
4. Cantell, K., and Pyhälö, L., *J. Gen. Virol.* (in press).
5. Cantell, K., Strander, H., Hadhazy, G., and Nevanlinna, H. R., *in* "The Interferons" (G. Rita, ed.), pp. 223–232. Academic Press, New York, 1968.
6. Connolly, J. H., Dick, G. W. A., and Field, C. M. B., *Brit. Med. J.* **1**, 1315 (1962).
7. De Clercq, E., and Merigan, T. C., *J. Gen. Virol.* **5**, 359 (1969).
8. Falcoff, E., Falcoff, R., Fournier, F., Chany, C. C., and Galliot, B., *Ann. Inst. Pasteur, Paris* **111**, 562 (1966).
9. Finter, N. B., ed., "Interferons and Interferon Inducers," Amer. Elsevier, New York, 1973.
10. Haase, A. T., Johnson, J. S., Kasel, J. A., Margolis, S., and Levy, H. B., *Proc. Soc. Exp. Biol. Med.* **133**, 1076 (1970).
11. Havell, E. D., and Vilcek, J., *Antimicrob. Ag. Chemother.* **2**, 476 (1972).
12. Herrmann, E. C., *New Physician* **18**, 270 (1969).
13. Ho, M., and Postic, B., *Nature (London)* **214**, 1230 (1967).
14. "Interferons and Interferonogens." Institute of Poliomyelitis and Viral Encephalitis, Moscow, 1967.
15. Janis, B., and Habel, K., *J. Infec. Dis.* **125**, 345 (1972).
16. Jones, B. R., Galbraith, J. E. K., and Al-Hussaini, M. K., *Lancet* **1**, 875 (1962).

16a. Jordan, G. W., Fried, R. P., and Merigan, T. C., *J. Infec. Dis.* **130** 56–62 (1974).
17. Jordan, W. S., Hopps, H. E., and Merigan, T. C., *J. Infec. Dis.* **128**, 261 (1973).
18. Merigan, T. C., *Cancer Chemother. Rep.* (in press).
19. Merigan, T. C., Fried, R. P., and Jordan, G. W., unpublished observations (1973).
20. Merigan, T. C., Reed, S. E., Hall, T. S., and Tyrrell, D. A. J., *Lancet* **1**, 564 (1973).
21. Oie, H. K., Gazdar, A. F., Buckler, C. E., and Baron, S., *J. Gen. Virol.* **17**, 107 (1972).
22. Postic, B., and Fenje, P., *Appl. Microbiol.* **22**, 428 (1971).
23. Scientific Committee on Interferon, *Lancet* **1**, 833 (1962).
24. Scientific Committee on Interferon, *Lancet* **1**, 505 (1965).
25. Skreko, F., Zajac, K., Bahnsen, H. P., Haff, R. F., and Cantell, K., *Proc. Soc. Exp. Biol. Med.* **142**, 946 (1973).
26. Soloviev, V. D., *in* "The Interferons" (G. Rita, ed.), pp. 233–245. Academic Press, New York, 1968.
27. Stevens, D. A., Jordan, G. W., Waddell, T. F., and Merigan, T. C., *N. Engl. J. Med.* **289**, 873 (1973).
28. Stevens, D. A., and Merigan, T. C., *J. Clin. Invest.* **51**, 1170 (1972).
29. Strander, H., Cantell, K., Carlström, G., and Jakobsson, P. Å., *J. Nat. Cancer Inst.* **51**, 733 (1973).
30. Stringfellow, D. A., and Glasgow, L. A., *Infec. Immunity* **6**, 743 (1972).
31. Tommila, V., *Acta Ophthalmol.* **41**, 478 (1963).
32. Tyrrell, D. A. J., and Reed, S. E., in "'Non-Specific' Factors Influencing Host Resistance" (W. Braun and J. Ungar, eds.), pp. 438–442. Karger, Basel, 1973.
33. Worthington, M., *Infec. Immunity* **6**, 643 (1972).

DISCUSSION

Dr. Tyrrell: We have followed up one of the points which Dr. Merigan mentioned, the need for an inducer that would avoid the problems of manufacturing large amounts of human interferon. Dr. Matthews and Dr. Deed have tested the LEV4 vaccine–live echovirus 1 vaccine of Dr. Voroshilova as a possible inducer of interferon and resistance to respiratory infection in man. The virus was given by mouth and multiplied well in the gut. Little circulating or local interferon was found. There was no reduction in the amount of virus produced after an intranasal challenge with rhinovirus Type 4, or in the frequency of clinical symptoms and signs.

Dr. Kaufman: I think that some of the studies with eye disease may give a little bit of light and highlight some of the points made by Dr. Merigan and Dr. Tyrrell. First of all, it is clear in the treatment of very good animal models of eye disease that interferon seems to be an excellent prophylactic agent, and at best a modest therapeutic agent. There is a clear-cut difference between prophylaxis and therapeusis, and the available antivirals—idoxuridine (IDU), adenine arabinoside, cytosine arabinoside, trifluorothymidine—have a reverse effect. For example, interferon in small doses will prevent infection and will prevent recurrences of herpes in rabbit models quite effectively, whereas the therapeutic effect is slight—ara A, IDU, and trifluorothymidine will not prevent recurrences. With the antivirals, the recurrences come, are small, and go away rapidly, but there is a real disassociation between prophylaxis and therapy which is very clear in these models.

Second, in these models and others, it is clear from our studies as well as those of others that there is an enormous species difference, and that humans and monkeys are relatively unresponsive, if not almost totally unresponsive, to polyinosonic–polycytidilic acids (poly-IC) and tilerone. We need also to keep in mind that Dr. Tyrrell studies an infection precisely timed for interferon prophylaxis. For prophylaxis we need to have interferon there before the infection starts, so that if inducers are to be useful, inducers that do not have a prolonged hyporesponsive state are necessary. At present it is not clear that such inducers are available for man, which is why human interferon is being used. Human interferon in monkeys, for example, will very effectively prevent infection with herpes, just as interferon will prevent recurrences of herpes keratitis in rabbits. Once a human has had herpes keratitis, the odds are about 50% that he'll get another attack within 2 years. This is a potentially devastating disease once it is in the eye. It is clear in monkey systems that as little as a drop of interferon in the morning and a drop at night of moderate titer, no more than 15,000 units/cc human interferon, will effectively prevent infections in these monkeys. And this is the basis on which the double-blind controlled human studies are being undertaken. I think that there is every reason to expect that they may well be fruitful. It is a totally different kind of principle than that on which Dr. Merigan is basing his studies, which is early therapeutic effect. Merigan's studies with cytosine arabinoside, as he pointed out, used immunoincompetent or least markedly suppressed hosts with a drug (CA) that further suppresses the host's immune response. This is likely to be different from the normal host, but there also may be a big difference, it seems, between therapeutic drugs. Studies in our laboratory indicate that adenine arabinoside in comparable antiviral concentrations has a much smaller and probably negligible immunosuppressant effect compared with cytosine arabinoside. And this might be a very important factor in considering the choice of antiviral agents.

Dr. Warren: I would like to make a general comment on introducing foreign proteins into the upper respiratory tract and attributing subsequently appearing viral interference as necessarily being due to the induction of interferon. In 1965, Drs. Jensen, Takano, and I found that yeast RNA, a single-stranded nucleic acid, administered by nasal instillation or exposure to an aerosol, would protect mice against a subsequent challenge with large amounts of encephalomyocarditis (EMC) or influenza viruses. However, no interferon was detected in the circulation of mice that were completely protected by yeast RNA [Takano, K., Warren, J., and Jensen, K. E., *J. Bacteriol.* **90,** 1542–1550 (1964]. The point I wish to make is that before we can ascribe protection against influenza and other respiratory viruses to the instillation of a substance as relatively crude as human leukocyte-derived interferon, it will be essential to determine whether such induced resistance is actually due to the action of interferon alone or to some type of competitive RNA incorporation [Bishop, D. C., and Abramoff, P., *Proc. Soc. Exp. Biol. Med.* **120,** 378–383 (1965)].

Dr. Merigan: We controlled for the amounts of protein in our leukocyte interferon in the placebo. I think a very interesting kind of support for our contention that the interferon was producing the antiviral effects we observed comes from Dr. George Jackson's recent studies, and I hope he can comment on them. It appears as though the levels that we produced in the nasal washings by spraying in exogenous interferon are very nearly identical to the levels that seem to be effective in his recent studies with an inducer stimulating production of human nasal interferon. Hence, it appears that, if you can produce levels in the respiratory tract of some 200 or so reference units of interferon, then you can produce resistance to rhinovirus challenge. Whether it is sprayed in or induced, you seem to get this result.

DR. TYRRELL: In our situation we had to use all the interferon that there was in Britain to produce a significantly sized study in order to show that there was an effect. We didn't feel like destroying half of it to deal with a possible control that we might never need to use. And so that is the reason that it was not done. I agree that formally we have not shown that this happens. I agree with Dr. Merigan that the presumption at the moment is that this is so. Because we put very much cruder stuff in years before and produced no effects at all, we have tested the effect of "rubbish" already and we know it is not much good.

DR. JACKSON: The problem that Dr. Warren has referred to of the time-related temporal association of interferon appearance and subjective and virologic improvement in infected hosts is a very difficult one to unravel. I think the full explanation of these associations will need to come from Dr. Revel and his *in vitro* studies. As Dr. Merigan has suggested, however, our studies using inducers as prophylaxis in man establish a clear relationship between relief of clinical illness and the interferon level in nasal secretions. Now that can still be a temporal association, not cause and effect. The only ways one has to dissociate these factors is to (1) administer preformed human interferon, (2) give a chemical inducer, or (3) induce interferon as a cell-mediated immunological product. These will permit different mechanisms for induction of interferon ahead of the time of the viral challenge. In doing so, everything seems to fall into the pattern showing that as one preinduces interferon by one or more different methods, or uses preformed interferon, both the symptomatic and the virologic responses correlate with the time of appearance and the interferon titer. We are still investigating what the minimal effective titer is as well as the temporal relationships. However, the titer, as we measure it, is that excreted into the nasal wash; we have no measure of what the titer might be at the interstices of the cells or the cell surface. As Dr. Merigan indicated, he had to use huge doses of preformed interferon applied topically. From the titration of nasal washes after a topical drug inducer, clinical response is not observed until one exceeds about 10 units/ml of interferon; above that, between 10 and 100 units/ml, the temporal relationship and presumably the cause-and-effect relationship can be demonstrated.

DR. FJELDE: Why did you choose human leukocyte cultures as the means for making interferon? We have been interested for many years at Roswell Park in using cell cultures for therapeutic programs. All cultures have some advantages and some disadvantages.

DR. MERIGAN: It's simply a question of practicality as leukocytes can be obtained as explants in sufficient quantity for a reasonable price to produce the units that we required. Considering the costs of cell propagation on a cost-per-unit basis, leukocyte interferon was the most economical at that time.

DR. CASALS: Among the viruses that you considered as possible targets for trials with human interferon you have listed arboviruses and arenaviruses. I presume that under arboviruses you had in mind those that cause serious illness, particularly the equine encephalitis viruses and a number of Group B agents that also cause encephalitis. The problem that one would be confronting, however, is that the majority of persons clinically infected with these viruses do not develop encephalitis; one would not contemplate to treat all the patients that may have a mild febrile illness. On the other hand, when encephalitis has developed, it may well be that the virus is no longer accessible to interferon, unless the latter can freely penetrate the blood-brain barrier; or, the virus may even be declining in titer and what causes the signs and symptoms are the lesions.

With respect to certain other arboviruses that cause serious illness, such as dengue viruses causing dengue shock or dengue hemorrhagic fever, there is, of course, some evidence that implicates an immune component to account for the syndrome, i.e., superinfection of a previously sensitized person. The same pathogenesis seems to be considered in the case of serious infection with (LCM) virus.

With respect to other arenaviruses, Junin and Machupo, their epidemiology is such that it seems to me that preventive vaccination would be the logical direction in which to orient our efforts.

DR. KAUFMAN: In relation to the tissue culture interferon, there was a real worry above and beyond practicality as to whether the primary leukocyte interferon that is made directly from removing leukocytes from donors would actually be the same *in vivo* as interferon produced by a stable diploid line. The leukocyte interferon was used in our human studies not only because of its availability but also because the basic *in vivo* work was done with a leukocyte interferon. It is clear that there are chemical differences between the leukocyte- and culture cell-produced interferon and there was reason to fear that there might be a difference in *in vivo* effect. Studies in our laboratory with interferon obtained by Drs. DeSomer and Eddy indicate that, in fact, the diploid cell interferon does work exactly the same *in vivo* as the leukocyte interferon. This knowledge and some evidence that lower titers can be used on the cornea than on mucous membranes is encouraging as to future practical use.

DR. LEHMANN-GRUBE: I wish to enlarge a point which has already been raised by Dr. Casals. Looking at the table in which Dr. Merigan has listed potential candidates for the therapeutic use of interferon, one is startled to discover diseases for which not even the etiology, let alone the pathogenesis, is well established; in others we have reason to believe that the infectious process has come to a standstill long before clinical signs first made their appearance. Let us clearly distinguish between virus infection and virus disease. Interferon can only be expected to inhibit the former; it is not likely to alter one of the still rather dark pathways that lead from infection to disease.

DR. MERIGAN: I believe Dr. Lehmann-Grube has helped to underscore one of my main points. That is, of course, that interferon or any other antiviral should only be used when we understand enough of the disease pathogenesis to believe the host still at risk for further difficulties resulting from continuing viral replication. I'm sure our preceding comments allow others to perceive clearly the various priorities within Table 5 of our paper.

CHAPTER 15

Interferon Cell Receptor Interactions

C. Chany
H. Ankel*
M.-F. Bourgeade

The genetic analysis of the interferon system using somatic monkey–mouse hybrid cells demonstrated the discordant segregation of chromosomes involved in interferon synthesis and the antiviral state (*2–4*).

I. CELL SPECIES SPECIFICITY OF INTERFERON

We report arguments in favor of the hypothesis that only the cell membrane receptor system is responsible for the cell species specificity of the interferon molecule. We postulate that all the subsequent events leading to the antiviral state are unspecific. The arguments are the following:

1. In somatic monkey–mouse hybrid cells, primate and mouse receptors can be distinguished because of their different sensitivity to trypsin. In monkey cell lines and in monkey–mouse hybrids, trypsin destroys only monkey receptors as seen when the freshly trypsinized cells are kept in the presence of primate interferon for 4 hours (Table 1). Mouse receptors are not destroyed under the same experimental conditions. The addition of 2% calf serum to the medium rapidly restores full inter-

* On leave from the Department of Biochemistry, The Medical College of Wisconsin, Milwaukee, Wisconsin.

TABLE 1. ANTIVIRAL EFFECT OF PRIMATE AND OF MOUSE INTERFERON[a]

Cell type	Dispersing agent	Human white blood cell interferon		Mouse interferon	
		No serum	2% Calf serum	No serum	2% Calf serum
CV-1	Trypsin	0.07	3.42	N.D.	N.D.
Clone M	Trypsin	0.49	2.06	2.76	2.86
MKS-BU-100	Trypsin	N.D.	N.D.	1.13	1.75
L	Trypsin	N.D.	N.D.	2.80	2.97
CV-1	EDTA	2.86	3.07	N.D.	N.D.
Clone M	EDTA	2.23	2.25	N.D.	N.D.

[a] Antiviral effect of primate and mouse interferon in trypsinized or EDTA-dispersed parental and clone cells. Numbers represent $\log_{10}$ inhibition of VSV yield in interferon-treated cells as compared to contact cells subjected to the same experimental conditions but incubated with media containing no interferon. N.D., not done.

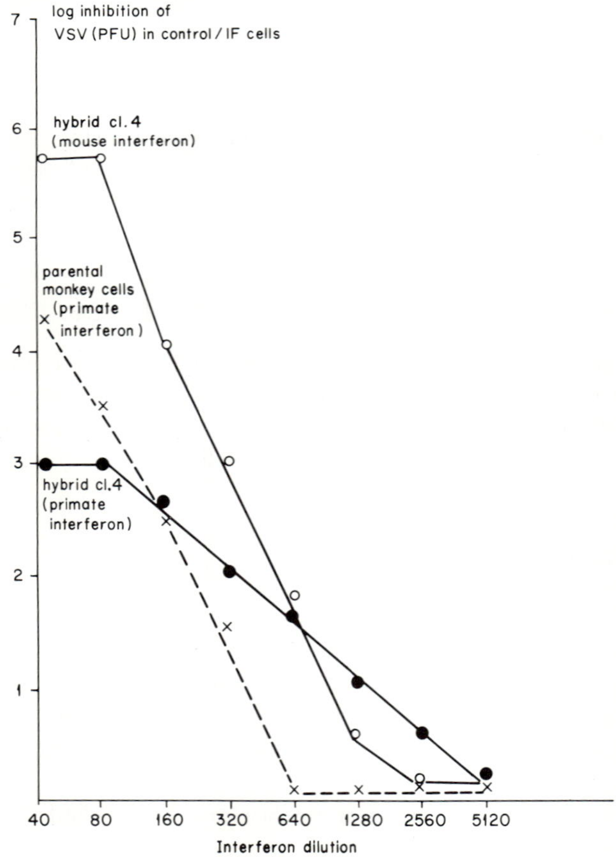

Fig. 1. Dose-response curves (log 10 inhibition) in hybrid clone 4 (ninety-second passage). VSV, vesicular stomatitis virus; IF, interferon.

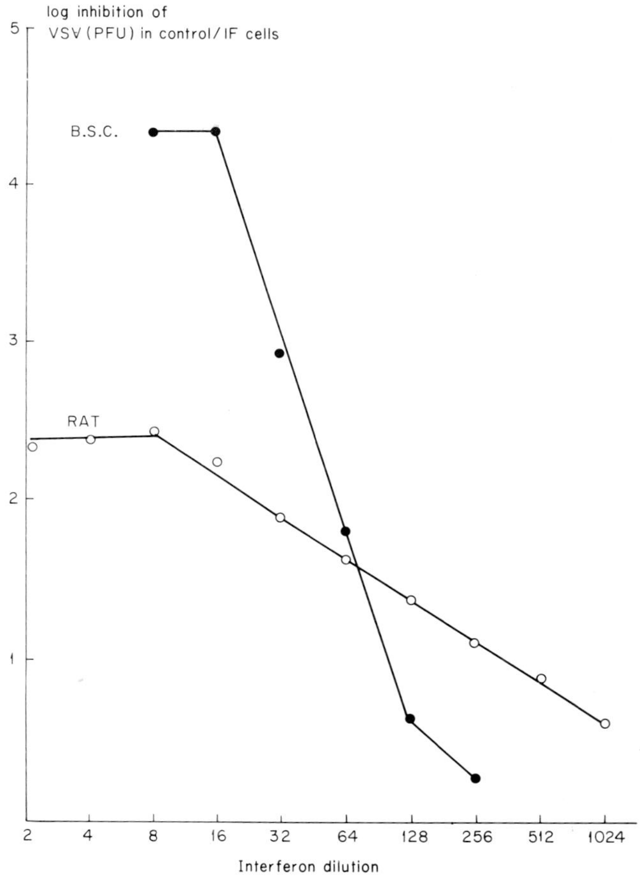

Fig. 2. Dose–response curves of interferon (IF) from human amniotic membrane. In BSC-1 cercopithecus monkey cells (●); in rat embryonic fibroblasts (○). VSV, vesicular stomatitis virus.

feron sensitivity to primate interferon. When EDTA (200 μg/ml) is used instead of trypsin, similar effects are not observed.

2. The dose–response curves in normal and hybrid cells for mouse interferon are sigmoidal. For monkey interferon, however, the dose–response curves vary from one hybrid clone to another and tend to become hyperbolic. Apparently their character depends on the amount or on the quality of monkey antigens present in the cell (Fig. 1). In most instances the dose–response curve crosses the sigmoidal dose–response curve observed in the parental CV-1 cells. Similar results are obtained when the antiviral effect of human amniotic interferon is assayed in homologous primate cells and heterologous rat embryonic fibroblasts (Fig. 2).

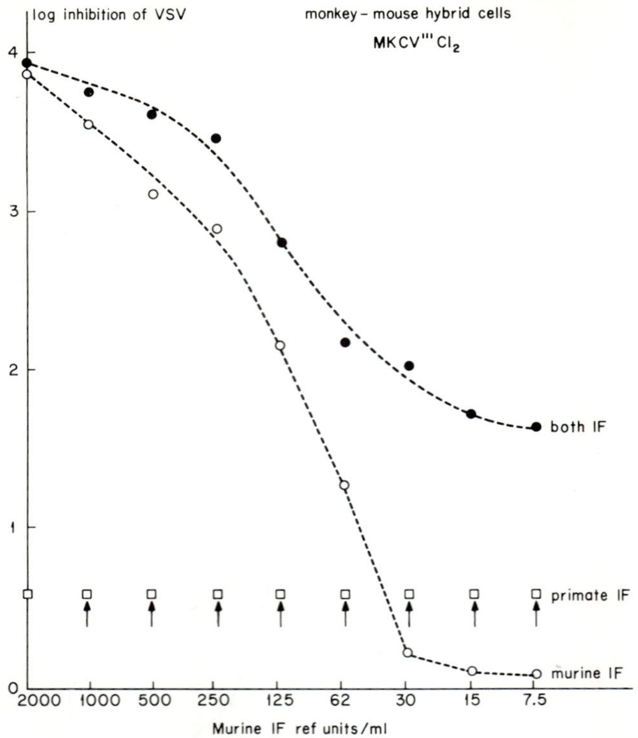

Fig. 3. Dose–response relationship between the amount of interferon (IF) is expressed in reference units as the log of the viral yield in control/the log of the viral yield in interferon-treated cells. (●) Cells treated with both murine and primate interferon; (○) cells treated with murine interferon; (□) cells treated with a constant amount of primate interferon. VSV, vesicular stomatitis virus. From Chany *et al.* (5).

3. When monkey and mouse interferons are added simultaneously to the hybrid cells, their combined antiviral effect is not simply additive, but increases 50–100-fold (Fig. 3). No such potentiating effect is observed when monkey cells are incubated with a corresponding mixture of human and monkey interferons (C. Chany, M.-F. Bourgeade, unpublished data).

II. INDUCTION OF ANTIVIRAL STATE BY INTERFERON

The interferon molecule acts on the receptor-activator system located in the cell membrane by contact, as previously shown (*1*). Interferon covalently bound to Sepharose beads induces the antiviral state. The number of beads per cell needed to induce the antiviral state is surprisingly small (Fig. 4). This could be explained either by detachment

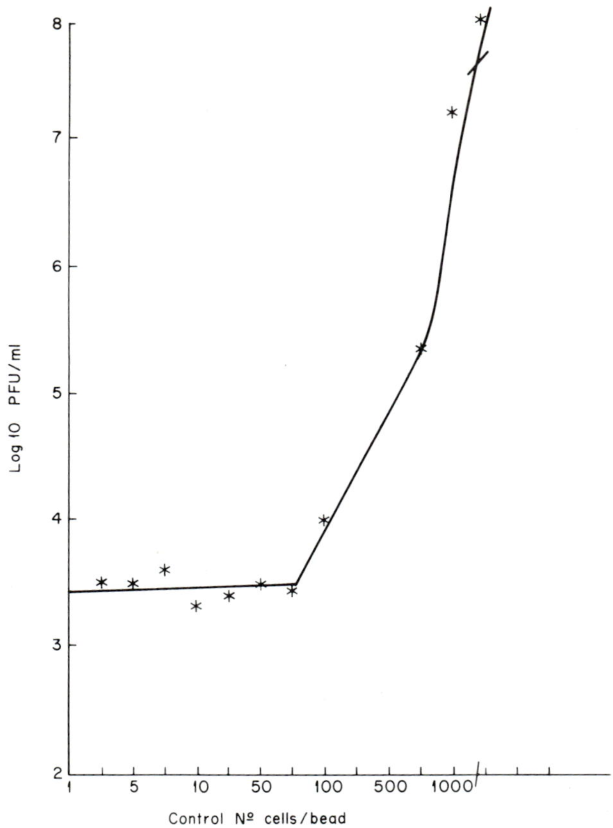

Fig. 4. Dose–response relationship between the number of interferon–Sepharose beads per cell and the antiviral state induced. Decreasing numbers of interferon-bound Sepharose beads were incubated with sensitive L cells grown in Eagle's medium in 35-mm plastic petri dishes. After 22 hours of contact, the beads were removed; the cells were washed twice with medium and infected with VSV at a multiplicity of infection (m.o.i.) = 0.1. After incubation for 16 hours at 37°C, the viral yield was determined using a routine plaque technique in L cells. From Chany *et al.* (5).

of small amounts of interferon molecules or by repeated contacts between the interferon-treated beads and the receptor sites of the cells.

In favor of this latter hypothesis are the following experiments:

1. When the number of beads is small, unable to induce the antiviral state under static conditions, a significant antiviral effect can, however, be observed if the beads are agitated during the incubation period (Table 2).

2. On the contrary, when the movements of the beads are restricted by containing them in the interior of a plastic ring placed on the surface

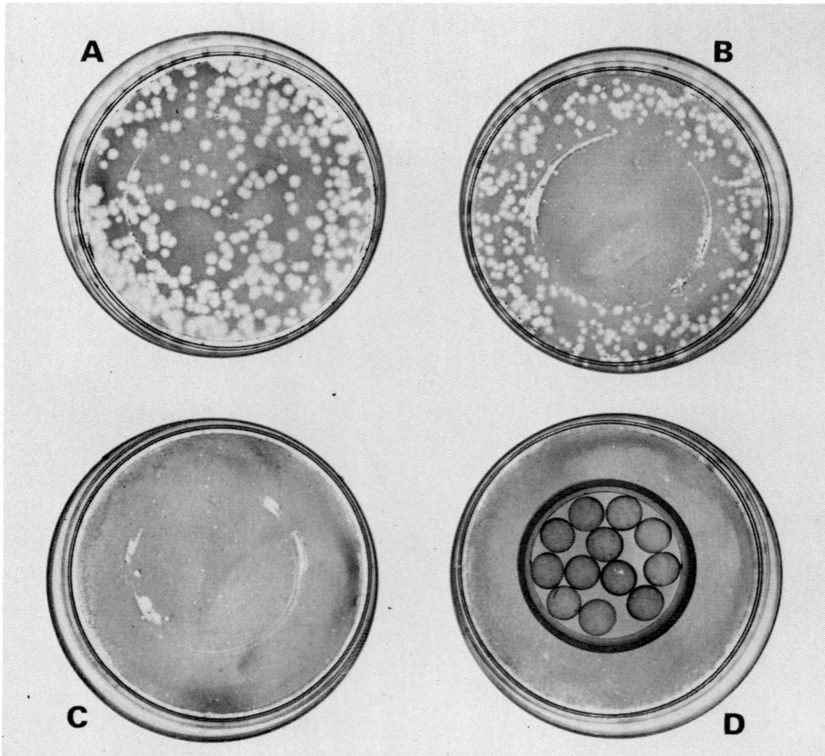

Fig. 5. A polyethylene ring was placed in the center of a 60-mm plastic petri dish containing a monolayer of mouse L cells. In one set of experiments, interferon–Sepharose beads (1 bead/10 cells) were placed in the center of the ring, which kept them from escaping into the surrounding area (B). In another set, free interferon (8×10^4 reference units/ml) was added to the center of the ring (C). After 22 hours the beads or free interferon were removed and about 400 infectious units of VSV were added. In the control not treated with interferon (A), about 390 plaques were counted, of which about 70 were found in the area corresponding to the interior of the ring used in B and C. In B, about 310 plaques were counted outside of the ring and only small, hardly visible microplaques inside of the ring. In B and C, the imprint of the ring is easily detectable. Thus, it is evident that antiviral material freely diffused outside of the ring when free interferon was used (C), but remained restricted to the interior of the ring when Sepharose-bound interferon was applied (B). For the purpose of illustration of the experiment, glass beads of distinguishable size, enclosed by the same type of plastic ring as used in B and C, are shown in D. From Chany *et al.* (5).

of the tissue culture sheet, the antiviral effect is only observed in the interior of but not on the outside of the ring (Fig. 5).

3. The amount of interferon in contact with cell receptors is of critical

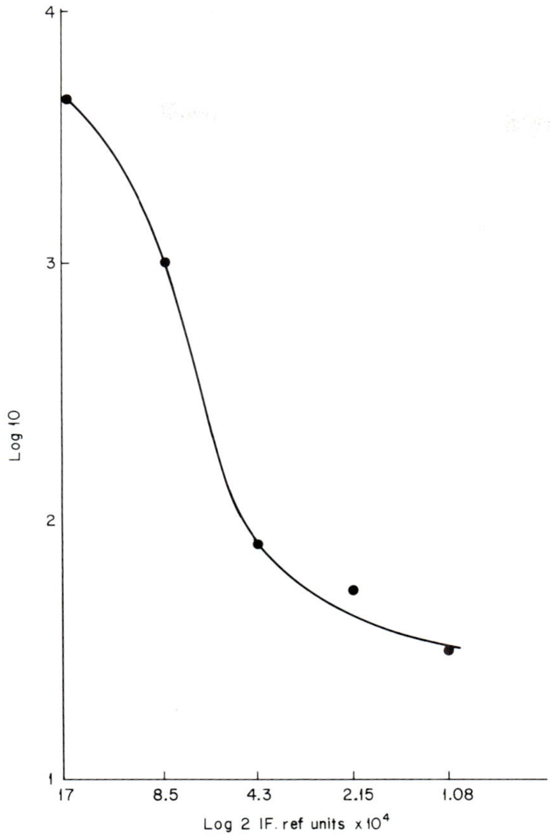

Fig. 6. Serial concentrations of murine interferon (IF) fixed on CNBr-activated Sepharose beads (15 ml/1.5 gm Sepharose dry weight). Before coupling, the protein concentration for each dilution was adjusted to 1.3 mg/ml, using purified bovine albumin. Interferon–Sepharose beads (1 bead/10 cells) were then added to 10^6 L cells grown in Falcon plastic (35-mm) petri dishes and incubated for 24 hours at 37°C. The beads were then removed, the cells washed and challenged with VSV at a m.o.i. = 0.1 PFU/cell. Viral yield was titered after a 16-hour incubation period at 37°C. The results are expressed as log 10 VSV yield in control/treated cells. From Chany *et al.* (5).

importance. When serial two-fold dilutions of interferon are bound covalently to a constant number of Sepharose beads, the antiviral effect decreases in a sigmoidal fashion (Fig. 6).

III. CONCLUSION

From the data presented here the following working hypothesis can be proposed. In many respects interferon can be compared to hormones.

TABLE 2. EFFECT OF AGITATION ON ANTIVIRAL STATE INDUCED BY SEPHAROSE–INTERFERON[a]

No. of cells/bead	With agitation		Without agitation	
	(PFU)	(HA)	(PFU)	(HA)
1000	4.7×10^7	1024/2048	3.3×10^8	2048
2000	4.5×10^7	≥ 2048	6×10^8	≥ 2048
Control	6.8×10^8	≥ 2048	6×10^8	≥ 2048

[a] Interferon–Sepharose beads were incubated with L cells grown in 35-mm plastic petri dishes either with or without agitation. The results are expressed in plaque-forming units (PFU) or in hemagglutinating units (HA). Hemagglutinations of EMC virus were performed using human group O erythrocytes. From Chany *et al.* (*5*).

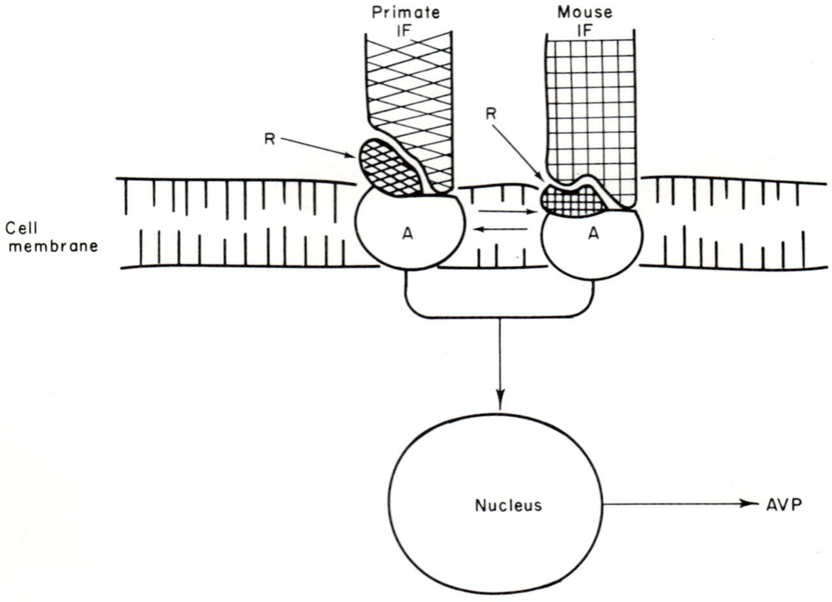

R–Receptor site; A–Activator site; AVP–Antiviral protein; IF–Interferon

Fig. 7. Model of the receptor-activator system in somatic monkey–mouse hybrid cells. Both receptors are closely connected to activator sites. The receptor site for primate interferon could become extruded from the cell membrane and, thus, accessible to trypsin. Both primate and mouse interferon are able to interact only with their own receptor sites and trigger activator sites. Each individual receptor-activator system could affect neighboring sites, thus cooperating with each other and spreading their effect all over the cell membrane. The subsequent events follow a common pathway, explaining the potentiation of mouse by primate interferon and vice versa. This can only be obtained in hybrid cells and stipulate the integrity of the two receptor systems.

Like insulin, interferon diffuses from the site of its production through the bloodstream of distant sites. Like some of the peptide hormones, interferon seems to interact with cell receptors located on the cell membrane and induces from there the series of events that lead to the antiviral state. The receptor–interferon interaction is probably of a cooperative type, which means that interferon induces in the receptor configurational changes leading to activation. Thus the receptor system consists at least of two operational sites: the binding and the activator sites. Cooperative activation between receptor systems located at different sites on the cell membrane is also among the possibilities (Fig. 7).

The data seem to indicate that interferon does not participate as such in the changes it induces other than at the receptor–activator site. It cannot be excluded, however, that small fragments detaching from the active molecule do penetrate into the cell. It also cannot be excluded that soluble interferon molecules penetrate entirely into the cell. This could be, however, unrelated to antiviral or other biological actions and might be related only to the degradation of the interferon molecule.

It can be predicted that the model presented here, even if only partially true, will provide interesting developments, not only from the theoretical point of view, but also for practical applications.

REFERENCES

1. Ankel, H., Chany, C., Galliot, B., Chevalier, M.-J., and Robert, M., *Proc. Nat. Acad. Sci. U. S.* **70**, 2360–2363 (1973).
2. Cassingena, R., Chany, C., Vignal, M., Suarez, H., Estrade, S., and Lazar, P., *Proc. Nat. Acad. Sci. U. S.* **68**, 580–584 (1971).
3. Chany, C., Grégoire, A., Vignal, M., Lemaitre-Moncuit, J., Brown, P., Besançon, F., Suarez, H., and Cassingena, R., *Proc. Nat. Acad. Sci. U. S.* **70**, 557–561 (1973).
4. Tan, Y. H., Tischfield, J., and Ruddle, R. H., *J. Exp. Med.* **137**, 317 (1973).
5. Chany, C., Ankel, H., Galliot, B., Chevalier, M.-J., and Gregoire, A., *Proc. Soc. Exp. Biol. Med.* **147**, 293–299.

CHAPTER 16

Inhibition of Enveloped Viruses in a Model of Virus-Induced Hyperlipemia*

Sidney E. Grossberg
Frank E. Frerman
Ernest E. Lengle

I. INTRODUCTION

Approaches to antiviral chemotherapy have been directed principally to the inhibition of *de novo* biosynthesis of viral nucleic acids and/or enzymic or structural viral proteins. Chemical agents that affect lipid metabolism have not been utilized to any extent, probably because no conceptual basis existed for their use and no direct experimental evidence suggested their possible value. Nearly 50% of cellular and viral membranes can be lipid (*1, 14*). The profound changes that may occur in cellular membranes during the replication of viruses, especially those that contain lipids, imply important changes in host lipid metabolism. A conception of the formation or assembly of membranes requires an understanding of how their components are formed. Our approach has been to try to understand the biochemical basis for the changes in host metabolism, particularly those involving lipids, in an experimental model in which lipid metabolism is markedly abnormal following infection with certain small RNA lipid-containing viruses. In our studies of this model of virus-directed host response, a low molecular weight compound,

* These studies were supported by grants from the National Institutes of Health, the American Heart Association, and the Wisconsin Heart Association.

currently utilized clinically to reduce levels of circulating lipid in several hyperlipemias of different possible etiologies, was found to reduce the levels of circulating lipid and virus (*8*). An appreciation of how such drugs might affect the replication of enveloped viruses requires an explanation of the experimental model.

II. THE PHENOMENON OF VIRAL LIPEMIA

During studies of interferon and antiviral host resistance in the chicken embryo, it was noted that plasma became progressively cloudy during the course of infection with Japanese encephalitis virus. Shown in Fig. 1 is the daily increase in opalescence of plasmas of chicken embryos infected with Japanese encephalitis virus in comparison with plasmas taken from uninfected control embryos. By the third day after infection the plasmas of embryos infected with this virus looked like heavy cream. Solvent extraction, gravimetric analysis, and thin-layer chromatography established that there was indeed increased lipid, mostly triglyceride, in these plasmas (*6*).

III. SPECIFICITY OF LIPEMIA

It was essential to determine whether this hyperlipemia was specific. Specificity was manifest in two ways. First, the total amount of serum lipid in each infection attained a maximum level characteristic of that

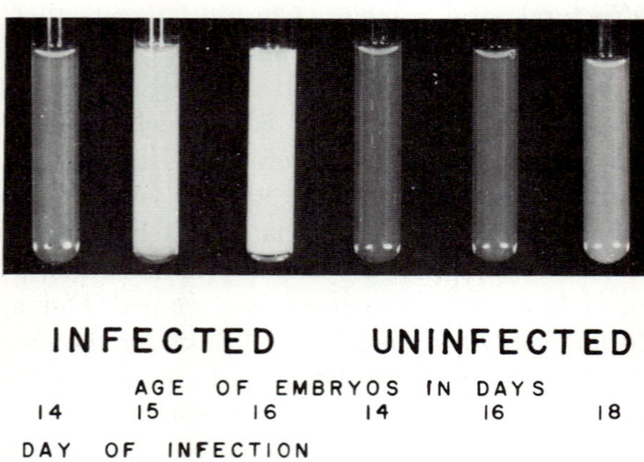

Fig. 1. Lipemic response in chicken embryos infected with Japanese encephalitis virus in comparison with plasmas taken from uninfected embryos of the same ages. [Reproduced from *Nature* (*London*) (*6*) with permission.]

infection (9). For example (Fig. 2), we tested three Group B togaviruses that were closely related immunologically. In comparison with normal values of about 10 mg/ml, the maximum amount of lipid produced by West Nile virus by day 4 was in the range of about 60 mg/ml and a smaller maximum amount of about 30 mg/ml for St. Louis encephalitis virus; intermediate values were obtained after infection with Japanese encephalitis virus.

The second way specificity was manifest was in the changes in lipid classes. A systematic examination of possible alterations in lipid classes was undertaken in infections by several viruses of different chemical and immunological types with and without envelopes; so far, the lipids in fourteen infections have been studied. Figure 3 summarizes the quantitative data in several representative infections. The amounts in each major lipid class were quantitated directly on chromatograms by a fluorometric method developed for measuring nanomole amounts of nonpolar lipids (*15*).

Infection with a virus that contains no structural lipid, such as encephalomyocarditis virus (a picornavirus that grows to high titer in embryonic fluids and organs) provokes no alteration in either total lipid content or distribution of major lipid classes in plasma. However, with the closely related immunological subgroup of the Group B togaviruses (West Nile, Japanese encephalitis, and St. Louis encephalitis viruses), there were significant increases in total lipids with very similar changes in major lipid classes, especially increased triglyceride and decreased polar lipids. With yellow fever virus, which is not so closely related to the aforementioned immunological subgroup, there was no significant increase in total lipids, although there was a change in the distribution of major lipid classes. Another arbovirus, California virus, immunologically un-

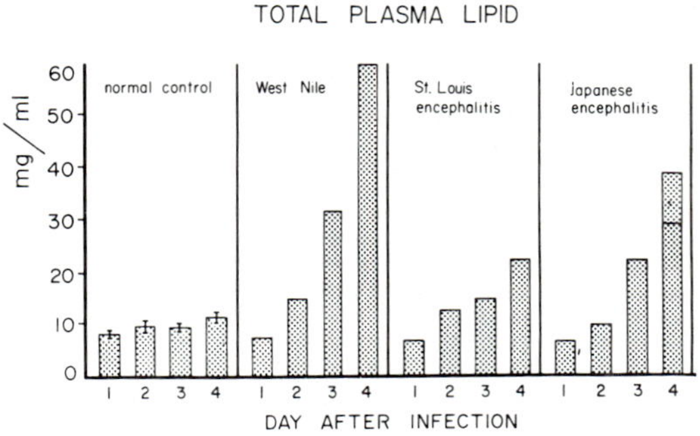

Fig. 2. Total extractable plasma lipids from chicken embryos infected with three closely related Group B togaviruses. The vertical bars show the range of values obtained in normal plasmas of individual embryos. [Reproduced with permission (*5*).]

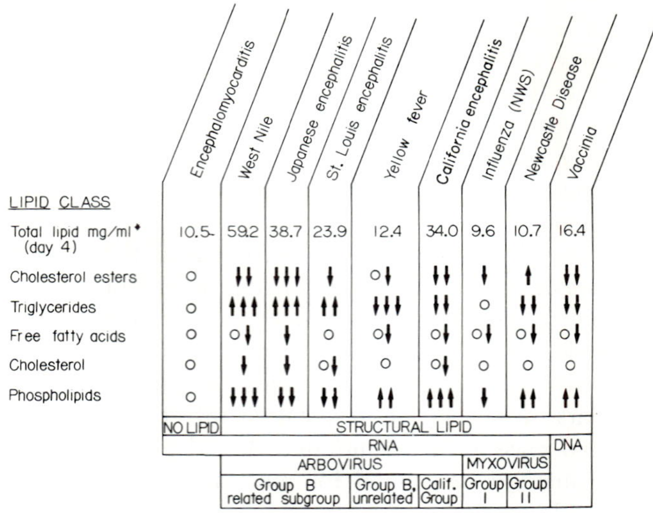

Fig. 3. Summary of changes in circulating lipids during infections with viruses of different chemical and immunological categories. The number of arrows indicates the degree of change, and the direction of arrows indicates an increase or decrease compared to the normal values simultaneously obtained for each lipid class. Zero indicates no change from the normal.* Normal total lipid day 4 = 11.3 (10.5–12.2) mg/ml.

related to the others, also elicited a hyperlipemia, but the pattern of the lipid distribution was quite different, with major increase in circulating polar lipids. Other enveloped viruses, e.g., Newcastle disease or influenza viruses, did not provoke increases in total lipid, but lipid class distributions were abnormal and distinctive (7–9).

IV. MECHANISM OF LIPEMIA

Since our use of an antilipemic agent resulted from attempts to determine the mechanism of viral-induced hyperlipemia, it may be helpful to discuss briefly the possible basis for the markedly increased amounts of circulating lipids observed. Our studies to date have concentrated on West Nile virus, which produces a lethal infection in the embryo and the greatest lipemia. Virus inoculated into the allantois subsequently invades the blood and is disseminated to brain and liver (Fig. 4). Most embryos are dead by the fifth day. The most significant lipid findings are the decrease in phospholipids and massive increase in triglycerides (see Fig. 3). Circulating glycolipids, not yet identified, are increased twofold in this infection. Pre-β- and β-lipoproteins, commonly associated with the transport of endogenous triglyceride, are also markedly increased.

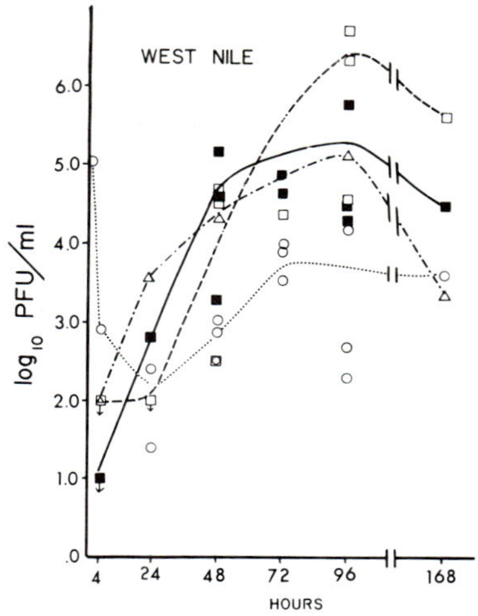

Fig. 4. Growth and distribution of West Nile virus in chicken embryos inoculated into the allantois. Organs made up as 10% suspensions. □, brain; ■, blood; ○, allantoic fluid; △, liver. [Data taken from Frerman and Grossberg (2).]

Our study of the metabolic events in this model suggests that carbohydrate metabolism is importantly linked to lipid metabolism during viral infection. Several considerations led us to study this in liver. (*a*) In the older avian embryo, like most animal species later in their development, the liver is the major site of neutral lipid synthesis for export (*4, 16, 17*). (*b*) Cellular and tissue changes in liver are noted by histological and electron-microscopic examination. (*c*) The distribution of lipids in infected liver are abnormal—triglyceride was elevated as found in plasma, but in contrast to plasma, polar lipids were also increased; these findings might be expected if cellular membranes were being formed at an increased rate during infection. (*d*) Finally, a survey of enzymes released into the circulation (9) showed that certain enzymes that are abundant in liver, such as glutamic oxalacetic transaminase and lactic dehydrogenase, increased in blood as these enzymes decreased in liver (Fig. 5). Such enzyme losses were not detected in other major organs, including brain (5).

Because of the observed losses of such enzymes, we considered that the problem in liver might involve regulation of carbon flow in lipid biosynthesis. We therefore began two series of investigations, still in progress, first to measure the activity of hepatic regulatory enzymes along metabolic pathways that direct carbon flow toward or away from lipid biosynthesis, and, second, to determine the relative efficiency of incorporation of radiolabeled precursors into tissue lipids. Our plan was to

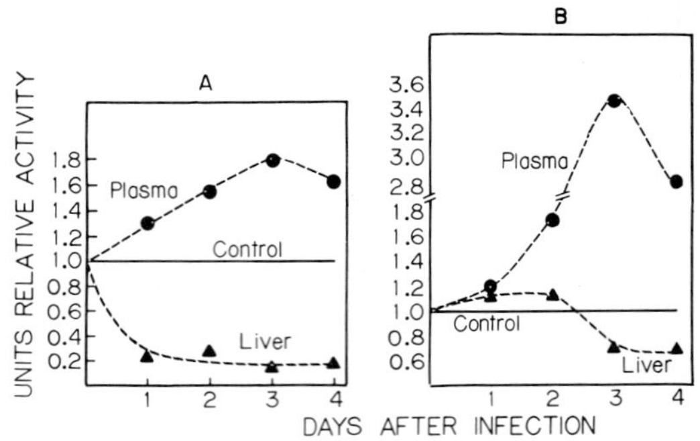

Fig. 5. Balance studies of glutamic oxalacetic tansaminase (A) and lactic dehydrogenase enzymes (B), showing increase in plasma and decrease in liver during West Nile virus infection.

study enzymes and carbon precursors in the major metabolic pathways that might influence the biosynthesis of lipids (Fig. 6)—the Embden-Meyerhof, Krebs, and pentose phosphate pathways.

When glucose is catabolized via the pentose pathway, 5-carbon sugars, such as ribose, are provided for nucleic acid synthesis. In addition, reducing power is generated as NADPH which is essential for fatty acid synthesis. This pathway also provides carbons that can reenter the Embden-Meyerhof pathway as 6-carbon or 3-carbon sugar phosphates. When pyruvate is oxidized to acetyl-CoA in the mitochondrion, the 2-carbon fragment can condense with oxalacetate to form citrate in the

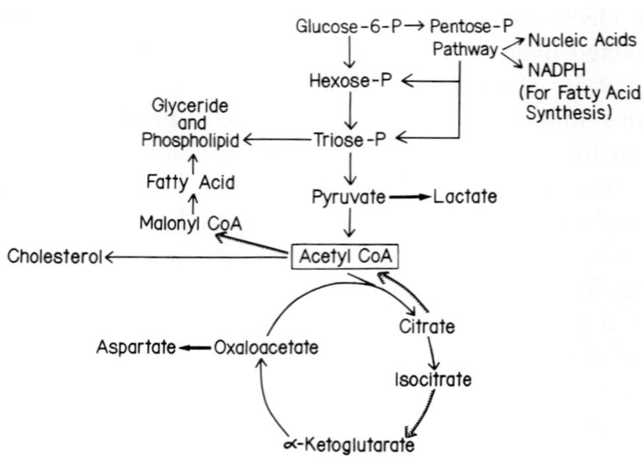

Fig. 6. Representation of some relationships between the Embden-Meyerhof pathway for the production of carbons that enter the Krebs cycle for energy production to the pentose phosphate pathway.

Krebs cycle. The citrate that enters the cytoplasm can be cleaved by citric acid lyase to provide acetate for the formation of acetyl-CoA. Cytoplasmic acetyl-CoA carboxylase fixes CO_2 to acetyl-CoA to form malonyl-CoA for the production of fatty acid. The key enzyme that catalyzes this step in cytoplasm is acetyl-CoA carboxylase, the rate-limiting and modulated step in fatty acid synthesis.

A. Enzymic Changes

Several regulatory enzymes that are importantly involved in carbon metabolism were studied in liver of West Nile virus-infected embryos. The enzymes are shown in Table 1 grouped according to their effects. The first group is composed of three enzymes that contribute to the flow of carbon fragments to acetyl-CoA; their activities were significantly increased. Most striking was the great increase in activity of citric acid lyase, which generates acetate from the cleavage of citrate to form cytoplasmic acetyl-CoA for cytoplasmic fatty acid synthesis. Activity of a second group of enzymes, those in the pentose phosphate pathway, were also increased; this pathway provides NADPH, for the reducing power essential to fatty acid synthesis, as well as ribose for RNA synthesis. In contrast to the first two groups, the activity of enzymes of the third group that might divert carbon flow away from acetyl CoA remained unchanged or decreased (*12*).

TABLE 1. RELATIVE ACTIVITIES OF ENZYMES INVOLVED IN CARBON METABOLISM IN LIVER OF WEST NILE VIRUS-INFECTED EMBRYOS

Enzyme[a]	Fold increase in activity in infected liver over control levels
Enyzmes contributing to increased acetyl-CoA formation:	
Citric acid lyase	21.0
Phosphofructokinase	3.3
Malic enzyme	2.5
Pentose pathway providing NADPH needed for fatty acid synthesis:	
Pentose phosphate dehydrogenases	4.5
Glucose-6-phosphate dehydrogenase	2.3
Enzymes diverting carbons away from acetyl-CoA formation:	
Malic acid dehydrogenase	0.75
NAD—isocitric acid dehydrogenase	1.2
NADP–isocitric acid dehydrogenase	1.2

[a] NAD, nicotinamide adenine dinucleotide; NADP, NAD phosphate; NADPH, reduced NADP.

The controlling enzymic step for fatty acid biosynthesis is acetyl-CoA carboxylase. The activity of this enzyme was increased twofold in homogenates of infected liver (Fig. 7). The enzyme has the characteristics of carboxylase reported by others, namely, its sensitivity to avidin, dependence on adenosine triphosphate (*17*), and inhibition by the antilipemic drug, clofibrate (*13*), (see below). Low molecular weight activators of the enzyme, citrate and isocitrate, were found to be increased 30% in West Nile virus-infected liver. These allosteric effectors appeared to be responsible for the abnormally high activity inasmuch as carboxylase activity returned toward normal levels following dialysis or chromatography with Sephadex G-25 (Table 2) (*2*). Thus, increased activity, and not new synthesis of carboxylase appears to be responsible for the increase in enzymic activity. These enzymologic studies provided strong circumstantial evidence that during viral infection the regulation of acti-

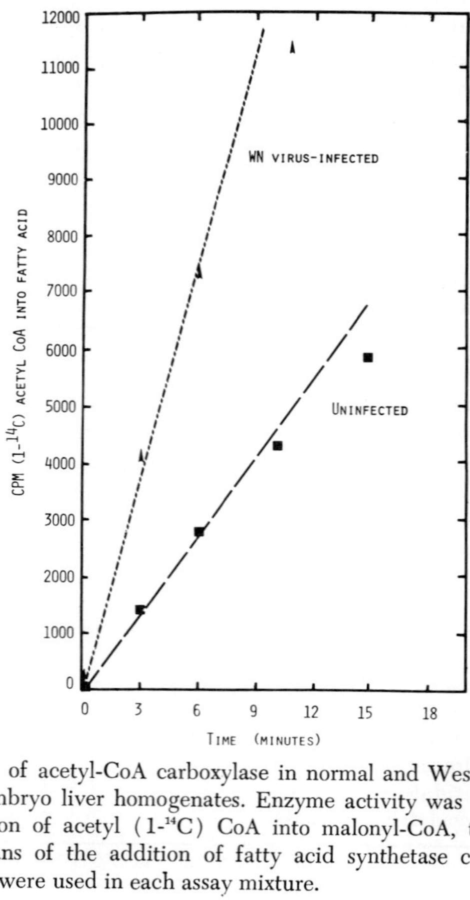

Fig. 7. Activities of acetyl-CoA carboxylase in normal and West Nile (WN) virus-infected chicken embryo liver homogenates. Enzyme activity was measured by coupling the incorporation of acetyl (1-^{14}C) CoA into malonyl-CoA, to the formation of fatty acids by means of the addition of fatty acid synthetase complex. Equivalent amounts of protein were used in each assay mixture.

TABLE 2. SPECIFIC ACTIVITY OF LIVER ACETYL-CoA CARBOXYLASE IN WEST NILE VIRUS INFECTION

Test	nmoles min^{-1} mg^{-1} Protein	
	Crude	Dialyzed
Control	0.52 (± .05)	0.31 (± .08)
Infected	1.05 (± .05)	0.32 (± .02)

vities of enzymes in pathways linking carbohydrate and lipid metabolism were deranged so as to provide carbons and activators for new lipid synthesis.

B. PRECURSOR STUDIES

The incorporation of different carbon precursors into lipids was studied during West Nile virus infection. For these experiments, liver fragments taken from previously infected embryos or from uninfected controls were incubated with different precursors, and the amount of incorporation into cellular lipid was measured during a 60-minute interval. Acetate was a relatively poor precursor for lipid synthesis in infected or normal liver. However, as shown in Table 3, carbons from carbohydrates, such as glucose labeled either in the 1- or the 6-carbon position, pyruvate, or lactic acid were converted more readily into lipid in infected embryos than in controls. Carbon from lactic acid, although incorporated to a greater extent in infected liver lipids than in the normal, was not metabolized into lipid as much as pyruvate or glucose presumably because of the loss of lactic dehydrogenase from liver (refer to Fig. 5). Oleic acid, an 18-carbon monounsaturated fatty acid, was incorporated twice

TABLE 3. EFFECT OF WEST NILE (WN) VIRUS INFECTION ON INCORPORATION OF LIPID PRECURSORS IN CHICKEN EMBRYO LIVER FRAGMENTS

Precursor	Microatom carbon incorporated/mg cellular lipid	
	Control	WN virus-infected
($1-^{14}C$) Glucose (9.75 m*M*)	0.515	0.837
($6-^{14}C$) Glucose (9.75 m*M*)	0.552	0.799
($2-^{14}C$) Pyruvic acid (10.6 m*M*)	2.533	4.587
($2-^{14}C$) Lactic acid (9.9 m*M*)	0.154	0.213
($1-^{14}C$) Oleic acid (14 μM)	0.514	1.043
($6-^{14}C$) Glucose (9.75 m*M*) and oleic acid (1.42 μM)	0.346	0.725

as well in infected liver as in the control. This important finding indicates that there is no defect in the ability of infected liver to take up fatty acid.

Finally, when excess unlabeled fatty acid was added to liver fragments given radiolabeled glucose as precursor, there was inhibition of glucose incorporation into lipids of normal liver; such inhibition was not observed in infected liver in which carbon from glucose, despite the presence of exogenous fatty acid, continued to be incorporated into lipid at the same rate as glucose alone (*5, 12*). Thus, during infection there is a loss of control by end-product inhibition.

In the above experiments the release of $^{14}CO_2$ was also measured. The failure to observe any change in the rate of formation of CO_2 from oleic acid indicates that fatty acid oxidation was unchanged during infection. Comparison of ratios of CO_2 from glucose labeled in the 1- or 6-carbon position (*11*) showed that the pentose pathway was being used preferentially (*5, 12*).

Studies of incorporation of labeled glucose or triglyceride into liver fragments incubated with infected plasma suggest that there may be a hormonal and/or viral factor(s) in infected plasma that enhances the incorporation of precursor carbons into cellular lipids (*5, 12*).

V. ANTIVIRAL EFFECTS OF ANTILIPEMIC AGENTS

Another approach to understanding the underlying metabolic abnormalities in this model was to evaluate the effects of antilipemic agents, the mechanism of action of which had been determined, at least in part, in other systems. The drug clofibrate (Fig. 8), or Atromid-S (Ayerst Laboratories), lowers circulating lipids in man after oral administration of the esterified form. Its deesterified salt, sodium chlorophenoxy methyl-

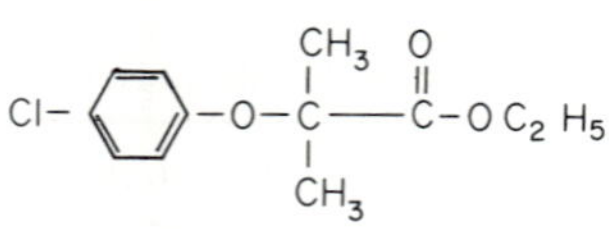

Ethyl 2-(*p*-chlorophenoxy)-2-methylpropionate

Clofibrate (Atromid-S)

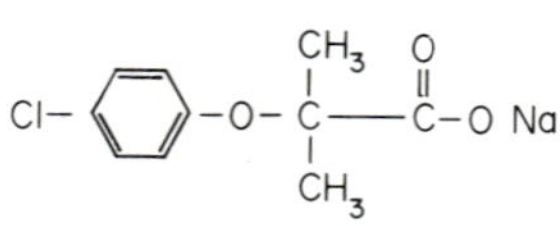

Sodium clofibrate

Fig. 8. Clofibrate and the sodium salt of the active deesterified form.

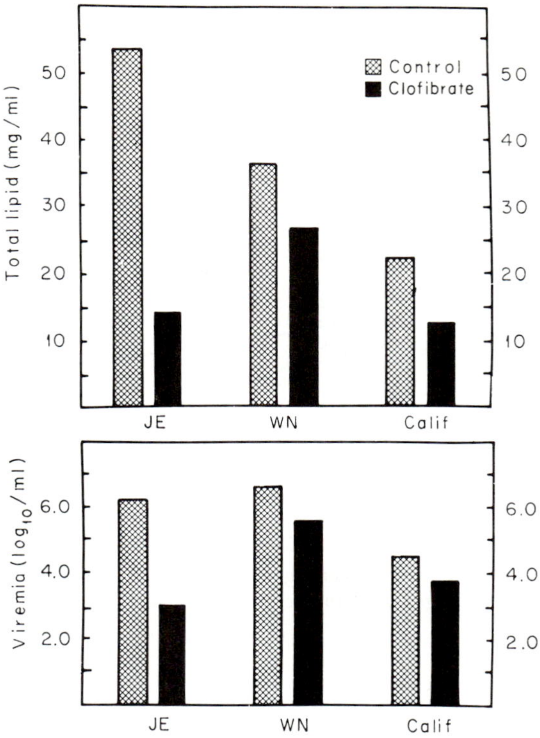

Fig. 9. Antilipemic and antiviremic effects of clofibrate during infections by the viruses of Japanese encephalitis (JE), West Nile (WN), and California encephalitis (Calif). The titer of Japanese encephalitis in liver was reduced by 90%.

propionate (mol wt 237) has been shown to be antilipemic and antiviremic after parenteral administration to chicken embryos (8). Its antilipemic and antiviral effects in three different infections are summarized in Fig. 9. In these experiments, 10 μg was injected repeatedly 2–3 times a day beginning several hours after virus was injected. The most striking effect was seen in Japanese encephalitis (JE) virus infection where the lipemia was nearly completely corrected and the viremia was reduced about 2000-fold; titers of JE virus in liver were reduced ten-fold. The effects of the drug on lipemia and viremia were less striking but significant, nonetheless, in West Nile and California encephalitis virus infections. In all three infections circulating triglyceride was most significantly reduced by clofibrate.

Shown in Fig. 10 are dose–response curves relating the dose of clofibrate to its *in vitro* inhibition of acetyl-CoA carboxylase from livers of control and West Nile virus-infected embryos. Clofibrate has been

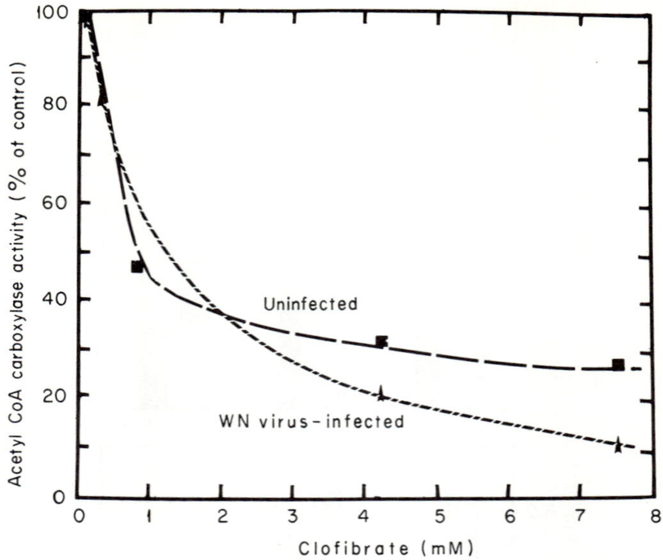

Fig. 10. Dose–response curves relating clofibrate dose to inhibition of acetyl-CoA carboxylase in liver homogenates from control and West Nile (WN) virus-infected embryos.

shown by Maragoudakis and Hankin (*13*) to act as a competitive inhibitor of citrate, the allosteric activator of acetyl-CoA carboxylase. Although such a mechanism of action would suggest that clofibrate's activity is exerted by regulating the formation of lipids either in the viral envelope or in host membranes necessary for viral development, it is possible that the drug may have other actions as well (*19*). We have observed paradoxical effects by administering very large doses of the drug where the lipemia is accentuated. Since it is thought that one effect of clofibrate is displacement of triglyceride from serum lipoproteins, by analogy it is conceivable that the drug may act by displacing virions from carrier serum lipoproteins. Thus, it is possible that the effect observed as reduced circulating virus titers merely represents the displacement of virions from blood back into the tissues. In that case, the viral infection should become more severe, but it does not. Certainly, Japanese or California encephalitis viruses, which produce inapparent infections in the embryo, when treated with clofibrate, do not die and virus titers are not higher than in untreated embryos. In West Nile infection there is no change in the pattern of death after clofibrate treatment; however, the metabolic derangements are so grossly abnormal that the embryo may, nevertheless, die a metabolic death. Since clofibrate is only a competitive inhibitor, the fundamental derangements in carbon metabolism are not corrected.

Two diphenyl analogs of clofibrate, (I.C.I. 54,856 and I.C.I. 55,897) (*18*) supplied by Imperial Chemical Industries, England, were also tested. One of these analogs, a diphenyl aryl ether (I.C.I. 54,856) having a methylene group interposed between the ring and the oxygen, was antilipemic in West Nile virus infection. Both analogs were effective in the JE infection.

VI. IMPLICATIONS AND DISCUSSION

Considerable evidence obtained in these studies would suggest that the regulation of host lipid metabolism may be coupled to the replication of enveloped viruses. Certainly, clofibrate, which competitively inhibits acetyl CoA carboxylase, reduced levels of circulating lipid and virus in the chicken embryo model. An agent that reduces circulating virus titers may be useful. Viremia plays a critical role in the ecology of arbovirus infections in nature as well as in the dissemination of virus within a given host. It is possible that by combination with another agent that reduces virus titers in target organs, even greater antiviral effects can be achieved.

In JE viral infection of the chicken embryo, clofibrate reduced viral titers in liver as well as in plasma. At present, it is not possible to explain why the reduction in viremia with West Nile or California encephalitis viruses is not paralleled by a reduction in virus titers in organs, especially liver or brain. It is hoped that the kinds of studies presented here may suggest new approaches to antiviral chemotherapy of lipid-containing viruses. Thus, it may be possible to utilize metabolic inhibitors of host physiological processes whch might be coupled to viral replicative events to result in the attenuation of the production of virions.

Our experimental observations in the chicken embryo may be relevant to those made in higher animals. Suckling rats double their circulating lipids during West Nile virus infection. Horses infected with equine infectious anemia virus, an enveloped virus, develop hyperlipemia (3), and volunteers injected with sand-fly fever virus, an arbovirus, develop abnormalities in their lipoprotein and lipid patterns, especially an initial decrease followed by an elevation in serum triglycerides (*10*).

VII. SUMMARY

1. Certain lipid-containing viruses of different chemical and immunological types induce hyperlipemia.

2. Lipemic responses are qualitatively and quantitatively specific for the individual viral infection.

3. Regulation of host lipid metabolism is markedly affected, especially that integrating carbohydrate and lipid metabolism, since in liver (a) carbohydrates are more readily converted into lipids; (b) the pentose phosphate pathway is used in preference to the Embden-Meyerhof glycolytic pathway; (c) there is loss of feedback inhibition by fatty acid of liponeogenesis from glucose; (d) activity of enzymes contributing to acetyl-CoA formation is increased; and (e) activity of acetyl-CoA carboxylase is increased.

4. Clofibrate, which competitively inhibits acetyl-CoA carboxylase, reduces levels of circulating lipid and virus.

ACKNOWLEDGMENTS

We are indebted to Mrs. Lee Ann Roch, Mr. W. Ted Hron, and Mrs. Mary Dixon for their expert technical assistance.

REFERENCES

1. Davis, B. D., Dulbecco, R., Eisen, H. N., Ginsberg, H. S., and Wood, W. B., "Microbiology," 2nd ed. p. 35, Harper, New York, 1973.
2. Frerman, F. E., Lengle, E. E., and Grossberg, S. E., *J. Biol. Chem.* (submitted for publication).
3. Gainer, J. H., Amster, R. L., Needham, J. W., and Schilling, K. F., *Amer. J. Vet. Res.* **27**, 1611–1621 (1966).
4. Goodridge, A. G., *Biochem. J.* **108**, 655–661 (1968).
5. Grossberg, S. E., Frerman, F. E., Lengle, E. E., Roch, L. A., Hron, T., and Dixon, M., *RES, J. Reticuloendothel. Soc.* **15**, 334–341 (1974).
6. Grossberg, S. E., and O'Leary, W. M., *Nature (London)* **208**, 954–956 (1965).
7. Grossberg, S. E., and Roch, L. A., *Bacteriol. Proc.* p. 201 (1971).
8. Grossberg, S. E., Roch, L. A., and Frerman, F. E., *Progr. Immunobiol. Stand.* **5**, 289–293 (1972).
9. Grossberg, S. E., Roch, L. A., Tompkins, R., and Cichanski, K., *Antimicrob. Ag. Chemother.* pp. 501–506 (1969).
10. Lees, R. S., Fiser, R. H., Jr., Beisel, W. R., and Bartelloni, P. J., *Metab., Clin. Exp.* **21**, 825–833 (1972).
11. Lehninger, A. L., "Biochemistry, the Molecular Basis of Cell Structure and Function," p. 361. Worth Publ., New York, 1970.
12. Lengle, E. E., Grossberg, S. E., and Frerman, F. E., *Fed. Proc., Fed. Amer. Soc. Exp. Biol.* 33, 1543 (1974).
13. Maragoudakis, M. E., and Hankin, H., *J. Biol. Chem.* **246**, 348–358 (1971).
14. Robertson, J. D., *in* "Cell Biology in Medicine" (E. E. Bittar, ed.), p. 6. Wiley, New York, 1973.
15. Roch, L. A., and Grossberg, S. E., *Anal. Biochem.* **41**, 105–115 (1971).
16. Romanoff, A. L., "Biochemistry of the Avian Embryo. A Quantitative Analysis of Prenatal Development," p. 64, Wiley (Interscience), New York, 1967.
17. Ryder, E., *Biochem. J.* **119**, 929–930 (1970).
18. Thorp, J. M., *Proc. Eur. Soc. Study Drug Toxicity* **13**, 98–109 (1972).
19. Thorp, J. M., Cotton, R. C., and Oliver, M. F., *Progr. Biochem. Pharmacol.* **4**, 611–617 (1968).

DISCUSSION

DR. PLOTKIN: I would like to point out the possible significance of Dr. Grossberg's interesting work to Reye's syndrome in which there is massive fatty infiltration of the liver with resultant hepatic encephalopathy. Epidemiologic association of the syndrome has been made with influenza B and varicella, both lipid-containing viruses.

To determine the possible relevance of his experiments to Reye's syndrome, I would like to ask Dr. Grossberg four questions:

1. Reye's syndrome (RS) follows the prodromal illness and is not contemporaneous with it. What is the relationship of the viremia in the chick embryos with the hepatic changes?
2. Survivors of RS show complete return of the liver's functions. What happens in the chick embryos?
3. Reye's syndrome is not seen in adult patients and is only a disease of children. Do postnatally infected chickens show the same metabolic changes?
4. Does clofibrate given after the disease has begun in the chick embryo ameliorate the lipid changes or does the drug have to be given from the beginning?

DR. GROSSBERG: May I take the questions up in reverse order. The clofibrate effect is not very great, and we did not want to reduce the chances of showing an effect by delaying treatment more than a few hours. We avoided adding it at the same time that the virus was given because of the possible binding of the compound to the virus itself. However, we have not tried to delay the administration of clofibrate until late in the infection.

We have tried to induce lipemia postnatally. We have demonstrated a doubling in the amount of circulating lipid in chicks infected at 1 or 2 days of age. I cannot tell you about the effects on older chickens.

In response to your question about the return of what is once a fatty liver to normal, the histological changes in the liver are minimal to begin with, in contradistinction to the major infiltration of lipid which I understand to occur in RS. The lipemia and the viremia in Japanese encephalitis, St. Louis encephalitis, and California encephalitis infections tend to subside. The embryo without any immunological mechanism—we believe because of the production of interferon—manages to survive those infections; as the virus tends to disappear, so too do the serum lipid values tend to return to normal. Nonetheless, the chicks are unable to hatch for lack of energy. We understand better now why they lack that energy to get out of the shell; they may peck a hole and may begin to cheep. We have tried "Cesarean" delivery by taking the shell away, but they are too weak to stand. Possibly because of some avian reflex mechanism if they do not stand, they will not take nourishment. We have not tried lifesaving parenteral therapy or gavage, but we would like to sustain them because it would be interesting to have a model of pediatric atherosclerosis. In West Nile infection, it is clear that the infection progresses and the embryos die.

Your first question has to do with the lipemia in relation to viremia. An observation we find curious is that on the first day of infection with at least three viruses, circulating lipids decrease just as virus begins to circulate. Then, as the virus titers go up throughout the embryo, so too do the serum lipids. We have no explanation for that early decrease, but that is about the only discrepancy between lipemia and viremia that we have noted.

DR. WELLER: Could you give me a short lesson in the lipid content of the yolk sac? Because it seems to me as viremia gets under way, there must be tremendous capillary damage at the endothelial lining between the vascular system and the yolk sac and there might be selective spillage from the yolk into the bloodstream and that this might, in part, explain some of the phenomena that you have observed. Have you put any radioactive materials into the yolk sac and observed their uptake to see whether this is accelerated in your viremic embryos?

DR. GROSSBERG: We have injected labeled triolein into the yolk sac and the uptake into the circulation is very poor. That does not mean that the lipemia may not be related to the kind of mechanism that you suggest. We suspect that there was very poor mixing of the labeled material with the yolk. However, there is very little histologically visible arteriolar or capillary damage in the yolk sac membrane or elsewhere. It is remarkable how little pathology there is. We think that the processes of transport of yolk sac material certainly continue to go on—fatty and other materials continue to be transported across the yolk sac membrane and be taken up into the circulation. The uptake and oxidation of fatty acids and other lipids by liver continues to go on normally; if that did not continue the embryo would probably die quickly. The yolk sac is the ultimate source of the carbons for energy and other metabolism.

DR. TOKUMARU: Our experiment showed that pseudorabies virus infection (a member of herpes group) in rabbits produced diffuse fatty liver degeneration with high lipidemia after subcutaneous inoculation. This was dissimilar to that seen with herpes simplex virus which forms directly necrotic foci in the liver. It has been reported in Eastern European literature that acetyl-CoA activity was found to be increased in such instances. This may be an additional animal model for disturbances in lipid metabolism to that of chick embryo systems.

DR. GROSSBERG: Is there necrosis in the liver? There is none in our model.

DR. TOKUMARU: No.

DR. SCHER: What is the comparison of the patterns of specific lipids in the plasma and in the different viral envelopes?

DR. GROSSBERG: That is a very serious question to which one day we hope to be able to give a definitive answer. As you know, each host in which enveloped viruses grow may change the distribution of lipids in the envelopes of the progeny virus. This means that in order to make a valid comparison, we must grow these viruses in chicken embryos; chicken embryo cells cultured in serum would not be acceptable. Titers are not high in chicken embryos. For example, we would need to collect many liters of allantoic fluid in order to get enough virus to analyze. However, we must do it.

DR. BRUNELL: 1. Does clofibrate reduce the titer of virus if they are incubated together? 2. Do the same enzymic changes occur when the virus is propagated in tissue culture?

DR. GROSSBERG: We have no evidence that the compound inactivates virus *in vitro*. In fact, so far at least, in single-cycle yield experiments, we have not demonstrated any effect on viral growth in chicken embryo cell cultures infected with several different enveloped viruses. In response to your first question, we have not looked at enzyme changes in the same way in cell culture as we have in chicken embryo liver.

DR. BRUNELL: If you make dilutions of virus with your compound, will this reduce the titer of your virus?

DR. GROSSBERG: We have not done that.

DR. NEVA: As to the specificity of the phenomenon, does the hyperlipemia occur in other hosts such as mice, in which embryonic metabolism is not at issue. Also, to rule out other causes, can the effect be inhibited by specific antiserum?

DR. GROSSBERG: Antibody inhibits it. If the virus does not replicate, the phenomenon does not occur. We have demonstrated virus-induced hyperlipemia in suckling rats. Since the phenomenon occurs in the rat, a species distant from the chicken, we think the phenomenon may have some general validity. There is so much to study in this single model that we have not looked further for new ones.

CHAPTER 17

Photodynamic Inactivation of Herpesvirus

Joseph L. Melnick
Craig Wallis

I. INTRODUCTION

The treatment of herpes simplex lesions on the lips, skin, and especially the genitalia has proved frustrating for clinicians. No effective means of either relieving the symptoms or preventing recurrences have been available. However, we have been pleased by the promising results that a number of our colleagues have been obtaining in the past few years by applying our laboratory findings to the clinic. The treatment utilizes photodynamic inactivation of herpesvirus. This involves photosensitization of the virus by a heterotricyclic dye (Fig. 1).

II. PHOTOSENSITIZATION OF VIRUS WITH DYE

Proflavine and other heterotricyclic dyes can attach to the double-stranded DNA genome of the virus. Then, when light energy is added, it is absorbed and contributes to an oxidation reaction that disrupts the structure of the DNA so that it is no longer functional. The more light, the greater the loss of function.

In Fig. 2, the mechanism of the dye–light therapy is shown. On the left is the intact viral DNA with paired bases of thymine (T), adenine (A), guanine (G), and cytosine (C). The dye intercalates between the base pairs, as shown by the black horizontal bars. If the dye is present in the cell during the replication of the virus, the dye is readily incorporated into the architecture of the virus and the virus becomes hypersensitive to light. For proflavine, the most effective light is that

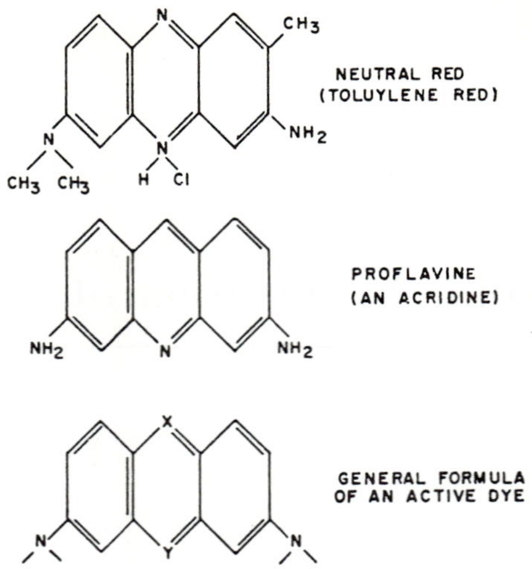

Fig. 1. Structural formulas of photodynamically active dyes.

which is emitted at 450 nm (blue), which is where proflavine absorbs most. When, as shown in the center section, light hits the virus–dye complex, guanine molecules are excised from the DNA, leaving gaps in the base sequence (right). Thus, the virus is no longer able to replicate and infect other cells.

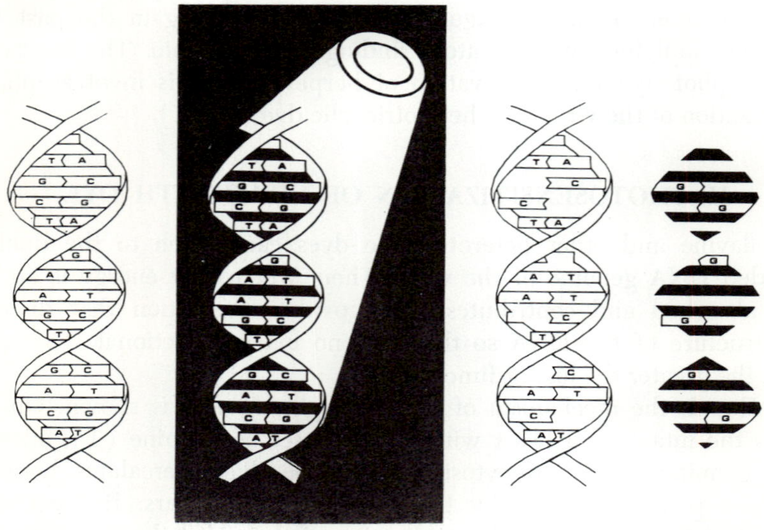

Fig. 2. Mechanism of photoinactivation. Left—viral DNA; center—dye (horizontal bars) intercalated between base pairs and the virus–dye complex being exposed to visible light; right—fragmented DNA from which guanine has been excised.

We have been engaged in photodynamic studies of a number of viruses over the past 15 years, and it was from this background of experience (*2, 16–21*) that our clinical colleagues were led to apply the technique. Actually, this work had its origin in studies going back to the turn of the century, when the phenomenon of photodynamic inactivation was discovered by Raab (*10*) in his observation that paramecia suspended in an acridine solution were killed by exposure to ordinary daylight. Three decades later viruses were shown to be photosensitive (*9, 15*). However, assay methods in the 1930s were crude, and no quantitative results were reported. In 1958, Yamamoto (*25*) reported the first quantitative studies on photodynamic inactivation of bacterial virus, and in 1960, Hiatt *et al.* (*5*) extended this work to a number of animal viruses (pox-, adeno-, herpes-, and papovaviruses) but found that enteroviruses were resistant to photosensitization. Poliovirus, naturally photoresistant, can be made photosensitive if the virus is grown in cells maintained with medium containing proflavine, neutral red, or acridine orange (*2, 12, 23*). During replication of the virus, the photoreactive dye becomes incorporated within the virus structure, a finding that has been confirmed in later studies (*13, 24*). However, in the investigations cited, completely photosensitized harvests could not be obtained. Wallis and Melnick *et al.* (*16, 17, 19–22*) found ways to bind dyes irreversibly to nonenteroviruses and even to render enteroviruses sensitive to heterotricyclic dyes.

A. Photoinactivation of Enteroviruses

The parameters for the system were worked out with some of the enteroviruses. Some data are shown for echovirus 7. This virus is actually resistant to photodynamic inactivation if the dye is added to the virus in a test tube. However, if the dye is added to the cell cultures in which the virus grows, it can be photosensitized. In the experiment shown in Table 1, we see the concentrations of dye, neutral red in this case, that were used. The virus was allowed to remain in the cell for 24 hours, harvested, kept in the dark, and then titrated in the presence of weak light. Aliquots of each harvest were exposed to white light for 15 minutes, and the rapid fall in titer was obtained. Dilutions of dye up to 1:80,000 produced fully photosensitive virus. Even a concentration of 1:160,000 produced photosensitive virus with over 99.9% of the virus inactivated.

Using a concentration of neutral red of 1:40,000, we were interested in determining how long we had to treat the cells with dye before photosensitive virus developed. As shown in Table 2, it took 1 hour for dye to penetrate the cell and be at the location of viral replication,

TABLE 1. EFFECT OF NEUTRAL RED CONCENTRATION ON DEVELOPMENT OF PHOTOSENSITIVE ECHOVIRUS 7

Concentration of dye	Log PFU/ml Dark	Log PFU/ml Light (15 min)
None	7.5	7.6
1:10,000	5.2	0.0
1:20,000	7.3	0.0
1:40,000	7.5	0.0
1:80,000	7.4	0.0
1:160,000	7.7	3.2

where it could be incorporated into the virus particle, making it photosensitive.

The next experiments were set up to answer the question as to how long cells that were exposed to the dye—the dye was then removed and the cells washed—would retain the ability to produce photosensitized virus. At periods of 24 to 120 hours after exposure to dye, virus was added; the cultures were harvested 24 hours later and examined for photosensitized virus. The results in Table 3 indicate that the cells retained the ability to produce photosensitized virus for as long as 72 hours after they were exposed to the neutral red.

B. PHOTOINACTIVATED ENTEROVIRUS AS A VACCINE

We next carried out studies to determine if photoinactivated virus could serve as a vaccine. Our experiments were set up not so much to prepare a vaccine for echovirus 7 as to determine whether or not the protein capsid was in any way denatured by the photoinactivation

TABLE 2. EFFECT OF DURATION OF PRETREATMENT OF CELLS WITH NEUTRAL RED ON DEVELOPMENT OF PHOTOSENSITIVE ECHOVIRUS 7

Duration of treatment of cells with 1:40,000 neutral red prior to virus inoculation (hours)	Log PFU/ml Dark	Log PFU/ml Light
0	7.7	7.6
0.5	7.5	4.7
1.0	7.3	0.0
2.0	7.6	0.0
4.0	7.3	0.0

TABLE 3. PRODUCTION OF PHOTOSENSITIVE ECHOVIRUS 7 FROM CELLS INOCULATED WITH VIRUS AT VARIOUS INTERVALS AFTER 2-HOUR PRETREATMENT WITH NEUTRAL RED

Time interval between dye treatment of cells and virus inoculation (hours)[a]	Log PFU/ml	
	Dark	Light
None (inoculated immediately)	7.5	0.0
24	7.3	0.0
48	7.7	0.0
72	7.2	0.0
96	7.5	2.5
120	7.1	4.1

[a] Virus harvested 24 hours after inoculation.

procedure. As shown in Table 4, three types of vaccine were prepared from the same harvest and then tested in rabbits, for echovirus does not replicate in these animals. An unmodified live virus vaccine served as a control and produced a mean antibody titer of 1070. When formalinized vaccine was prepared, it still produced antibodies but the titer was reduced to 90. The photoinactivated vaccine produced an antibody with a titer of 640, markedly higher than the formaldehyde vaccine and close to that produced by the unmodified live virus. Thus, the photo-

TABLE 4. ECHOVIRUS TYPE 7 VACCINES (V)

Virus treatment	Log PFU/ml		
	Live virus vaccine	Formalinized vaccine	Photoinactivated vaccine
Untreated (dark)	7.6 (V)	—	7.4
Exposed to light (15 min)	7.5	—	0.0 (V)
5 days at 37°C	4.2	0.0 (V)	—

Bleedings (days)	Serum titers for rabbit no.: 1	2	3	5	6	7	8	9	10	11	12
0	0	0	0	0	0	0	0	0	0	0	0
10th	10	640	160	0	0	0	0	160	40	10	40
20th	160	1,280	320	0	80	0	0	320	320	160	320
30th	320	2,560	320	40	160	80	80	1,280	640	320	320
Mean titers (30th day)		1,070			90				640		

inactivation destroys the genome but has no deleterious effect on the external capsid antigen of the virus.

III. DYE–LIGHT THERAPY OF HERPESVIRUS INFECTIONS

A. Herpesvirus-Induced Keratitis in Rabbits

Photodynamic inactivation of virus was next tested as a potential therapeutic tool for herpesvirus-induced keratitis, which is a leading cause of blindness and disability (8). Experimental keratitis produced in rabbits by infection with herpesvirus isolated from a human periorbital lesion was treated with proflavine and light. The herpetic keratitis could be rapidly resolved by the dye–light treatment, whereas improvement was not seen in untreated rabbits, nor in those treated with light alone or dye alone.

Serially diluted herpesvirus was inoculated into scarified eyes, and the subsequent degree of keratitis was scored. Resulting encephalitis due to virus infection was also recorded. The results in Table 5 show that the severest keratitis obtained without fatalities was with a dose of 10^4 PFU per rabbit cornea. Therefore, the following experiments were carried out with this virus concentration. Rabbits scarified and inoculated with virus (10^4 PFU per cornea) were treated with dye and light.

A stock of 1% proflavine sulfate was made in physiological saline and sterilized by filtration through a 0.22-μm membrane (Millipore Corp.). The sterile product was stored in amber bottles. Rabbits were treated with 0.1 ml of the dye 3 times daily at intervals of 1 hour. Two hours after the last dose, eyes were exposed to light (150-W incandescent

Table 5. Titration of Herpesvirus in Rabbits

Virus dose (PFU)/rabbit[a] (dose each eye)	Severity of keratitis[b] on day:				No. dead on 10th day/ total no. inoculated
	1	2	3	4	
10^6	+	++++	++++	++++	3/3
10^5	+	++++	++++	++++	1/3
10^4	±	++++	++++	++++	0/3
10^3	0	++	++++	++++	0/3
10^2	0	0	0	+	0/3
10^1	0	0	0	±	0/3

[a] Scarified corneas were inoculated with herpesvirus as described in the text.
[b] Scores: 0 = No detectable lesions; ± = occasional lesion; ++ = moderately severe lesions with 50% opacity; ++++ = severe and generalized keratitis.

bulb, at a distance of 3 ft) for 5 minutes; 2 hours later, a repeat light treatment was performed as described. Dye and light treatment was carried out on the second day when severe herpetic keratitis was manifest, and was repeated on the following day. Four different concentrations of dye were used, and 0.1% dye was found to be the highest dilution that gave striking therapeutic relief.

The corneas of 54 rabbits were scarified and inoculated with virus; 48 hours later, 48 rabbits (96 eyes) manifested severe herpetic keratitis. The other rabbits failed to manifest bilaterally severe lesions and were not used. The 48 rabbits were randomized and treated by the procedures described in Table 6.

Of 24 corneas treated with dye and light, 22 were completely cured on the tenth day. The improvement and cure of keratitis in these rabbits was rapid, starting on the third day after treatment; by the sixth day, 15 rabbits were either cured or almost cured. On the other hand, rabbits treated only with dye and no direct light exposure showed a total of 6 cures of 24 corneas on the tenth day. Similarly, corneas treated with light and no dye or untreated corneas (no dye or light) failed to improve in significant numbers.

B. Herpes Simplex in Humans

Following favorable results in tissue culture studies and in animals, preliminary clinical trials of dye-light treatment to inactivate herpes simplex virus Types 1 and 2 were undertaken. A double-blind study was then conducted in a selected group of 32 patients who had been

Table 6. Photodynamic Treatment of Herpes Keratitis

No. of corneas infected	Treatment[a]	No. of corneas cured[b] on posttreatment day indicated:							
		2	3	4	5	6	7	8	10
24	Dye + light	0	1	3	7	15	22	22	22
24	Dye + dark	0	0	0	0	1	3	4	6
24	Light only	0	0	1	2	3	5	7	8
20	Dark only	0	0	0	0	0	1	2	2
4	Totally dark (sewn)	—	—	—	—	—	—	0	0

[a] Corneas were scarified and infected with 10^4 PFU of herpesvirus. On the second and third days after inoculation, all corneas with ++++ lesions were treated with 0.1% dye or light or both. Dark represents housing in conventional animal quarters with normal lighting. Sewn designates that the eyelids of rabbits were surgically sealed and opened on the eighth day.

[b] Cures indicate that corneas have returned to normal or show less than 10% opacity. [From Moore *et al.* (*8*), with permission.]

experiencing at least four episodes of recurrent herpes simplex per year (lip, skin, or penis) for a minimum of 2 years. The following is from the report of Felber *et al.* (3).

Twenty patients were treated with neutral red dye plus light and were followed for 18 to 30 months, whereas 12 control patients, followed for 12 to 24 months, received exactly the same type of treatment, except that a similarly red-colored compound (phenol sulfonphthalein) was applied; this control substance has no known photoreactive or virucidal activity. Neither the patient nor the physician who administered the dye–light treatment and conducted the followup examination knew which dye (experimental or placebo) was used. At the time of admission to the study, each patient had an acute early episode of recurrent herpes simplex and a history of recurrence at approximately the same site in each episode. A detailed history was obtained from each patient and, in all cases, diagnosis was verified either by smear or by virus isolation.

The technique of treatment consisted of rupturing and unroofing early vesicular lesions with a sterile 25-gauge needle. Aqueous neutral red (0.1% solution) or the control dye was applied liberally to the base of each ruptured vesicle. The lesion was then exposed to an ordinary cool white 15-W fluorescent light for 15 minutes at a distance of 6 in. Each patient was instructed to reexpose the lesion for 15 minutes to fluorescent light, or to incandescent light, 1–6 hours later. The patients were then examined at 1- or 2-day intervals to ascertain symptomatic improvement and healing time. If new lesions developed within 1 to 3 days after treatment, the patient was treated again. The patients were emphatically instructed to return for repeat treatment if new lesions developed at a future date. The date of each new recurrence was noted, and the location of the lesion was recorded on a schematic anatomical diagram.

This study was designed to answer three questions: (1) Could treatment for each recurrence over a period of time reduce the frequency of recurrences? (2) Could new lesions be prevented from recurring at a treated site? (3) If the lesions were treated at a prevesicular or early vesicular stage, could symptomatic improvement and reduced healing time be achieved? The most objective of the criteria used for effectiveness of the treatment was the frequency of recurrences. Among the 20 patients receiving the experimental treatment, the average annual recurrence rate was decreased by 50% or more in 17 (85%); a few patients, who had been experiencing recurrences almost monthly, required treatment of three or four separate episodes before a decreased rate of recurrence was noted. By contrast, among 12 control patients treated with the placebo dye, only 2 (17%) experienced a significant decrease in frequency of recurrences. Of 18 patients who previously had recurrences to the same site, only 2 (11%) experienced a recurrence to exactly the same site after treatment with neutral red. The new recurrences, when observed in this group, were in a closely adjacent area. In contrast, among 11 placebo-treated patients with previous same-site recurrences, 9 (83%) continued to experience recurrences to the same location. The results suggest that by photoinactivation the virus in the site treated may be eliminated or altered, rendering it incapable of surviving to reenter a latent stage from which it otherwise could erupt once more after some inciting stimulus.

Supporting evidence of effectiveness was also observed in the patients' reports of symptomatic improvement and in the observations of reduced time required for healing. Of the 20 patients treated with neutral red plus light, 18 (90%) noted symptomatic improvement superior to that obtained with any mode of therapy previously used (these included administration of idoxuridine, smallpox vaccination, herpes "antigen," cryotherapy, and topical agents). Within the first few hours after treatment, patients noted a decrease in pain, tingling, and burning; within 24 hours they were virtually asymptomatic except for a painless crust. All of these 18 patients were noted also to have a 50% or greater reduction in healing time. Six patients were treated in the early erythemic prevesicular stage, and in 4 of them, the lesions resolved in 24 to 36 hours rather than progressing, suggesting that in some instances very early therapy may entirely abort the course of the lesion. A placebo effect—comparable to that observed in other therapeutic studies of herpes simplex—was also noted in these more subjective criteria: among the control patients, 6 (50%) also reported symptomatic improvement and some shortening of healing time. These data, however, are useful as supporting evidence when seen in conjunction with the more objective criteria described above.

In this clinical study, herpes genitalis due to the Type 2 virus appeared to be even more sensitive to photoinactivation than the Type 1 virus.

Further evidence to support the effectiveness of neutral red treatment of herpes vulvitis has been reported by Friedrich (*4*) and by Lefebvre and McNellis (*7*). At Baylor, Kaufman *et al.* (*6*) have recently published preliminary results obtained in treating women with proflavine. They reported as follows.

Fifty-seven women with acute herpes genitalis were treated with proflavine. In 26 cases the diagnosis was primary herpes genitalis and in 31 it was the recurrent type of infection. The diagnosis was made on the basis of clinical history and/or the results of serial antibody studies and cultures from vulva and cervix. Patients with no evidence of antibodies to the infecting virus at the time of the infection who subsequently developed serum antibodies were classified as having primary infection. Patients with evidence of antibodies at the time of initial examination were classified as having recurrent infection. In about 50%, the diagnosis was confirmed on the basis of positive viral cultures. In the remaining patients the diagnosis was made on the basis of clinical symptomatology and/or smears demonstrating typical changes of herpesvirus infection.

All of the patients when first seen presented with superficial ulcers characteristic of herpesvirus infection. Very few of the patients treated were seen in the vesicular stage of the disease. The herpetic ulcers were painted with 0.1% proflavine and the dye was allowed to soak well into the tissues. Following this, the vulva was exposed to incandescent (150 W) or fluorescent light at a distance of 6 to 8 in. for a minimum of 10 minutes. Dye was then reapplied to the ulcers and surrounding tissue, and the vulva again was exposed to light.

Table 7 outlines the results of treatment in both primary and recurrent infection. In patients with primary disease, relief within 4 hours of treatment was noted in 12 women. Eight women reported marked improvement within 24 hours, and in 3 women improvement was noted within 48 hours. In the patients with recurrent infection, 19 reported relief of symptoms in 4 hours; 2, in 1 day; and 1, within 48 hours. In 6 women the results were reported as being poor.

One of the features noted in treating the primary infection was that, whereas many patients reported rapid relief of symptoms, the objective clinical course of the disease did not seem to be altered. There was persistence of ulcers, and in most women new ulcers continued to develop despite treatment. As the new ulcers developed and were treated, symptoms were again rapidly relieved. In many women, rapid pain relief was noted, but with the development of new lesions there was return of pain. In the recurrent infection, once symptomatology was relieved, the ulcers rapidly dried up and disappeared, resulting in the shortening of the total clinical course of the disease.

TABLE 7. PHOTOTHERAPY OF HERPES GENITALIS: RAPIDITY OF PAIN RELIEF[a]

Time	Primary herpes	Recurrent herpes
4 Hours	12	19
1 Day	8	2
2 Days	3	1
>2 Days	3	9
	26	31

[a] Data of Kaufman *et al.* (*6*).

In 7 patients who entered the study with primary infection and in 12 of the women with secondary disease, there were recurrent episodes of herpes genitalis (Table 8). In 15 patients no notation was made regarding the location of the reinfection relative to the treated infection. In 4 women with recurrent infections after treatment, notation was made that new lesions occurred in sites different from the ones originally treated.

As Kaufman points out, the preliminary study was undertaken to obtain some insight into the relative effectiveness of proflavine plus light in alleviating symptoms, shortening the clinical course, and preventing recurrences of herpes genitalis. The results suggest that this form of treatment offers potential in the treatment of this distressing disease. Admittedly, this is not a well-controlled double-blind study, and the results do nothing more than suggest that further study would be worthwhile.

Even more favorable results would have been expected if the patients had been treated a second time with light about 18 hours following the application of dye. The immediate application of light inactivates preformed virus that has taken up the dye. On the following morning, any new infectious virus that had developed would have the dye incorporated into its structure. Such virus is exceedingly photosensitive and would be destroyed even if located intracellularly. Currently, a special light tube emitting a nearly monochromatic light is being developed which can be used by the patient at home as well as in the physician's office. The light tube would allow for intravaginal and cervical photodynamic inactivation of the virus.

We have continued to carry out studies in the laboratory with proflavine and neutral red, and I would like to report on some of these now, particularly in view of the recent report by Rapp *et al.* (*11*) that herpesvirus inactivated by neutral red plus light could transform hamster cells. We have found that proflavine is markedly more effective than neutral red—at least 35 times more so on a molar basis (Fig. 3)—and it is proflavine we are using at the moment for our photoinactivation studies. Also, commercially available proflavine is relatively easy to purify

TABLE 8. RECURRENT EPISODES OF HERPES GENITALIS[a]

Treated disease	Total	Recurrences[b]
Primary	26	7
Recurrent	31	12

[a] Data of Kaufman *et al.* (*6*)

[b] In 4 patients about whom notations were recorded, lesions occurred in sites different from those originally treated.

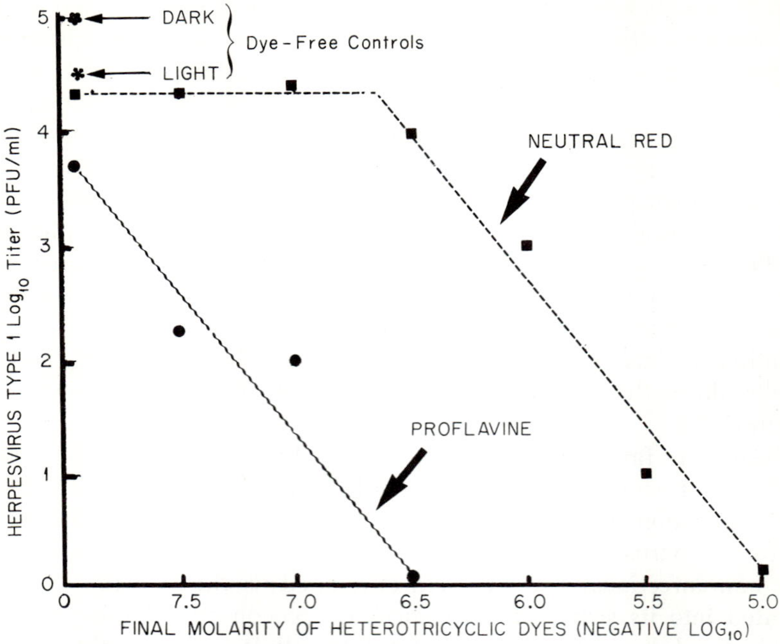

Fig. 3. Comparison of neutral red and proflavine action on herpesvirus. Herpesvirus was diluted 100-fold in glycine buffer to contain 10^5 PFU/ml. Representative samples, all at pH 9, were treated with dye to attain the final concentration indicated in the figure. After 2 hours sensitization at 37°C in the dark, aliquots were exposed to 5000 $\mu W/cm^2$ of 450-nm wavelength light for 15 seconds.

and neutral red is not. A large fraction of commercial neutral red is made up of impurities.

Another technique we are looking at currently is the use of a more nearly monochromatic light source that emits at the band where proflavine absorbs. In this way there is much greater light effectiveness than using ordinary white light.

Also, I think it should be kept in mind, that, as with almost every inactivating procedure, if one adds more of the inactivating agent or increases the time of inactivation, one can get more inactivation. Seemayer *et al.* (*14*), using not herpesvirus but SV40, in hamster cells, which are nonpermissive for replication but permissive for transformation, were able to measure transformation with a degree of accuracy that far exceeds what we can do with herpesvirus. As they increased the amount of radiation of toluidine blue-treated virus, they obtained a decrease in infectivity of 90 to 99 to 99.99% as they doubled the amount of light (Table 9). Similarly, the transformation capacity also decreased

TABLE 9. SV40 PHOTOINACTIVATION[a]

Irradiation (min)	Percent decrease in capacity	
	To infect	To transform
4	90	20
8	99	50
16	99.99	100

[a] Toluidine blue data of Seemayer *et al.* (*14*).

from 20 to 50% and finally to 100% destruction of the transforming capacity of the virus. So, there is no evidence for a residual bit of the genome remaining resistant to light. If enough light is added to the system, one can get complete destruction of the transforming principle of the genome.

However, the critical answer here is not what happens with SV40 but what happens with herpesvirus. We have used two strains of herpesvirus Type 2 that were isolated in our laboratory. Both of these have been sent to Dr. Rapp, and strain 333 is the Type 2 strain that he actually used in his studies. We also used neutral red, as had Rapp *et al.* (*11*). The results in Table 10 indicate various doses of light used, some to permit cytopathic effects and most to permit none at all. We had to use a permissive hamster cell in this particular case because we know of no system in which the virus will not replicate but will only transform. A number of foci of clumped cells appeared in the hamster cell cultures, especially those carried for periods of a month or 2. Foci were picked and passed, but no continuous cell lines were obtained. There is only one cell line that is now limping along in its seventh passage. The others all faded out fairly quickly, as did the 120 controls that were tested in the first experiment. In a second experiment the results have so far again been negative. In contrast, the same two strains, when inactivated with ultraviolet light, produced transformed cultures that are now beyond the fiftieth passage. At the moment, whatever transformation occurred in Hershey with photoinactivated virus has been difficult to reproduce, at least as far as our own laboratory is concerned.

C. SIDE EFFECTS OF PROFLAVINE

Is proflavine carcinogenic? Tests at the National Cancer Institute in Fisher rats and in hybrid mice that are highly susceptible to carcinogenic chemicals have been negative. Albert in his authoritative monograph (*1*) states that "no cancer-producing acridines are known." (Proflavine is 3,6-diaminoacridine.) Furthermore, proflavine has long been used, partic-

TABLE 10. TRANSFORMATION BY HERPESVIRUS TYPE 2

Strain	Treatment	Light exposure $\mu W/cm^2$ (450 nm)	Light exposure Min	CPE[c]	No foci passed	No. continuous lines obtained
Baylor 186	Neutral red[a]	400	6	4/4	—	—
		800	6	1/4	41	0
		1,200	6	0/4	50	0
		3,600	6	0/4	50	0
		10,800	6	0/4	50	(1)
Baylor 333	Neutral red	400	6	4/4	—	—
		1,200	6	4/4	—	—
		2,400	6	0/4	49	0
		3,600	6	0/4	50	0
		10,800	6	0/4	50	0
Controls (no virus)	With and without neutral red[b]	100 – 10,800	6	0/12	120	0
Baylor 333	Neutral red	2,000	5	3/4	33	0
		2,000	10	1/4	51	0
		2,000	20	0/4	18	0
		2,000	30	0/4	42	0
Controls (no virus)	With and without neutral red[b]	2,000	1–30	0/12	115	0
Baylor 186	Ultraviolet[d]	—	8	0	32	2 (>50 passages)
Baylor 333	Ultraviolet[d]	—	12	0	46	3 (>50 passages)
Control				0	40	0

[a] Neutral red incorporated into virus during infective cycle.
[b] Cultures with dye and those without are grouped together.
[c] Cytopathic effect.
[d] S. Kimura, V. L. Flannery, B. Levy, P. A. Schaffer (1975), *Int. J. Cancer* (in press).

ularly in Great Britain, as an antibacterial agent in the treatment of suppurative wounds, burns, otitis media, mastoid infections, and infections of the mouth and vagina.

In going through the literature, we noticed that some have reported an antitumor effect with some of the acridines, so we tested proflavine in this regard. We took two tumor cell lines—HeLa and a transformed SV40 line—and compared their sensitivity in monolayer cultures with BSC-1, a continuous monkey line, primary monkey kidney, and primary rabbit kidney. The interesting finding is that the tumor cells are highly susceptible to proflavine and light, whereas the normal cells are relatively resistant.

As shown in Table 11, with no dye or with dye diluted to 1 in 16

TABLE 11. EFFECT OF PROFLAVINE AND LIGHT ON TRANSFORMED AND NORMAL CELLS[a]

Dye concentration	Cultured cells[b]				
	HeLa	SV40 Tr	BSC-1	Primary monkey kidney	Primary rabbit kidney
1:500,000	++++	++++	+	+	++
1:1,000,000	++++	++++	0	0	0
1:2,000,000	++++	++	0	0	0
1:4,000,000	+	++	0	0	0
1:8,000,000	+	+	0	0	0
1:16,000,000	0	0	0	0	0
None	0	0	0	0	0

[a] Scores: 0 = no deleterious effects observed in five serial transfers for HeLa, BSC-1, and SV40 Tr cells, in two transfers for PMK and PRK cells; + = no immediate visible effect but cells could not be passed; ++ = during 5 days of observation, cells showed rounding, ballooning, and syncytial formation; ++++ = immediate rounding of cells after irradiation.

[b] Exposed to light for 15 seconds at a wavelength of 450 nm.

million, no changes in quality were observed in any of the cultures nor in their passages (5 for HeLa, SV40-transformed and BSC-1 cells; and 2 for the primary cultures used). However, the cancer cells and transformed cells were more sensitive to the dye–light treatment; the HeLa and the SV40-transformed cultures lost the ability to be passed after dye treatment at 1 in 8 million concentration; in these two lines the effects increased markedly with increasing dye concentrations, which at 1 in 1 million produced the maximum effect scored—immediate rounding of the cells after irradiation. No effect was observed in the normal cell cultures or in their passages after the dye treatment at any concentration from 1 in 16 million to 1 in 1 million. Only after treatment using the highest dye concentration tested, 1:500,000, was there an observable effect on the normal cells; after this treatment, the BSC-1 cells and the primary monkey kidney cells could not be passed, and the rabbit kidney cells in addition showed rounding, ballooning, and syncytium formation during the 5-day observation period. It may be that, because these cells have a higher rate of metabolism or DNA turnover, they become more sensitive to the effects of proflavine.

This is where our studies currently stand. Regardless of what we say or do in the laboratory now, we think the cat is out of the bag. Based on their results to date, our clinical colleagues continue to insist on using this treatment for patients who wish to receive it. Naturally, the judgment regarding the use of the treatment rests partly on the degree of discomfort and disability produced by the disease. At Baylor the clinical investigators keep accurate records on their patients and continue to believe they have something which is doing their patients some good. We hope that other clinicians who are so inclined might also use this treatment under carefully controlled conditions and record their observations for everyone to see.

REFERENCES

1. Albert, A., "The Acridines," 2nd ed. Arnold, London, 1966.
2. Crowther, D., and Melnick, J. L., *Virology* **14**, 11–21 (1961).
3. Felber, T. D., Smith, E. B., Knox, J. M., Wallis, C., and Melnick, J. L., *J. Amer. Med. Ass.* **223**, 289–292 (1973).
4. Friedrich, E. G., *Obstet. Gynecol.* **41**, 74–77 (1973).
5. Hiatt, C. W., Kaufman, E., Helprin, J. J., and Baron, S., *J. Immunol.* **84**, 480–484 (1960).
6. Kaufman, R. H., Gardner, H. L., Brown, D., Wallis, C., Rawls, W. E., and Melnick, J. L., *Amer. J. Obstet. Gynecol.* **117**, 1144–1146 (1973).
7. Lefebvre, E. B., and McNellis, E. E., *J. Amer. Med. Ass.* **224**, 1039 (1973).
8. Moore, C., Wallis, C., Melnick, J. L., and Kuns, M. D., *Infec. Immunity* **5**, 169–171 (1972).
9. Perdrau, J. R., and Todd, C., *Proc. Roy. Soc., Ser. B* **112**, 228 (1933).
10. Raab, O. Z., *Z. Biol. (Munich)* **39**, 524 (1900).
11. Rapp, F., Li, J. H., and Jerkofsky, M., *Virology* **55**, 339–346 (1973).
12. Schaffer, F. L., *Virology* **18**, 412–425 (1962).
13. Schaffer, F. L., and Hackett, A. J., *Virology* **21**, 124–126 (1963).
14. Seemayer, N. H., Hirai, K., and Defendi, V., *Int. J. Cancer* **12**, 524–531 (1973).
15. Shortt, H. E., and Brooks, A. G., *Indian J. Med. Res.* **21**, 581 (1934).
16. Wallis, C., and Melnick, J. L., *Virology* **21**, 332–341 (1963).
17. Wallis, C., and Melnick, J. L., *Virology* **23**, 520–527 (1964).
18. Wallis, C., and Melnick, J. L., *Photochem. & Photobiol.* **4**, 159–170 (1965).
19. Wallis, C., and Melnick, J. L., *J. Bacteriol.* **89**, 41–46 (1965).
20. Wallis, C., Sakurada, N., and Melnick, J. L., *J. Immunol.* **91**, 677–682 (1963).
21. Wallis, C., Scheiris, C., and Melnick, J. L., *J. Immunol.* **99**, 1134–1139 (1967).
22. Wallis, C., Trulock, S., and Melnick, J. L., *J. Gen. Virol.* **5**, 53–61 (1969).
23. Wilson, J. N., and Cooper, P. D., *Virology* **17**, 195–196 (1962).
24. Wilson, J. N., and Cooper, P. D., *Virology* **21**, 135–145 (1963).
25. Yamamoto, N., *J. Bacteriol.* **75**, 443 (1958).

DISCUSSION

Dr. Melnick: I would like to comment on the significance of hamster cell transformation *in vitro,* since the issue continues to be raised. First, I do not know any data that allow us to transfer hamster findings to human findings. Adenoviruses are highly oncogenic for the hamster. Not only do they produce transformation in hamster cells, which after transformation can be passed and produce tumors in the hamster, but also the viruses themselves produce tumors in the hamster. I do not know the significance of this for the human host, since despite intensive efforts no evidence has been obtained for oncogenicity of adenoviruses in man.

As regards the alleged production of greater transforming ability in the virus by photodynamic inactivation, this is open to question. We cannot as yet measure quantitatively the transforming capacity of herpesviruses in the absence of the proper cells for assay. The cells that are used are permissive for herpesvirus, so the infective component of the virus must first be inactivated before we can determine whether the virus can transform. In a situation like that with SV40, where transformation can be measured accurately, the virus does not become enhanced for transformation when it is treated with dye and light; instead, it actually decreases in its transforming ability. Perhaps the photodynamic treatment of herpesvirus, by removing the infectious component, allows the detection of any transforming components that remain.

Those of us who are in the game of working with herpesviruses are not interested in working with them on the basis of merely trivial infections. There are data from a number of laboratories that show an association between herpesviruses and certain types of cancer. One then has to evaluate the situation of being able to eliminate herpes lesions as soon as they become visible, early in life, as against allowing the virus to remain present for decades, to produce periodic recurrences of active lesions, and to conduct a continuous assault on cells not only with virions that replicate but also with defective virions. Every time herpesvirus replicates, it produces perhaps 1,000 defective particles that do not replicate. How many of these defective particles are capable of transforming cells?

If we look at the total amount of transforming herpesviruses that a person would be subjected to in a lifetime and compare this with what might be the situation in the case of a successful early treatment that eliminates the virus from the body, we find an enormous excess of potentially transforming herpesviruses that would continue to plague the individual over a lifetime, in contrast to the situation following a successful treatment.

So the question has to be answered—whether a person is better off as far as cancer is concerned (if herpesvirus is related to cancer) with having the oncogenic insult removed early in life, or should this insult be allowed to remain for long periods of time with a continuous assault of defective, presumably oncogenic, viruses month in and month out.

Dr. Oxman: In view of the observations that between recurrences herpes simplex remains latent in the sensory ganglia, at a distance from the skin, how do you think that local therapy of skin lesions can reduce the incidence of recurrences?

Dr. Melnick: I think it is clear from recent data of Baringer and of Stevens that the virus is latent in the ganglia. When recurrences occur it must come down from the ganglion into the skin. However, from the beautiful electron micrographs of Baringer, it appears that the virus that is latent in the ganglion is present as a whole virus. In careful autopsies that he has done, he found this to be present in virtually 100% of the cases, indicating that during recurrences something happens to the virus—it seems to move to the skin and becomes active. If it can be destroyed at the time when it is replicating in the skin, then I think that we may interrupt its return to the ganglion when it again becomes latent.

Dr. Nahmias: I'm sorry but I think that Dr. Melnick misquoted Dr. Baringer. The number of ganglia that release viruses is in the range of 50 to 60%, not 100%, and Dr. Baringer had to search thousands of slides to be able to find one or two cells that showed a virus particle. So I'm not sure that one can say that the virus is in the ganglia as an infectious particle. The point that Dr. Melnick brings up is a very crucial one, that is, how does the virus, starting, let's say, from herpetic gingivostomatitis, get to the ganglia in the first place? Why doesn't it go further back into the brain? Also, if virus in some form is in the ganglia, how many of the cells are affected? Does it go back into the skin as a full virus or as viral DNA perhaps? And does it then go back and forth several times? I'm leaning currently toward the view that there must be traffic back and forth between ganglia and external sites.

DR. MELNICK: I believe that if Dr. Baringer finds virus by his technique in 50 to 60% of cases it must be present in close to 100%.

DR. SABIN: I think it is entirely premature and unwarranted to assume that, just because some virus has been demonstrated in sensory ganglia, it is not also present in the skin site in which recurrent herpes occurs. The main difference, I would say, is that with any virus that recurs in the complete infectious form to destroy more cells, obviously the whole genome remains but not necessarily the whole virus. The assumption that it is only in the ganglion, I think, is unwarranted at the present time. Moreover, I would say that, from all the points that Dr. Melnick made about the other problem, obviously we have to learn more; but I think just because we have to learn more, it is important to wait and not to jump too soon.

DR. FJELDE: Is there any difference in the viruses that are treated with low doses of irradiation compared with those that are treated with high doses? Do you see abnormal viruses (empty cores, etc.) in EM studies?

DR. MELNICK: There is no visible effect of the radiation of the virus.

DR. FJELDE: If you give them high doses or low, you have the same effect?

DR. MELNICK: We haven't done much in the way of quantitative work in this area, but in the little we have done, there has been no obvious change in the morphology of the virus.

DR. POLLARD: Through how many cell layers will visible light penetrate?

DR. MELNICK: I refer you to a paper from Baylor by Drs. Bean, Felber, and Knox [*Arch. Dermatol.* **107**, 204 (1973)], who studied proflavine staining of skin as demonstrated by fluorescent microscopy.

CHAPTER 18

Studies on Adenine Arabinoside—A Model for Antiviral Chemotherapeutics

Richard J. Whitley
Lawrence T. Ch'ien
Robert A. Buchanan
Charles A. Alford, Jr.

I. INTRODUCTION

The research of the previous decade on the biology and biochemistry of viral replication and function has heralded a coming era of antiviral chemotherapy. Each year numerous new compounds are synthesized but because of toxicity fail to go beyond the stage of *in vitro* experimentation. Most of the antiviral agents with acceptable toxicities that have been studied to date effect one or more of the intracellular events responsible for both host cell and viral replication and viral assemblage. Until the events unique to viral and host cell replication are more clearly delineated, antiviral chemotherapy will remain nonspecific. At the present there are several compounds such as benzimidazole, guanidine hydrochloride, and rifampicin or its derivatives believed to be specific for viral function. However, these agents have not been studied nearly as completely as some of the other available agents. Current interest of viral and cancer chemotherapists has come to focus on the pyrimidine

* Portions of the work summarized here have been supported by the grants from National Institute of Child Health and Human Development (HD-01687), a training grant (HD-00413), from the National Cancer Institute (CA-131148), and the Robert E. Meyer Foundation.

and purine nucleosides. Included among these agents are the pyrimidine analogs, 5-iodo-2′deoxyuridine (IDU) and 1-β-D-arabinofuranosylcytosine (ara-C), and the purine analog 9-β-D-arabinofuranosyladenine (ara-A). Of this triad of compounds attention will focus on ara-A, the drug most familiar to our own laboratory. Where appropriate, comparisons will be drawn to its two pyrimidine counterparts. From the extension of the studies of not only our own laboratory but those of others in the field and with a greater understanding of viral replication, new, more specific antiviral agents most surely will be developed.

II. EARLY INVESTIGATIONS OF ARA-A AND ANALOGS

Historically, ara-A or adenine arabinoside was first synthesized in 1960 by Lee *et al.* (*7*) as a potential anticancer agent. Shortly thereafter, in 1964, Privat de Garilhe and De Rudder (*9*) reported the drugs antiviral activity *in vitro* against herpes simplex virus and vaccinia virus. Three years later, in 1967, Sidwell *et al.* (*11*) noted the drug's ability to inhibit the replicative cycle of cytomegalovirus in tissue culture. Thus, a broad spectrum of antiviral activity for ara-A was readily established through the employment of similar *in vitro* systems. This *in vitro* spectrum was most marked for the classic DNA viruses—namely, herpes simplex, herpes samari, herpes B, pseudorabies, vaccinia, varicella zoster, and cytomegalovirus. The potential antiviral activity is much analogous to ara-C and IDU. This antiviral activity spurred numerous animal model experiments which led to reports of therapeutic activity in the treatment of herpes simplex keratitis (*5*), and herpes simplex (*10*) and vaccinial encephalitis (*10*).

Concomitant with the investigations into the antiviral activity of ara-A, various laboratories attempted the elucidation of the drug's mechanism of action—particularly as an antineoplastic agent. These efforts were paralleled by similar and even more detailed studies on cytosine arabinoside and idoxuridine. The biochemical structure of these compounds is shown in Fig. 1. Although the precise intermediaries responsible for and sites of inhibition necessitate further clarification, certain fundamental principles have been excepted. First, like ara-C and IDU, ara-A is readily phosphorylated to its mono-, di-, and triphosphate derivatives (Fig. 2). These intermediaries show little inhibitory effect against the purified phosphorylating enzyme system as is seen with ara-C but in contrast to IDU which does inhibit the thymidine kinase system. The triphosphate derivative, adenosine triphosphate arabinoside (ara-ATP), then, has been shown by York and LePage (*12*) to be a noncompetitive inhibitor of DNA polymerase. Later work has suggested mixed order kinetics in the inhibition of the reaction catalyzed by DNA polymerase

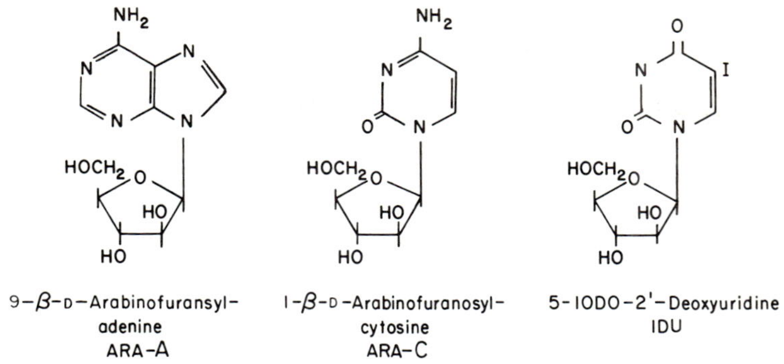

Fig. 1. Structural comparison of ara-A, ara-C, and IDU.

(2). There is but one report as of the present to suggest the incorporation of ara-A into DNA (personal communication, A. Kornberg). Thus, the primary site of inhibition remains DNA polymerase. The translation of these cellular events into biochemical time-sequence studies of virus-infected cells has not yet been performed or reported to our knowledge.

Cytosine arabinoside most nearly resembles the inhibitory pattern of ara-A with its primary site of action being the inhibition of DNA polymerase. However, it has also been suggested that ara-C can be incorporated into the 5′ end of the growing DNA chain and inhibit the

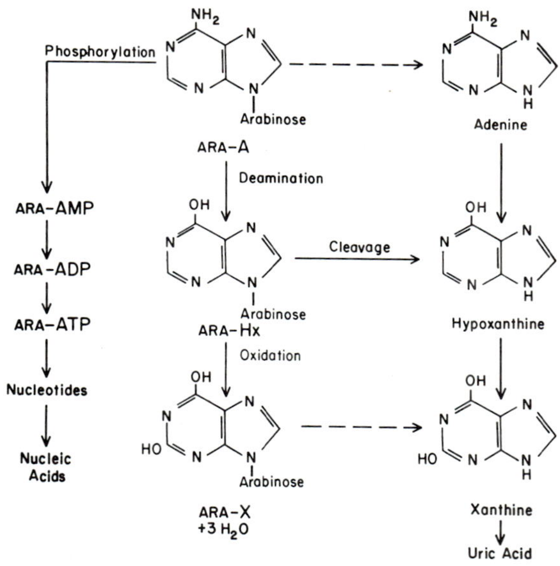

Fig. 2. Metabolic pathways of phosphorylation, deamination, and oxidation for adenine arabinoside (ara-A).

nucleoside diphosphate reductases. On the other hand, IDU possesses the capability of interfering with the thymidine kinase system and can also be incorporated into DNA. From these patterns of cellular and viral interference, a clearer understanding of both the spectrum of antiviral activity as well as potential toxicities can be appreciated.

A. Toxicity of Ara-A in Animals

The early animal models testing therapeutic activity were extended to pursue both the clinical pharmacology and toxicity of ara-A. It was readily noted, as predicted, that ara-A was much less toxic in animals and did not appear to suppress the immune response to the same degree as ara-C or IDU. Similarly, ara-A had little effect on rapidly dividing cells in contrast to the pyrimidine analogs. Among species of animals, however, there was a marked difference in their tolerance to ara-A. Lower animal forms, namely, mice, rats, rabbits and dogs, could tolerate immensely higher (100-fold and greater) parenteral doses than could the higher animal forms such as the simians. These doses were tolerated with minimal and reversible clinical signs of toxicity and essentially no organ histopathology. In monkeys there were signs of anorexia and weight loss at far lower doses. Experiments with tritium labeling at the C-2 position of ara-A helped to explain these striking differences. Lower animal forms very rapidly oxidize ara-A at the C-2 terminal into volatile inactive components, notably xanthine and heavy water, which, in turn, are rapidly excreted in the urine. Only 15% of the original dose is removed as the nonvolatile hypoxanthine derivative. In contrast, higher animal forms metabolize the drug far more slowly via deamination to the primary metabolite of arahypoxanthine (ara-Hx) which accounts for 85% of the excreted product. Of particular interest to viral chemotherapists is that this primary metabolite of ara-A has significant although less antiviral activity than the mother compound. This effect remains in direct contrast to the metabolites of ara-C and IDU which possess no such activity.

B. Clinical Pharmacological Trials in Man

With this background and in order to understand the dosage, route of administration, toxicity, and drug disposition as well as the species differences, it was necessary to undertake similar clinical pharmacological studies in man. Dr. Larry Ch'ien of our own laboratory and the chemists from Parke-Davis under Dr. Anthony Glazko (*3*) proceeded with a metabolic disposition study of ara-A in 7 human volunteers with incurable and imminently fatal disease. A single administered dose of ara-A at 1 mg/kg and tritium labeled so that no patient would receive more than

100 μCi was given intravenously. The radioactive labeling technique was employed in order to avoid the unnecessarily large doses of drug that would have been required for chromatographic determinations. Serial plasma, urine, and fecal determinations were performed over 10 days. These plasma levels are graphed in Fig. 3. The nonvolatile tritium represents the ara-A, the hypoxanthine derivative, or other adenosine metabolic end products. The volatile tritium represents products from further oxidation at the C-2 terminal with the relase of a single hydrogen to form heavy water.

Following the single intravenous infusion, plasma levels of 1.1 to 1.5 μg/ml were noted at 30 minutes with a rapid fall occurring over the next 8 hours to 0.1 to 0.2 μg/ml where they remained for the duration of the study. The biologic half-life was subsequently calculated as 1.5 hours. On the other hand, ara-C has a much shorter initial half-life of 30 minutes. Over the first 4 hours the nonvolatile radioactivity approximated the total plasma radioactivity. The volatile fraction of radioactivity does not become significant until approximately 8 hours after dosing. This suggests that the primary metabolic pathway of ara-A in man is

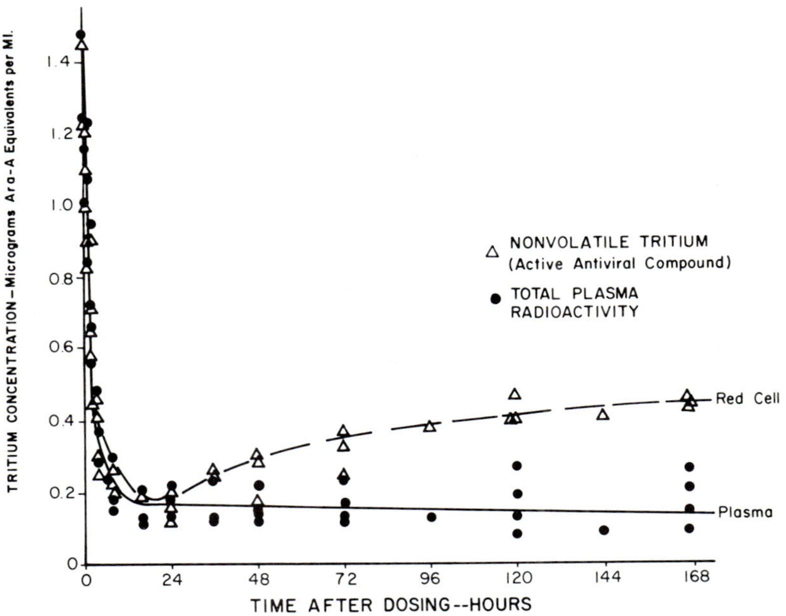

Fig. 3. Total tritium in plasma and nonvolatile tritium in red cells following a 1-mg/kg dose IV of ara-A in 4 subjects. [Reproduced by permission; Ch'ien *et al.* (*1a*).]

one of deamination to ara-Hx as monitored by the nonvolatile fraction. Thus, man most closely resembles the monkey in the metabolism of the drug. As noted previously, ara-Hx is a unique nucleoside metabolite with antiviral activity and, therefore, provides a distinct advantage over ara-C and IDU which have inactive metabolites.

The kidney represents the primary site for clearance of the drug in man with nearly 52% of the drug appearing in the urine within the first 24 hours of dosing. Fecal excretion remains negligible. Figure 4 demonstrates the urinary excretion pattern of nonvolatile and volatile fractions over the first 24 hours. As can be readily noted, there is rapid renal clearance of 16% of the original dose of ara-A and ara-Hx over the first 3 hours all as the nonvolatile component. After the first 3 hours, lower excretion rates are attained reaching 0.1% of the dose per hour in the 12–24 hour period. With decreasing rates of excretion, there is a progressive increase in the oxidized metabolites or volatile tritium fraction which remains consistently less than the nonvolatile. The earlier and more pronounced appearance of volatile tritium in the urine may well reflect both the site of excretion of ara-A and, perhaps, a local contribution by the kidney to oxidation of the drug.

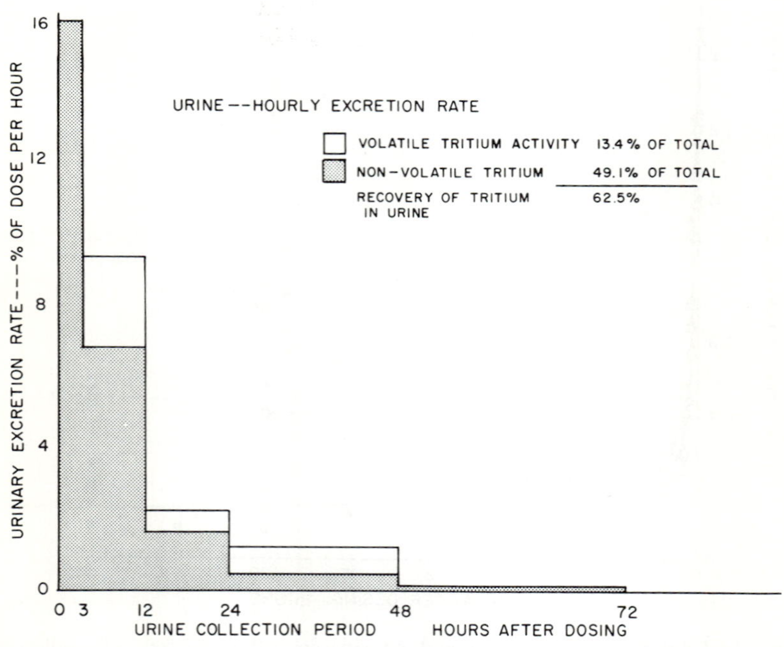

Fig. 4. Urinary excretion of tritiated nonvolatile ara-A and its metabolites compared with volatile degradation products.

Four of the 7 patients were similarly studied for red cell concentrations of the nonvolatile tritium fraction, predominantly ara-Hx, following a single intravenous dose of drug. Figure 3 demonstrates the progressive increase in red blood cell concentrations with a plateau (of 0.4 μg/ml) occurring approximately 5 days after dosing. This is in contrast to total plasma radioactivity corresponding to 0.18 to 0.20 μg/ml. These levels persisted for the duration of the study. These data become of increasing importance in that the primary sites of action of ara-A and ara-Hx are intracellular. These red cell results are further corroborated by intracellular levels of labeled material obtained on 1 patient with terminal acute lymphoblastic leukemia who died 11 days after dosing. The highest levels were obtained from kidney, liver, and spleen (0.51–0.56 μg/gm tissue), whereas lower levels were detected in brain, skeletal muscle, and cerebral spinal fluid. A spinal fluid determination of the nonvolatile component was performed at 2 hours after medicating in an additional patient. This level was found to be approximately one-half of the plasma level.

These early clinical pharmacological trials of ara-A in man yielded several important answers while posing many new questions. First, ara-A's metabolism in man more closely resembles the monkey model than that of the lower animals. Future metabolic disposition studies may well, then, employ such a model. Second, due to the extremely rapid clearance of the drug by the kidney, ara-A should be able to be administered by a continuous drip without excessive drug accumulation. Third, and of particular pertinence to the therapy of CNS infections, ara-A freely, yet variably, crosses into the cerebral spinal fluid. Finally, and, perhaps, of most significance, the drug and its metabolites are capable of concentrating intracellularly as suggested by the red blood cell and necropsy analyses. As viral replication is an intracellular phenomenon dependent on host cell systems, drug concentrations intracellularly are imperative. At the same time, the shortcomings of the present studies came to light. Newer and better techniques that clearly differentiate ara-A from ara-Hx need to be developed. Concomitantly, drug disposition studies have to be undertaken at drug dosages replicative of human therapy trials. No less, the intracellular drug and metabolite levels need clarification for the appropriate dosage regimens. With this disposition foundation and the development of more refined analytic techniques, further metabolic studies will be undertaken.

Limited field trials of ara-A were initiated for the treatment of severe herpetic and pox infections. These trials employed varying dosages of ara-A in order to determine an appropriate medication range as well as preview clinical toxicity in man. The results of these preliminary studies, as will be detailed later, showed an appropriate dose range of 10 to 15 mg/kg/day. Thus, an expanded metabolic study was performed in 5 patients at the aforementioned dosage for 5 days. Each patient received ara-A by continuous intravenous infusion over a 12-hour interval in order to insure the uniformity of administration. A high-pressure liquid chromatography procedure was employed by Parke-Davis chemists under Dr. A. Kinkel (6) to assay the serial samples of plasma obtained on each of the patients under study. The precision of assay as based on replicate determinations was 7.99%.

Figure 5 graphs the plasma levels of ara-A and ara-Hx when determined serially. It should be noted that at the time of collection deaminase inhibitor was added to each specimen. This figure graphically depicts in further detail the earlier findings from the tritium studies. Namely, the levels of ara-A and ara-Hx have a uniform daily pattern. There is no accumulation of ara-A and an almost immediate conversion of ara-A to its first-step metabolite ara-Hx. On the other hand, ara-Hx does accumulate in the plasma during the period of infusion. At the end of the infusion period ara-Hx is rapidly cleared from the bloodstream. More importantly, these levels remain replicable over the 5-day period of study. It is crucial to recognize that these levels are only the plasma determinations and do not include intracellular drug concentrations. At the present, more precise intracellular determinations have

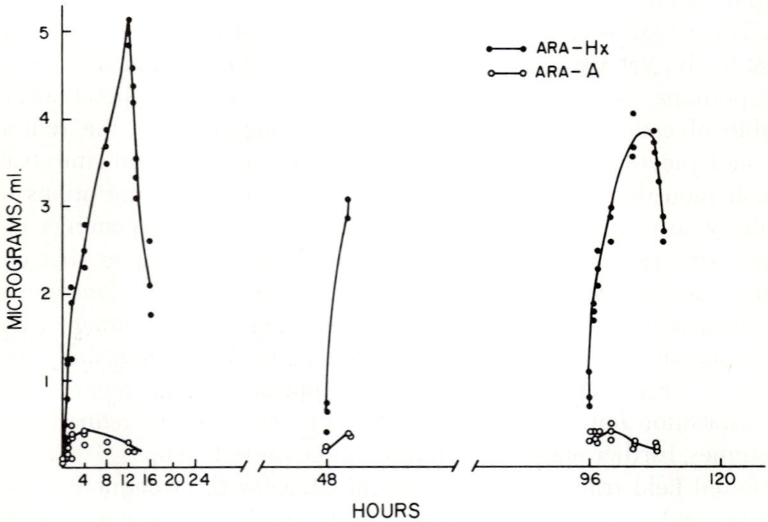

Fig. 5. Mean blood levels of ara-A and ara-Hx following daily 12-hour constant infusions (10 mg/kg). Summary data derived from study of 5 patients.

not yet been performed. Thus, one can only speculate on the real significance of these findings. Should there be a similar pattern of concentration of ara-Hx occurring intracellularly, then, perhaps, it is this metabolite that has the greater antiviral significance.

The maximum plasma levels of ara-A and ara-Hx are, respectively, 0.5 and 5.0 μg/ml. The sensitivities of the different herpesviruses have been tested against varying concentrations of ara-A in tissue culture systems by both our own laboratory and the one of Dr. James Luby (*4*). From these *in vitro* experiments, it was possible to demonstrate the inhibition of viral replication of herpesvirus hominis, type 1, with extracellular drug concentrations of both ara-A and ara-Hx at 20 to 40 μg/ml. Herpesvirus hominis (HVH), type 2, was less responsive to ara-A and responsive to ara-Hx within the same dosage range. Similar results have been demonstrated with IDU and ara-C. A perusal of the plasma drug levels readily shows that the peaks so obtained do not even approach those extracellular levels required for a suppression of replication *in vitro*. This is but one further piece of evidence pointing toward the necessity of intracellular qualitative and quantitative drug determinations.

III. TREATMENT OF HERPETIC INFECTIONS WITH ARA-A

These initial laboratory and metabolic studies paved the way for limited field trials of ara-A. The clinical and virologic studies have been performed primarily at our own institution, the University of Alabama in Birmingham. Those patients selected for treatment came exclusively from a population in which viral infections are recognized as leading to potentially serious complications. Thus, these patients included organ transplant recipients, newborns, and those with such underlying diseases as leukemia, lymphoma, or generalized malignancy, particularly those receiving immunosuppressive therapy with steroids, antimetabolites, or radiation. Table 1 reveals the number of patients treated under each herpetic disease category.

The response to therapeutic trials seems, in part, mediated by the localization or dissemination of the infection and, if disseminated, its duration. Therefore, the clinical data will be presented according to localized infections as HVH, type 1 and/or type 2, of adults and newborns and dermatomal herpes zoster or as to disseminated infections of herpes neonatorum, cytomegalovirus, and varicella zoster. Adults with progressive mucocutaneous herpetic infections were selected on the basis of documented HVH infection persisting for longer than 2 weeks, extending beyond the oral or vaginal cavity, associated with incapacitating

TABLE 1. SUMMARY OF HERPETIC DISEASES AND THE NUMBER OF PATIENTS RECEIVING ARA-A UNDER PHASE II PRELIMINARY TRIALS

Type of infection	Patients treated
Herpesvirus hominis	
Adult	
Mucocutaneous	14
Encephalitis	1
Neonatal	
Encephalitis	10
Localized	3
Varicella zoster	
Adult	
Localized	9
Disseminated	11
Cytomegalovirus	
Congenital	9
Mononucleosis	2
Generalized (immunosuppressed)	5
	64

pain and necessitating hospitalization. The patients received an ara-A intravenous infusion of 3.3 to 15 mg/kg/day over a 12-hour interval. The duration of therapy varied between 5 and 14 consecutive days.* The drug was mixed with a 5% dextrose and one-quarter normal saline solution not to exceed 0.7 mg of ara-A/cc. Concentrations greater than this are insoluble. Figure 6 depicts graphically the characteristic response of 8 patients having HVH, type 1, in contrast to 3 matched control patients. The virologic and clinical responses are graphed according to the duration and dose of ara-A therapy. Pretreatment virus titers ranged from 1000 to 100,000 infectious units/0.2 cc of swabbed material from clinical lesions. In all but one of the treated patients to date, there has been a 90% reduction of viral titer within 72 hours of the initiation of therapy. Virus was not detectable in most patients (6 of 8) 8 days after the commencement of therapy. The clinical responses paralleled the virologic clearing with healing occurring by 9 days. These results stand in comparison to our 3 untreated controls who demonstrate both persistence of the virus in higher titers as well as clinical signs and symptoms of disease. The excretion of HVH in low levels from the throat and saliva did return in approximately 1–3 weeks after the

* Adenine arabinoside was supplied by Parke-Davis and Co., Ann Arbor, Michigan, in concentrations of 50 or 200 mg/cc.

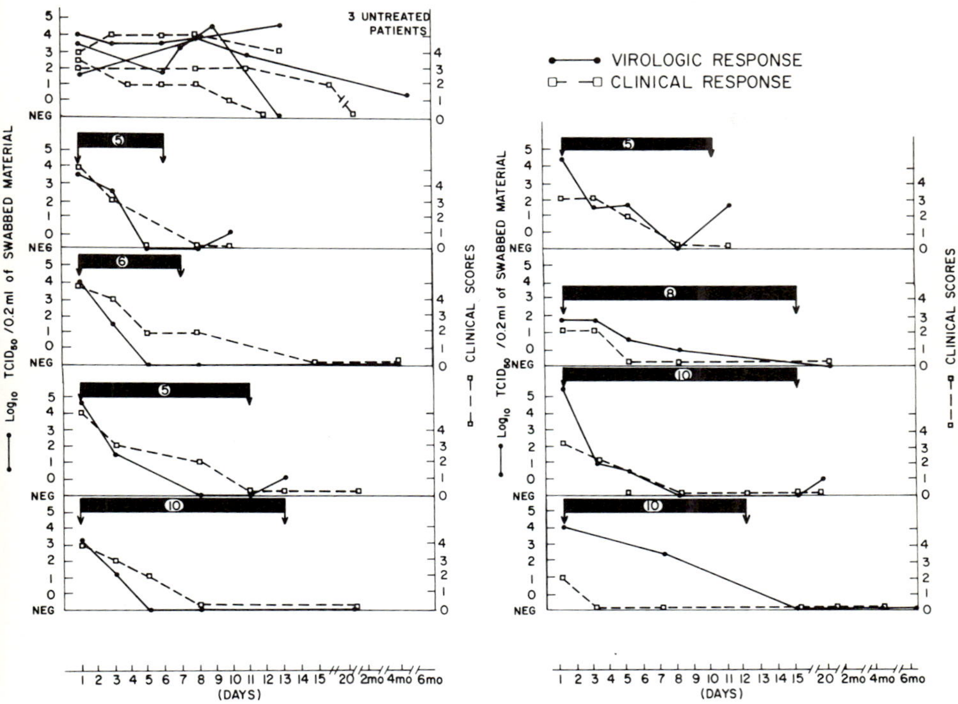

Fig. 6. Clinical and virologic response of Type 1 mucocutaneous herpesvirus hominis in 8 treated and 3 control patients. The numbers on the bars indicate the daily dosage (mg/kg) of ara-A while the length represents the duration of therapy. [Reproduced by permission; Ch'ien *et al.* (*1*).]

cessation of therapy in 50% of the patients. However, in no case was there associated clinical recurrence of disease (*1*).

The response of HVH, Type 2, to ara-A is less remarkable than Type 1. Those 3 patients having concomitant lesions from Types 1 and 2 HVH had a more protracted virologic and clinical resolution of their Type 2 disease. Figure 7 is a representative graph of the difference in response between infections in 1 patient.

The fact that there was virtual and near unanimity of response of HVH progressive mucocutaneous infections in spite of considerable variation in the duration of disease prior to therapy lends credence to the concept of drug responsiveness. The return of silent low-level viral excretion contributes to the proposal of viral responsiveness. Like the pyrimidine analogs, ara-A does not result in complete viral clearance but acts to suppress or reduce viral replication and, thereby, allow the altered host to bring the infection under control. Thus, if this theory is borne

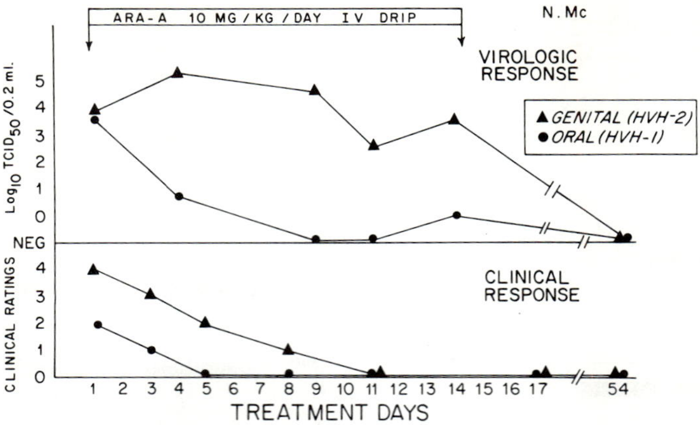

Fig. 7. Comparative clinical and virologic responses of genital lesions caused by herpesvirus hominis (HVH) Type 2 and HVH Type 1 oral lesions in one patient having concomitant infections. [Reproduced by permission; Ch'ien *et al.* (*1*).]

out, the viral load, the extent of dissemination, and the preexisting degree of immunosuppression will all influence the response of the infection to ara-A therapy as well as the other nonspecific viral inhibitors. An additional contributory factor to the outcome of therapy may well prove to be related to the antigenic differences between the strains of HVH 1 and 2. This was noted in the few patients with concomitant HVH Type 2 lesions where both virologic and clinical resolution was more prolonged. This has not only been witnessed with ara-A but has been suggested that ara-C and IDU *in vitro*.

Three newborns with herpetic infections localized to the skin and eye responded rapidly to therapy clinically and ceased to excrete virus within 5 days of the initiation of treatment. Similarly, 9 immunosuppressed adults were treated with ara-A for localized or dermatomal varicella zoster (VZ). These adults responded to therapeutic regimens as detailed earlier in a manner essentially identical to the progressive mucocutaneous HVH patients. Namely, new vesicle formation ceased within 48 hours of the initiation of therapy. Clear vesicles quickly clouded and, then, scabbed over. Thus, localized HVH and VZ show particular susceptibility to this form of antiviral chemotherapy.

The response of disseminated herpes neonatorum and early disseminating zoster appears to be intermediate between the therapy of localized progressive infection and fulminant disseminated disease. Ten neonates with disseminated neonatal herpetic infections were treated in a manner

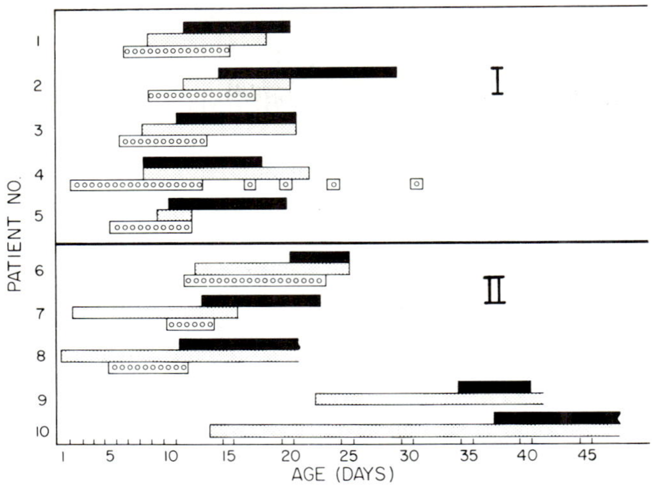

Fig. 8. Treatment categories for neonatal herpes infection. Group I consists of 5 neonates having disseminated herpetic infection and treated early. Group II consists of 5 having disseminated infection and treated late. Solid bars—therapy; stippled bars—neurological signs; bar with circles—skin vesicles.

nearly identical to the adults and three other newborns who had progressive mucocutaneous herpes. The significant differences included a dose range of 10 to 20 mg/kg/day with the duration of therapy ranging from 10 to 15 days (mean 12). In contrast to the adults with herpetic infections, neonatal HVH may be far more severe and life threatening with the potential for devastating neurological sequelae in the survivors. The disseminated infection has been reported to carry a mortality rate well in excess of 50% (8). As the prevalence of CNS involvement in disseminated disease demands adequate cerebral spinal fluid levels of ara-A, the dosage was nearly doubled from earlier trials. In our own treatment series of 10 neonates having disseminated disease with CNS and multiple sites of visceral involvement, 5 treated early (Fig. 8, Group I) within 4 days of the onset of neurological signs survived with no apparent neurological deficits detectable at 6 months and 1 year of age. In contrast, 5 were treated with ara-A some 8–21 days after the onset of neurological signs (Fig. 8, Group II). Of this group, 4 died and 1 was left with severe neurological sequelae. Thus, it appears that the outcome is directly related to the time of recognition of infection in relation to the time of initiation of therapy.

These results for the early treatment group of neonatal herpes are truly encouraging in light of the severity of the disease and in consideration of the high mortality. Of particular interest is the identification of the far greater incidence of type 2 HVH than type 1 in causing disease yet its apparent responsiveness to ara-A. This responsiveness appears to be greater in the newborn than the adult. However, further studies employing higher dosages of ara-A in the treatment of genital herpes are indicated before any conclusions can be reached.

Similarly, 11 patients with early disseminating zoster seemingly benefited from therapy with the cessation of new vesicle formation over a 72 to 96 hour time interval. The vesicles, in the majority of patients, clouded quickly and then scabbed over as in the therapy of localized zoster. Although encouraging, these results must be interpreted cautiously until the results of blind studies are complete.

IV. TREATMENT OF CYTOMEGALOVIRUS WITH ARA-A

By comparison, other disseminated viral infections pose particularly difficult treatment dilemmas. The patterns of response to therapy as will be demonstrated with the treatment of congenitally acquired cytomegalovirus (CMV) are significantly influenced by the spectrum of severity of the infection. Nine newborns and infants were treated for congenital CMV with the same drug regimen as employed with neonatal herpes and for the same reasons. These infections are particularly difficult to treat due to their intrauterine acquisition. In contrast, neonatal HVH infections are usually acquired intranatally and progress rapidly to systemic involvement. These 9 treated infants may be graded according to the severity of the viral load as monitored by the quantification of urinary virus excretion. Six mildly symptomatic babies had less than 100,000 infectious particles/0.2 cc of urine. Therapy of this group resulted in a prompt stepwise reduction of urinary virus to negative titers in all cases but 1. Ten to 15 days after cessation of therapy, urinary viral excretion returned in all the cases just as with the herpes infections. The results of six typical therapy trials are compared to a representative control infant in Fig. 9A. Those 3 neonates with acute fulminant congenital CMV (Fig. 9B) were treated with 20 mg/kg/day of ara-A for 14 days. Each baby had a urinary virus excretion of greater than 1,000,000 infectious particles/0.2 cc of urine. None of these newborns survived long enough for followup. All succumbed from overwhelming infection. Each of these newborns showed some reduction of urinary excretion of virus as witnessed in Fig. 9B. These differences are obviously based on the increased quantity of virus present in the later treatment group and, therefore, the stage to which the disease had progressed. Although the urine excretion was affected, the minimal levels achieved

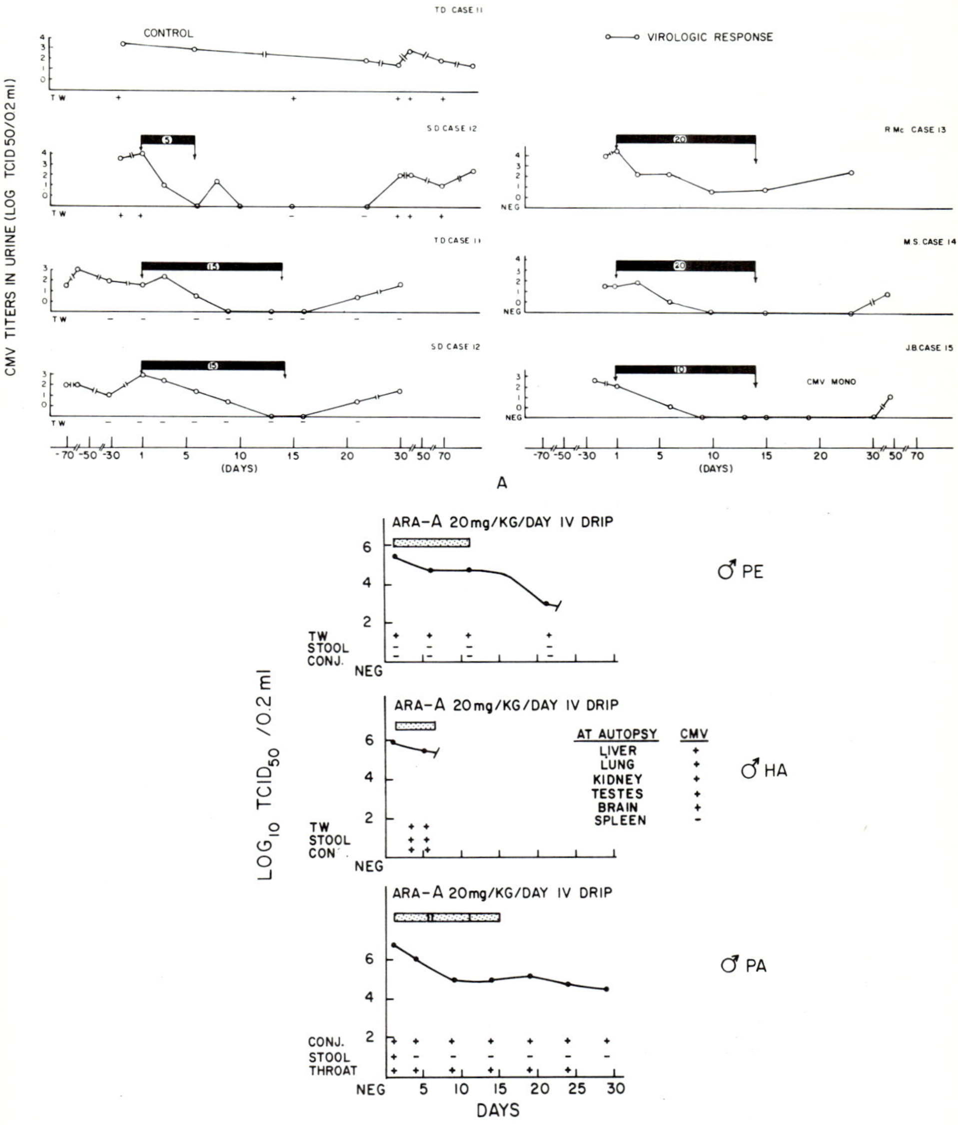

Fig. 9. (A) Urinary cytomegalovirus (CMV) excretion in untreated control as compared to 6 therapeutic trials of ara-A in infants. The dose and duration were varied as noted by the bars. Patients had minimal or "burned out disease." Note: lower levels of viral excretion in urine as compared with values shown in part B. (B) Therapy with ara-A in 3 neonates with acute fulminant CMV infection showing quantified urinary excretion and additional sites of viral isolation. The dose and duration of therapy are noted by the bars.

are not impressive in relation to those in the first group. The 3 neonates had virus persist throughout therapy in the stool, conjunctivae, and throat at high titers as well as remain easily retrievable from the white cells. Other factors that influence the effect of therapy include the time of initiation of therapy, the intracellular concentration of active antiviral metabolites, and the dosage of ara-A.

The results demonstrated with congenital CMV infections are supported by the data obtained from therapy of the immunosuppressed renal transplant patients with generalized CMV. Five transplant patients showed little change in their urinary excretion pattern with persistent multisite retrieval most closely resembling the acute fulminant form of disease.

The differences in response among CMV patients raised the possibility of excreted urinary metabolites inhibiting the replication of CMV *in vitro*. A biological assay, therefore, was performed with undiluted urines of patients on therapy for HVH against strains of CMV from the congenitally infected infants. These studies failed to show a decrease in the viral titer. A similar assay is presently being employed to study saliva and throat swab material from the HVH patients. It is anticipated that these results will similarly be negative.

V. THERAPEUTIC POTENTIAL OF ARA-A

The truly significant advantage of ara-A over its counterpart pyrimidine analogs is its virtual lack of acute toxicity. All patients treated with ara-A to date have been monitored serially for changes in biochemical and hematologic status. The only adverse reaction to the drug has been the appearance of nausea in 30% of the adult patients. Monitors of renal and hepatic function showed no alteration with therapy. Hematologic parameters showed only a transient decrease in the packed cell volume and platelet counts. These changes were of no clinical significance and did not necessitate transfusion with blood or blood products. Immunoglobulin production as well as the formation of neutralizing antibodies to HVH infections is not impaired by therapy. Of particular interest is the ability of ara-A to break chromosomes *in vivo*. The peak incidence of 25% breakage occurs within 18 to 24 hours after the initiation of therapy. A gradual decline is noted for the duration of treatment with only 9% breaks recorded 4 weeks after the cessation of therapy. Even with these toxicities, ara-A may be employed with general safety. The toxicities described, however, are all acute as we have not had a sufficient period of time to seek out signs of chronic toxicity.

The clinical implementation of ara-A remains in its early preliminary stages of employment. Like ara-C and IDU, ara-A remains a nonspecific inhibitor of viral replication acting intracellularly on host enzyme systems. The further development of antiviral chemotherapeutic agents requires more precise elucidation of the biology and biochemistry of viral replication. Even with such knowledge, careful intracellular metabolic studies will need to be performed. As has been shown from our own Phase II studies, early localized infection of HVH and VZ appears to be amenable to therapy, whereas the treatment of late disseminated infections has little to offer. The nature of the virus, even the viral strain, the viral load or quantity of virus present, and its duration as would be reflected by the localization, progression, and/or dissemination, play the most significant roles in determining responsiveness to therapy. These variables of themselves determine the dosage of the drug. Certainly, and of equal importance, these factors are modulated by the host falling on the spectrum from an intact immune system to one severely suppressed by either drugs or underlying disease. Even age will probably prove to be a significant variable. An understanding of these determinants most certainly will provide greater insight into the infectious states of latency and chronicity. With such knowledge, new and more specific antiviral agents will certainly be developed. Our own studies continue to teach us of the complexity of viral chemotherapy for herpetic infections. Nevertheless, the results of the studies for localized and early disseminated disease are encouraging and have led to a definitive large-scale multiinstitutional double-blind or Phase III study which is now underway to confirm or refute the suggested therapeutic efficacy of ara-A. However, the data from this study are insufficient for review at this time. Adequate Phase III studies have yet to be performed for ara-C and IDU; in spite of this they are in widespread use, especially the latter. Control studies are a perfectly normal sequence of events for assaying the development of a drug in man. Only through such an approach can efficacy be ascertained and this is absolutely essential given the toxicity of the purine and pyrimidine nucleosides.

ACKNOWLEDGMENTS

We thank Ms. K. G. Stubbs and Mr. R. Reynolds for technical assistance, Ms. Jan Holley for secretarial assistance. Drs. Arnold G. Diethelm, Charles W. Scott, Paul Bridgeford, and John R. Durant referred patients and provided helpful suggestions. The following physicians contributed neonatal cases: Dr. Richard Alper, Minneapolis, Minn.; Drs. David Carver and Susan Shurin, Johns Hopkins University, Baltimore, Md.; Drs. Roy Cioletti, W. S. Newman, Larry Frenkel, and Joseph Bellanti, Washington, D.C.; Drs. Gilbert Schiff and Calvin Linnemann, Cincinnati, Ohio. We also thank Drs. Robert A. Brackett, A. Kinkel, A. Glazko, and Ms. Sylvia Weiss at Parke-Davis and Company, Detroit, Michigan, for excellent laboratory support.

REFERENCES

1. Ch'ien, L. T., Cannon, N. J., Charamella, L. J., Dismukes, W. E., Whitley, R. J., Buchanan, R. A., and Alford, C. A., Jr., *J. Infec. Dis.* **128**, 658–663 (1973).
1a. Ch'ien, L. T., Schabel, F. M., and Alford, C. A., *in* "Selective Inhibitors of Viral Functions" (W. A. Carter, ed.), p. 241. CRC Press, Cleveland, Ohio.
2. Furth, J. J., and Cohen, S. S., *Can. Res.* **28**, 2061 (1968).
3. Glazko, A. J., unpublished results in personal communication (1972).
4. Jones, S. R., Luby, J. P., and McElligott, D., *Clin. Res.* **21**, No. 3, 603 (abstr.) (1973).
5. Kaufman, H. E., Ellison, E. D., and Townsend, W. M., *Arch. Ophthalmol.* **84**, 783 (1970).
6. Kinkel, A., unpublished results in personal communication (1973).
7. Lee, W. W., Benitez, A., Goodman, L., and Baker, B. R., *J. Amer. Chem. Soc.* **82**, 2648 (1960).
8. Nahmias, A. J., Alford, C. A., and Korones, S. B., *Advan. Pediat.* **17**, 185–226 (1970).
9. Privat de Garilhe, M., and De Rudder, J., *C. R. Acad. Sci.* **259**, 2725 (1964).
10. Schabel, F. M., *Chemotherapy (Basel)* 13, 321 (1968).
11. Sidwell, R. W., Arnett, G., and Dixon, G. J., *Progr. Abstr., Intersci. Conf. Antimicrob. Ag. Chemother., 7th, 1967* No. 64 (1967).
12. York, J. L., and LePage, G. A., *Can. J. Biochem. Physiol.* **44**, 19 (1965).

DISCUSSION

Dr. Kaufman: It's clear from all the double-blind control studies that idoxuridine (IDU) is active clinically when given topically. It is equally clear that IDU is not a very good drug for systemic administration. For systemic administration, adenine araboniside (ara-A) has great possibilities. Our group has worked with this drug for several years. We've been particularly interested in it not for the treatment of superficial eye disease, which can be treated by other superior topical agents, but for the treatment of potentially blinding, long-lasting, awful eye infections. These chronic deep infections, iritis and stromal disease, seem to be produced largely by a host hypersensitivity reaction compounded by continual production of virus, and probably depend on continual viral replication. The present treatment regimen, based on our findings on pathogenesis, is to treat this with corticosteroids to suppress the hypersensitivity component and to give topical antivirals to prevent the corneal ulcers that will otherwise develop when corticosteroids are used alone. Although this is effective, it doesn't really cure anything. Under this regimen the disease can go on for months and even years. In fact, it may not be possible to control it this way. Because of this, we have studied the use of ara-A in the treatment of deep viral infections. First we tried to use it in a way that would not require systemic chemotherapy. Topically, penetration was unsatisfactory. Even injecting it subconjunctivally gave us inadequate concentrations and provided little clinical benefit. We then tried using smaller doses of ara-A systemically giving 20 mg/kg IM, a different route than used in Dr. Whitley's study. Again, in an open study, no clinical effect was seen whatsoever. On the other hand, an open study employing 30 mg/kg of ara-A given IV seemed to show efficacy. Thus, we started a controlled, double-blind study of intravenous ara-A. It is requesting a lot of a patient to ask him to come in the hospital for a week with only a 50% chance of getting active drug. But we offered the patient the alternative that if he stayed the same or got worse, we would break the code and then give him the option of receiving known drug. The patient would know our previous results and then decide whether he wanted the drug or not.

So far, although our studies are not complete, there seems to be a clear-cut drug effect. Out of about 9 treated, 4 had a rather dramatic improvement. Five had improvement which was more modest, but none of the placebo controls improved. Although several placebo-treated patients remained the same, most of them got worse. On removing aqueous humor from the anterior chamber of the eye, we found drug levels that are very consistent with those reported by Dr. Whitley: ara-A concentrations were about 0.5 μg/ml, and hypoxanthine arabinoside (ara-Hx) concentrations were about 6 μg/ml, very similar to what you would expect in spinal fluid. This extends their studies to some extent with controlled double-blind results which they do not have.

DR. O'CONNOR: I wish to report briefly on some findings from a study of antiviral antibiotics conducted in collaboration with my colleagues Pauline Schiop-Stansly (National Cancer Institute, Bethesda, Md.) and V. Sagar Sethi, Ahmed Hadidi, and Paul Okano (Litton Bionetics, Inc., Bethesda, Md.).

Following the initial report of Dr. Maurice Green and his associates on the inhibition of the DNA polymerase activity of oncornaviruses with substituted rifamycins, several investigators have reported that a variety of three-substituted derivatives of rifamycin SV can inhibit oncornavirus polymerase activity and, in some instances, can also inhibit replication of oncornaviruses in cell culture and transformation of cell cultures by these viruses.

We now report that the parent rifamycin antibiotics—rifamycin B, rifamycin O, and particularly rifamycin S and rifamycin SV—do *not* inhibit an oncornavirus polymerase but, nevertheless, inhibit replication of oncornaviruses and transformation of cells by oncornaviruses.

We have employed Swiss mouse-3T3 cell cultures infected with murine sarcoma virus (MSV) and human fetal spleen cell cultures infected with feline sarcoma virus (FSV) as test systems. In addition, we tested the inhibitory activity of the compounds for the DNA polymerase isolated and purified from simian sarcoma virus I and for the following enzymes isolated and purified from 12 to 16-day-old Swiss mouse embryos: DNA polymerase I, DNA polymerase II, RNA polymerase II. None of these enzymes were significantly inhibited by the antibiotics listed. However, each of these compounds and particularly rifamycin S at a concentration of 4 μg/ml produced a 97% decrease in focus formation by FSV infection of human spleen cell cultures.

How do these antibiotics work? The failure of these materials to inhibit the enzymic activity of viral DNA polymerase or of cellular DNA or RNA polymerases renders a direct inhibition of these enzymes unlikely. We have examined the toxicity of both rifamycin S and rifamycin SV. At a concentration of 5 μg/ml these compounds do not show detectable toxicity for uninfected 3T3 cells, 3T3 cells transformed by MSV but not shedding virus (S^+L^- cells), and 3T3 cells transformed by and shedding MSV, as measured by growth rate of cells and incorporation of labeled precursors into DNA, RNA, and protein. These compounds, however, cause upwards of several hundred-fold decrease in virus yield from infected cells if added within 2 hours of infection of the cells but do not affect virus yield when added to previously infected cells that are shedding progeny virus. We conclude that these compounds act by blocking some undefined step following virus absorption and penetrating rather than by a cellular toxicity effect.

Rifamycin O, rifamycin S, and rifamycin SV were originally obtained by chemical alteration of rifamycin B but are now available as metabolic products of mutant strains of *Streptomyces mediternanei* (Lancini, G., personal communication). Our findings indicate that parent rifamycin antibiotics, just as parent streptovaracin antibiotics (secreted by *Streptomyces spectabilis*), can inhibit replication of oncornaviruses. Both of these groups of antibiotics possess ansamacrolide structure. A third antibiotic, streptonigrin, which has a quite dissimilar structure, has been reported to have both antitumor and anti(oncorna)virus activity. Chirigos and his associates [Chirigos, M. A. *et al., Cancer Chemother. Rep., Part* 1 **57,** 305–309 (1973)] have reported that this compound inhibits oncornavirus DNA polymerase and inhibits replication of murine leukemia and sarcoma viruses both *in vitro* and *in vivo* in mice. We observed negligible inhibition of simian sarcoma virus, DNA polymerase activity by streptonigrin at concentrations below 10 μg/ml, but we observed 50% inhibition of focus production by FSV in the human spleen cell cultures at concentrations as low as 0.1 ng streptonigrin. Our investigation of the mode of action of this compound is continuing.

The ability of three different antibiotics metabolized by three different species of *Streptomyces* to inhibit oncornavirus replication suggests that the occurrence of such antibiotics may not be uncommon. Such findings may stimulate the search for other active antiviral antibiotics, investigation of the scope and mode of operation, and perhaps synthesis of active analogs.

DR. BRUNELL: I wish to comment on the sort of studies that are going to be required to evaluate this drug in the infections of the newborn infant. I am particularly concerned about this because I am a pediatrician and I think this is a concern that all of you who are involved in this should share. One of the ways that you have attempted to evaluate efficacy is by measuring the effect of therapy on shedding of virus in the urine; you have shown us very clearly that this agent is excreted in the urine. I would ask if what you are, indeed, observing when you tell us that the virus is not detectable during therapy is that you are carrying this drug over into your assay system?

The other parameter that certainly is more important in terms of evaluating efficacy is the clinical state of your patient at the conclusion of your therapy. As you have shown, in your very elegant studies, and the group from Cleveland and in Pittsburgh have demonstrated, infection with cytomegalovirus (CMV) can produce a very variable clinical outcome. Most of the infections will be subclinical. Very few of the children who are infected will have devastating disease. This obviously requires that some of the patients would be treated and others kept as controls and that we would then observe clinically what happens to both groups of patients. I am particularly concerned as you know about the effect of DNA inhibitors on the developing infant. At this time there is rapid growth of a variety of organs requiring a burst of DNA synthesis. Adenine arabinoside seems to be a safe drug compared to some of the previous analogs in terms of immunosuppression. It still is a potent and nonspecific DNA inhibitor; it is, moreover, inhibiting host as well as viral DNA synthesis. Studies from our laboratory and from other groups testing a variety of DNA inhibitors in neonatal animals indicate that these drugs can have devastating effects on neurological functions in newborn animals. We have studied the use of cytosine arabinoside (ara-C) in the mouse. To summarize briefly we have produced almost total ablation of the cerebellum which is the most rapidly developing portion of the mouse brain in the newborn period. These animals developed ataxia. The migration of primordial cells in the outer granular layer to the inner granular layers was inhibited since we gave this drug at a very critical time in the development of this organ. Also we have produced profound effects on the retina. Both the retinal and cerebellar changes were irreversible in these animals. It is true that human brain postnatal development is not entirely comparable to that of the mouse. Dobbing [*Pediatrics* **52**, 317 (1973)], however, has commented on the vulnerability of the human brain and particularly the importance of myelinization and dendrite formation that occur during the early months of life.

It would appear that to assess DNA inhibitors, it would be necessary to determine their effects on normal brain development during the newborn period. Such a study would obviously be immoral and, therefore, impracticable. I would urge that, since this cannot be done, we observe treated infants, especially those who come to autopsy, for evidence of damage caused by DNA inhibition to organogenesis in these rapidly growing and developing newborn infants.

DR. WHITLEY: In response to the original question that there was biological activity of residual hypoxanthine and ara-A, I presume you would mean in the urine. First off, the drug levels in the urine would not be adequate enough to block a tissue culture system since we worked out what would be required in that system. We have tried the biological assay. The only monitoring that we have done with urine has been with CMV. We have checked out to see if the urine from patients receiving the drug would block much lower levels of CMV than are seen in these experiments. It will not. The monitoring for the herpes patients is from the throat. As you could see from the plasma levels, which is essentially what you would be dealing with from swabbed lesions, is far, far below anything that could possibly block herpes 1 and 2. The other possibility remains the methodology. These are all concomitant titrations which means that at a 10 million dilution of the original specimen would have to block herpes. In other words, as we don't pretreat the cells, we actually do a titration of the swab itself. Since some of these levels were as high as 10^6, in order to give you the stepwise reductions of virus, it would have to be active after dilution at that time of monitoring. Furthermore, you would need to leave the compound on the cells over a long period of time in order to demonstrate *in vitro* suppression, so I don't think that what we are measuring is residual drug.

Second, in response to your plea that we not treat the asymptomatic type of congenital CMV infections, I couldn't agree with you more. Our purpose in setting out on these studies was the realization that across the country, ara-C and IDU were already being employed without adequate evaluation. It is an ostrich phenomenon on our part to pretend that it is not being done. It is not ideal; but it is being done. And, hence, we undertook the study of this drug to make it available for the kinds of infections that were being used for the IDU and ara-C studies. Essentially we do not think that these drugs should be used on anyone unless they have what we would consider very severe debilitating disease or life-threatening disease. We do not attempt, for instance, to recommend its use in the normal host at the present time for the reasons you have already pointed out; namely that DNA inhibitors could be dangerous. We realize that perhaps this could be carried to the extreme by the clinicians. But in all honesty, in our seventeen-institution study, when we were confronted with those seventeen human use committees whose approval was mandatory, 50% of them said that we were immoral because IDU and ara-C were already shown to be essentially active and that we had no right to build a control into the study. The other half said there is no evidence that anything works at all and, therefore, you ought not to be performing such studies. In the meantime, the patients on all of their hospital wards were being treated with ara-C as a "proven" medication. So this is the situation at the present time. I absolutely agree with you that this should be monitored very carefully. In our control study we will attempt to determine what it is we can treat. We do not think that we are treating disseminated infections. The message of our report is that we do not feel that we are doing much for disseminated disease. As in the CMV data, we cannot see any clinical efficacy, only a virological effect, because of the difficulty in assessing with clinical accuracy that state.

DR. SAWICKI: Are you sure that the mucous secretions you titrated for virus never contained any residual drug, so that when you transferred it into tissue culture the virus titer might have been decreased by the amount of drug introduced at the same time as the virus?

DR. WHITLEY: We've tried to assess biological activity of our inoculum without any effect. In other words, we cannot show evidence of residual drug in the biological monitoring that we have done thus far. We would have to concentrate these materials to get any effect on herpes at all.

DR. SAWICKI: I'm referring to the studies of Dr. Pavan-Langston in Boston. She has shown that the drug, when used as a paste or subconjunctivally, persists in the eye for 24 hours. I would not exclude the possibility that you may have residual drug, if measured by physical methods as she did.

DR. WHITLEY: We think we do have persistence of the drug inside the cell. We do have persistence of something. We don't know what the fate of this drug is, but the radioactive tag most definitely did persist inside of cells. However, the inoculum that we measure by titration is apparently not the thing that is blocking the results.

DR. SIMON: I would like to indicate that we have reported on the synthesis and broad spectrum antiviral activity of a triazole nucleoside, specifically called ribavirin, 1-β-D-ribofuranosyl-1,2,4-triazole-3-carboxamide. Dr. Sidwell of our group has shown that it possesses activity against a wide variety of both RNA and DNA viruses *in vitro*, and specifically, has a potent *in vivo* activity against all strains of influenza, and in addition, is effective in treatment of herpes keratitis in the rabbit, as well as herpes tail lesions in the mouse. This compound, unlike ara-A, is reported to have less toxicity in simian hosts than in lower animals, and, in addition, our studies have shown that it is not incorporated into nucleic acids, although the mono-, di-, and triphosphate of this compound can be formed *in vivo*, as well as *in vitro*. Phase I clinical studies have been carried out, and in addition, we hope to begin evaluation of this compound in Phase II clinical studies shortly.

CHAPTER 19

Epilogue

Roy Stern*

There are a few moments in a person's life, such as birth, marriage, birth of children, and death, which are qualitatively unique in that they cause fundamental and irrevocable changes of direction in the lives of the people involved. Having just passed through one of those moments, namely marriage, I would like to speak about some of the ramifications of getting married and how they might be of use to those in this audience.

When I was about to get married, I vowed never to do it again, not from an emotional point of view, but from a practical one, because the experience of getting married and the effort involved was so physically and emotionally draining that only a person who was unaware of the implications would go through with it. But of all the things that this experience has taught me and is still teaching me, I think the most important lesson is that when people of two backgrounds are united, the union will lead either to great discord or to a product even greater than the source. What is crucial is the kind of people who are involved, for if of the two, one predominates, then there is an imbalance that does not have much potential, and if the individuals are self-centered and uncompromising, then there can be only discord. But if they are understanding of each other and provide aid in areas necessary to the other, then they can achieve a greater good.

By now you are probably asking yourselves what this all has to do with science and virology, but no field is isolated or unique, and there

* Intern, Babies Hospital, Columbia Presbyterian Medical Center, New York, New York.

are truths which are universal. The two truths which I alluded to previously are balance and mutual understanding. These are two factors that are often lacking in the relationship between, for example, medical doctors and veterinarians, or between basic researchers and clinical researchers. Too often one hears of the rivalry for resources and the debate as to which is more important. Too infrequently does one hear the idea expressed that, if each branch thinks of itself as an end rather than a means and tries to exist independently, then the ensuing discord will result in a great waste of time and talent and in the nonachievement of many attainable objectives. The arguments used are often very complex, but in reality what causes these difficulties is not the specific differences between fields, but human nature. That is why factors such as balance and mutual understanding are so universal and that is why it is possible to overcome such differences.

For this reason, one of the founding principles of the "Perspectives in Virology" symposia was to maintain a balance by allowing all those who had something to contribute to be heard and, at the same time, to create an atmosphere where such an intermingling of ideas would result in the achievement of a greater good. *Perspectives in Virology IX,* whose subject is "antiviral mechanisms" has the greatest potential yet, of attaining the ultimate synthesis of its kaleidoscopic group of participants and, if successful, will be doubly meaningful.

Subject Index

A

Adenine arabinoside, 315 (*see also* Chemotherapy)
  and herpes infection, 323
  history of, 316
  therapeutic efficacy of, 330
  toxicity of, 318
  trials in man, 318
Adjuvants
  built-in, 96
Antibiotics (*see also* Antiviral mechanisms, *and* Chemotherapy)
  antiviral, 333
Antibody
  anticollagen, 94
  antineuraminidase, 104
  core and surface, 75
  hepatitis, 9, 49
  radiolabeled, 118
  to fastidious agents, 38
Antigen (*see also* Assay Systems, *and* Viruses)
  Australia, 49
  competition of, 92
  conformation of, 93
  diversity of hepatitis, 54
  hepatitis B, 56
  hepatitis surface, 52
  Rous sarcoma, 217
  specificity of, 58, 92, 94
  synthetic, 91
Antigenemia, 71
Antiviral mechanisms (*see also* Chemotherapy, Immunotherapy, Interferon, Therapy, Vaccines, *and* Viral replication)
  and antilipemic agents, 280
  and cell replication, 215
  antibiotics, 333
  antineuraminidase, 99
  by allogeneic chimerism, 162
  by chemical agents, 315
  defective particles, 77
  immunological, 176
  in cultured cells, 207
  in host, 175
  in superinfection, 179
  interferon, 269
  lymphocytes in, 153
  photodynamic, 297
  strategies of, 100
  transfer factor, 135
  with congenital infection, 175
Assay systems (*see also* Tests)
  activation by concanavalin, 133
  electrical resistance pulse analysis, 45
  for influenza, 106
  immune electron microscopy, 9
  immunofluorescent, 131

passive hemagglutination, 50
plaque size, 132
radioimmune, 50, 63
transfer factor, 151, 156
Autoimmunity
and T-cell deficiency, 155

C

Cancer (*see also* Chemotherapy, Herpesviruses, Therapy, *and* Viruses)
and immune deficiency, 155
and oncogenic viruses, 157
and replicating viruses, 167
and transfer factor, 135, 146
and viruses, 130, 153, 165
human, 153
immunodeficiency in, 157
immunotherapy for, 146, 162
inactivated virus and, 308
inhibition of, 215
osteosarcoma, 158
RNA viruses, 165
Cell mediated immunity (*see also* Antiviral mechanisms, Cancer, Immunotherapy, *and* Vaccines)
and transfer factor, 135, 138
in herpes infection, 115
relation to cancer, 155
Chemical scissors, 93
Chemotaxis, 127
Chemotherapy, viral, 107 (*see also* Antiviral mechanisms *and* Therapy)
alpha-amaniten, 231
amantadine, 112
antiviral agents, 77
Ara A and C, 265, 315
clofibrate, 288
colchicine, 225
cycloheximide, 86
DNA inhibitors, 334
FANA, 109
for cytomegalovirus infection, 328
for enteroviruses, 299
for herpes, 323
hazards in, 334
idoxuridine, 265, 332
neutral red, 299
nucleosides, 316
proflavine, 297
ribavirin, 336
Rous sarcoma virus, 216
trifluorothymidine, 265
Clofibrate, 288
Concanavalin, 133, 204
Congenital infection
and gene inheritance, 192
and genetic transmission, 169
cytomegalovirus, 328
genetic transmission, 171
Coronaviruses, 14
Cytomegalovirus
treatment of, 328

D

Dane particle, 50, 57
Defective particles, 77
Determinants
antigenic, 92
Disease (*see also* Congenital infection, Epidemiology, *and* Cancer)
gastroenteritis, 9
hepatitis, 9, 33
herpes genitalis, 306
herpes keratitis, 302
immune complex, 89
influenza, 99
Lassa fever, 2
Marburg, 2
neurological, 264
Norwalk disease, 26
responding to transfer factor, 138
varicella zoster, 257
virus expression, 180
DNA, infectious, 76

E

Endogenous viral genomes, 197
Enzymes, viral (*see also* Viral replication)
DNA polymerase, 317
lipogenic, 285
polymerase, 60
QB polymerase, 247
Epidemiology
genetic control of spread, 185
genetic transmission of virus, 165
new diseases, 1
of hepatitis, 50, 66, 75
Epistatic inhibitor gene, 186

F

Freund adjuvant, 124

G

Gastroenteritis, viral, 9
Genetic control
and partial expression, 181
of host susceptibility, 176
of virus expression, 180

H

Hepatitis
active immunity against, 65, 71
epidemiology of, 50
identification of agents, 33
immune electron microscopy, 9
immunoprophylaxis 49, 61, 71
in marmosets, 46
in nonhuman primates, 61
passive immunity against, 63
tests for, 50, 71
venereal transfer of, 53
Herpesviruses (*see also* Transfer factor)
immunity to, 115
photodynamic inactivation of, 297
therapy for, 145, 297, 323, 332
Host contamination by genes, 193
Hybrid cells
monkey-mouse, 269
Hybridization
Rous sarcoma virus, 219
viral, 83
Hyperlipemia, virus-induced, 279

I

Infections, viral
and cancer, 153
and degenerative diseases, 4
and genetics, 7, 165
and transfer factor, 135
cellular immune response, 115, 135
nonspecific inhibition, 121
spread patterns, 116
Immune complex disease, 89
Immune electron microscopy
and fastidious viral agents, 9
and tobacco mosaic virus, 10
coronaviruses, 14
Norwalk disease, 26
rhinovirus detection, 11
Immunotherapy (*see also* Antiviral mechanisms)
and cancer, 153, 157
and transfer factor, 135
and viral spread, 116
augmentation of, 136
built-in adjuvants, 96
by interferon, 269
effect of pH, 133
nonspecific, 121
site of action, 129
Immunity, cellular
and cancer, 153
and viral infections, 115
Immunodeficiency
and cancer, 157, 160
and nutritional deficiency, 160
and transfer factor, 135
Immunogenetics, 7, 95, 176
Immunotherapy (*see also* Antiviral mechanisms *and* Cell mediated immunity)
by bone marrow cell transplants, 162
for cancer, 146
herpes, 115
with transfer factor, 139, 143
Infected cells
immune attack on, 117
Inhibitors (*see also* Enzymes)
antilipemic agents, 279
epistatic, 186
from NZB cells, 207
nonspecific, 331
of DNA synthesis, 216
of mitosis, 225
of multiplication, 204
of type C virus, 207
properties of, 208
Interference
by T particles, 86
model for host resistance, 189
molecular basis of, 82
to superinfection, 179
Interferon
and cellular functions, 246
and hormones, 275

and varicella zoster, 257
cell receptor interactions, 269
clinical applications, 263
clinical utilization, 249
exogenous, 249
in congenital infections, 175
LEV-4 vaccine inducer, 265
local applications, 251
mechanism of, 233
pharmacokinetics of, 258
properties of, 250
systemic application, 255
tRNA dependency, 244

L

Leukemia
immunotherapy for, 162
Lipemia
mechanism of, 282
virus induced, 280, 294

M

Marburg disease, 2
Multiple sclerosis
and transfer factor, 145

N

Neuraminidase
viral vaccine, 99, 101

O

Oncogene, 175

P

Particles
Dane, 50, 57
defective interfering, 77
defective interfering T, 78
plaque forming B, 78
Photodynamic inactivation
and transformation, 308, 310
by neutral red, 299
by proflavine, 297
by toluidine blue, 308
herpes genitalis, 306, 310
herpes keratitis, 303
of SV40, 309
of transformed cells, 311
of viruses, 297, 299
Proflavine
side effects, 309
Provirus, 230

Q

QB polymerase, 247

R

Replication of tumor viruses, 167
Reticulum cell sarcoma
immunotherapy for, 162
Reye's syndrome, 293
Rhinoviruses, 11
Rous sarcoma virus
inhibitors of, 216
model for replication, 228

S

Slow viruses, 3
Subacute sclerosing panencephalitis, 142
Synthetic vaccines, 91

T

Tests (*see also* Assay systems)
for transfer factor, 151, 156
hepatitis A, 33
immune electron microscopy, 9
Therapy (*see also* Antiviral mechanisms, Chemotherapy, *and* Immunotherapy)
antiviral agents, 77
by adenine arabinoside, 315
by antilipemic agent, 279
by transfer factor, 135
dye-light, 297, 302
herpes, 249, 306, 323
interferon, 233, 249, 252, 269
of transformed cells, 311
varicella zoster, 257
Tissue cultures
and viral genomes, 173
chick embryo fibroblasts, 217
Chinese hamster ovary, 78

continuous infection, 78
differentiating cells, 76
inactivation of, 227
synchronized, 229
Transfer factor, 131, 135
clinical tests of, 156
for herpes, 145
for measles, 144
for vaccina, 142
properties of, 139
specific activities of, 141
Transformation
by inactivated virus, 308
Translation RNA
preparation and purification of, 234
Tumor immunology, 131

V

Vaccines, 5 (*see also* Antiviral mechanisms)
actions of, 100
against hepatitis, 49, 61, 63, 65, 71
antihemagglutin, 101
antineuraminidase, 101
defective interfering, 77
ECHO virus, 7, 301
immunopotent determinants, 91
infection permissive, 103
influenza, 99
multivalent synthetic, 97
photoinactivated, 300
smallpox, 47
specific, 96
synthetic, 91
Vaticinations, 1
Viral replication (*see also* Enzymes)
activation by genes, 183
and cell inhibitors, 215
and Freund adjuvant, 124
and hyperlipemia, 279
and sensitized leukocytes, 126
autoinhibition, 89
chemical inhibition, 107, 216
defective interfering, 77
effect of pH on, 133
hepatitis, 59
model for, 228
multicycle, 111
neucleocapsid accumulation, 80
pathways of, 84
role of polymerase, 60
serum enhancement, 221
Virogenes, 173, 175
Viroids
and epigenetic inheritance, 6
Viruses
and transfer factor, 135, 143
coronaviruses, 14
cytomegalovirus, 328
echovirus, 301
encephalomyocarditis, 266
enteroviruses, 299
enveloped, 279
fastidious agents, 9
gastroenteritis, 19
hepatitis, 9, 49
herpes simplex, 115, 145, 297, 315
human cancer, 153
influenza, 99
Japanese B, 280
Lassa fever, 2
leukemia, 89, 167
lymphocytic choriomeningitis, 89, 133
Marburg, 2
measles, 144
mengo, 233
new agents, 1
noninfectious, 67
Norwalk, 21
oncogenic, 130
photoinactivated, 297
RNA tumor, 165
rhinoviruses, 11, 266
Rous sarcoma, 215
slow, 3
SV40, 308
tomato bushy stunt, 12
type C, 207
vaccinia, 142
varicella zoster, 257
vesicular stomatitis, 78, 133
West Nile, 281
xenotropic, 190, 203, 207
Virus expression
genetic control of, 180

X

Xenotropic viruses, 190, 203, 207